R01611 81526

4-7-92
1ref
79.00

DISCARD

A Technical History of the Motor Car

A Technical History of the Motor Car

T P Newcomb
*Reader, Department of Transport Technology,
Loughborough University of Technology*

and

R T Spurr
*formerly Principal Research Officer,
Ferado Ltd*

Adam Hilger, Bristol and New York

© IOP Publishing Ltd 1989

All rights reserved. No part of this publication may be reproduced, stored in a retrieval system or transmitted in any form or by any means, electronic, mechanical, photocopying, recording or otherwise, without the prior permission of the publisher. Multiple copying is only permitted under the terms of the agreement between the Committee of Vice-Chancellors and Principals and the Copyright Licencing Agency.

British Library Cataloguing in Publication Data
Newcomb, T.P.
 A technical history of the motor car.
 1. Cars, to 1980
 I. Title II. Spurr, R.T. (Robert Thomas), 1922–
629.2′222′09

ISBN 0-85274-074-3

Library of Congress Cataloging-in-Publication Data
Newcomb, T.P.
 A technical history of the motor car
 T.P. Newcomb and R.T. Spurr.
 430 p. 23 cm.
 Bibliography: 2 p.
 Includes index.
 ISBN 0–85274–074–3
 1. Automobiles—Design and construction—History.
 I. Spurr, R.T. II. Title.
 TL15.N38, 1989
 629.2′3′09—dc19   88-35090

Consultant Editor: **A E de Barr**

Published under the Adam Hilger imprint by IOP Publishing Ltd
Techno House, Redcliffe Way, Bristol BS1 6NX, England
335 East 45th Street, New York, NY 10017-3483, USA

Printed in Great Britain by J W Arrowsmith Ltd, Bristol

To our wives

'History is bunk'
Attributed to Henry Ford

Contents

Preface ix

Part I The General Background

1 The beginnings 3
2 Developments 1900–1917 35
3 Developments between the Wars 50
4 Developments since 1945 58

Part II The Technical Development of the Motor Car

5 The engine 75
6 Carburation 128
7 Ignition 159
8 Self-starters 177
9 Engine lubrication 181
10 The cooling system 200
11 Instruments 207
12 Clutches 215
13 Gear change mechanisms 228
14 Transmission and axles 262
15 Chassis frames 277
16 Suspensions 284
17 Steering 312
18 Brakes and braking systems 333

19	Wheels	356
20	Tyres	368
21	Bodywork	388
22	Lighting	418
Bibliography		423
Index		425

Preface

This book is divided into two parts. Part I is a general synopsis, and is intended to put the various developments, including major technical developments, into perspective, and Part II deals with the various systems and components in more detail. To keep the book within a manageable length we have kept to the main stream only of development. Many interesting things have therefore been omitted because they appeared on only a few cars, or were in production for only a short time. If you are the proud owner of a 49/55 Pumperley (described in *Punch* in the thirties) do not expect to read how, on your vehicle, the reflex action of the Yarper snoother lubricates the fubbing nut, or just how the wire-wove grorbles feed the gong-budger.

Some inventions are made years before their need becomes apparent, or suitable manufacturing facilities are available to exploit them. An invention may not be taken up, not because it is bad, but because the royalties demanded are too high. In addition, one firm may want to keep the monopoly and not license competitors, and so the invention appears on only a small number of cars, but when the patent expires the idea may be taken up by the whole industry. We have generally made little attempt to trace an idea to its source, and indeed in many cases this would be practically impossible. Nor do we necessarily mention the first user of a particular idea if little came of it at the time.

The authors when writing the book went back to contemporary sources whenever possible and worked through, for example, all the Proceedings of the Institution of Automobile Engineers, all the Proceedings of the Automobile Division of the Institution of Mechanical Engineers, all the volumes of the *Automobile Engineer*, read books and company histories that had not been opened for fifty years, and visited many museums in this country and overseas.

But just because a statement is in print it is not necessarily true, so we have talked to twenty five retired eminent engineers who had been actively engaged in designing and making motor cars, and discussed their work and times with them. The memories of some of these gentlemen went back to the nineties and we are very grateful for their help, and any errors we have made were made despite their assistance.

T P Newcomb
R T Spurr
September 1988

Part 1
The General Background

Chapter 1

The Beginnings

Horse Drawn Vehicles

Homo sapiens has flourished for the best part of a million years, and for practically all that time if he wanted to go anywhere he has had to go on his own two feet. For the last six or seven thousand years he has had the horse to help him, but for most of that time the horse was the prerogative of the aristocracy and the military.

Sleds of sorts have probably been used for millennia and it was probably not long before the horse was used to pull the sled. To shift very heavy loads megalithic builders used small tree trunks as rollers—the stone block was supported by a number of trunks and the one from the back was brought up to the front as the block was moved. The great stones of Stonehenge were probably moved like this. This may have sparked off in the mind of some genius the concept of the wheel, one of the greatest inventions of all times. Sledges had been mounted on wheels in Sumeria by 3500 B.C. (figure 1.1). Progress was then slow but steady, and vehicles were eventually developed that could carry passengers without too much discomfort and at considerable speed. The Egyptians had horse-drawn chariots in 1600 B.C. These had a pair of spoked wheels and a light body carrying a pole to which a pair of horses were yoked. The chariots were used for fighting and hunting. Greek and Roman chariots had the same general lines as the Egyptians. The celebrated British war chariots had their opening in front instead of at the back and had wicker work side and back, but again consisted of essentially a platform on wheels with a pole and yoke attached.

There were other forms of wheeled vehicle—heavy, cumbersome carts pulled by oxen rather than horses, but these were for the transport of goods rather than people.

The Romans improved travelling conditions by their system of well made roads but in post-Roman and mediaeval times conditions retrogressed, particularly in country areas. The rich rode, the poor walked, and goods were carried on horseback. Ladies travelled in heavy four-wheeled wagons requiring a whole team of horses to pull them.

The first coach made in England appears to have been supplied by Walter Rippon to the Earl of Rutland in 1555. Queen Elizabeth received a coach from Holland in 1560 and its donor William Boonen became her coachman. Coaches became fashionable with the more wealthy of the aristocracy but by 1625 hackney coaches were established and traffic jams recorded in London by 1650. The difference between a wagon and a coach was that the body of the coach was isolated from the axles by straps or chains hung from corner posts in order to reduce jolting. The word 'coach' came from the German 'Kutsche' which in turn was based on the name, Kovacs, of the Hungarian town where such vehicles were built.

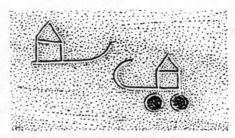

Figure 1.1 Sumerian pictograph used shortly before 3000 B.C. The addition of two circles to the sign of a sled (right) turns it into the sign for a wagon.

About the same time long wagons were used for the transport of goods and these would also carry passengers, though most people would have made better time by walking, for the 'stager' only did about 15 miles per day; indeed 'it is so tedious that none but women and people of inferior condition travel in this sort'. However the stager remained popular until late in the 18th century.

What was apparently the first stage coach ran from London to Chester in 1657—fresh relays of horses were supplied by inns along the route in ten to twenty mile stages. In 1669 a 'flying coach' took a day to travel from London to Oxford. There were, however, regular mail services from London to Oxford, Bristol, Colchester and Norwich by 1635.

In the 1660s Pepys used coaches much as we would taxis and he found it just as difficult to find a coach on a cold wet night when the theatres emptied as we do now to get a taxi. But it took a week to get from London to York in 1700 and the Fly coach took five days to reach Exeter. Travelling was uncomfortable even in London, for the coaches were unsprung and the body was slung on leather straps from four corner posts—so a fast drive was 'hell for leather'. Many people hated travelling by coach, and took a couple of days rest after a long journey to rest their tired and bruised limbs.

Coaches were built with a bent perch or backbone which supported a high box; the front axle which carried the shafts was pivoted under the box and so the front wheels had to be made much smaller than the rear wheels. This arrangement greatly reduced the turning circle of the coach.

Even in Pepys's day steel springs were being investigated and by the middle of the eighteenth century the four corner posts of the carriage were being replaced by laminated steel springs, the coach body being slung from the tops of the more or less perpendicular springs. By 1790 steel springs had come into general use and an industry to manufacture carriage springs came into existence. Steel springs helped bring about the coaching era. In 1760 the Manchester coach, for example, left London at 4 am on Monday and arrived in Manchester at 6 pm on Wednesday, while in 1767 the steel-sprung coach leaving London at 6 am Monday arrived in Manchester at 9 pm Tuesday.

In 1804 Obadiah Elliot mounted the coach body directly on to elliptical springs attached to the axles. The perch was no longer necessary and Elliot's invention revolutionised coach building.

Post-travelling was introduced into this country from France in 1743 by John Trull. The traveller supplied his own coach and hired his horses at the posting inns, changing horses in the same way as the stage coach. A light post-chaise was used; at first two-wheeled vehicles were drawn by one horse, later vehicles had four wheels and two or more horses. One horse of each pair was ridden by a postillion and so the post-chaise did not require a coachman.

Though the vehicles improved it was a long time before there was much improvement in the roads. Until the turnpike era country people were supposed to maintain the roads passing through their parish, but as they themselves made little use of these roads they did so very unwillingly, or not at all. The turnpikes came in at the beginning of the 18th century and on turnpiked roads the user had to pay a toll to travel on the road and this was supposed to be used to maintain the road surface. The much greater use of wheeled traffic damaged the roads and it was not until much later, and the time of Telford and Macadam, in the early nineteenth century, that good durable road surfaces were built.

The stage and mail coaches had their heyday between 1810 and 1837. Services ran to timetables with great regularity, and competition between the different proprietors helped to maintain and improve timings. Average speeds of nearly 14 mph could be reached and runs of up to 150 miles made in a day. The traffic was considerable; at one time 18 mail and 176 other coaches passed each day through Barnet on their way to and from the Midlands and the North.

Things were very well organised; the time required to change horses was incredibly short. When the coach was within earshot of the posting inn the guard would give a warning blast on his 'yard of tin' so that when the coach raced into the coach yard ostlers were standing by with the relief horses. The tired horses were unharnessed and the fresh ones slipped in and off the coach would race. Horses were changed on the crack 'Quicksilver', the Devon post mail, so quickly that the coach was on its way within 47 seconds! They did these things even better in France for the changes of horses on the malle postes was effected in 45 seconds! Stallions were used on the French Postes and many of the ostlers were women, who wore sharp pointed sabots on their feet and were not afraid to use them, so the stallions were not only fast but well behaved.

But all this was at a considerable cost, both directly and indirectly, typical fares were 2d per mile for outside and 4d per mile for inside seats. The horses had to be changed frequently, on the average one horse being used for every mile of the return journey; so the number of horses required to run these services was enormous—somewhere about the million mark. Each horse consumed the equivalent of the food for eight men, so the indirect cost to the country was high. It is not surprising therefore that, once the feasibility of the railway locomotive had been demonstrated, railways completely revolutionised transport in the UK.

Curiously, however, though the railways did away with the stage coach, they encouraged the private carriage among the well-to-do who moved out into the country and needed a conveyance between their house and the railway station.

The Victorians travelled in the big cities by train, but again people of rank and fashion used carriages. A lady would use a park phaeton for an airing in the park, a landau for making calls, and yet another carriage such as a brougham for an evening engagement with her husband. Many members of the professional classes and the middle class, however, could also now afford a carriage, though on much simpler scale. It cost about £100 per annum to keep a modest carriage in the 1870s when about 400 000 carriages were licensed, 40 000 carriages being made each year. There were just as bad traffic jams in London a hundred years ago as today; in some ways they were worse, for at least a waiting car does not empty its bowels in front of you. Congestion on the roads was so bad that the Underground was started; the Metropolitan line, running initially from Paddington to Farringdon Street, commenced operations in 1863.

But though the horse was the prime mover on the roads a number of people had tried over the years to replace the horse by a mechanical engine of some form or another, and as at first the only mechanical engine was the steam engine, attempts were made to propel vehicles with steam engines.

Steam Road Locomotives

The first road locomotive was that of Nicholas Cugnot which ran in Paris in 1769. It was a three-wheeled vehicle with the front wheel carrying the boiler and engine; the latter had a double cylinder and drove the wheel by two ratchet wheels and pawls which were moved by direct connection with the piston rods. The vehicle actually ran (into a wall on one occasion) but it was badly balanced and apparently overturned passing a corner near the Madeleine. Cugnot made a second larger vehicle (figure 1.2) in 1770/1 which is still in existence. The vehicles were intended to haul guns (Cugnot was a military engineer) and they did not lead to any further development.

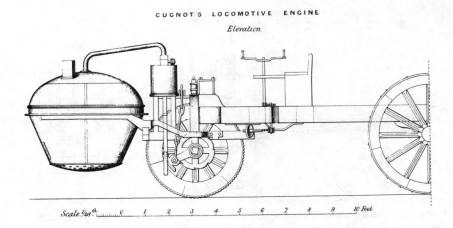

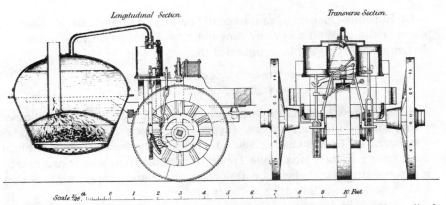

Figure 1.2 Cugnot's locomotive engine of 1771. By permission of The Council of the Institution of Mechanical Engineers (from *Institute Proceedings*).

The Swiss Isaac de Rivaz made a series of interesting vehicles during the late eighteenth and early nineteenth centuries. His first vehicle (1787) had a steam engine but later (about 1807) he used an internal combustion engine (see p 21). After Watt had invented a more efficient steam engine he considered installing such an engine on a vehicle and patented an arrangement, but only to stop anyone else from working on the same idea. His assistant William Murdock actually built an experimental steam-driven three-wheeled model (figure 1.3) in 1785, apparently against Watt's express orders. The model worked and travelling one night along a country road nearly frightened the wits out of the vicar of Redruth.

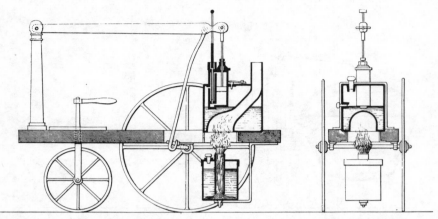

Figure 1.3 Murdock's three-wheeled model steam locomotive. By permission of The Council of the Institution of Mechanical Engineers (from *Institute Proceedings*).

Dr Erasmus Darwin, grandfather of the more famous Charles Darwin, was a friend of Watt and more sanguine than Watt. In his long poem *The Botanic Garden* (1789) he prophesied that:

> Soon shall thy arm Unconquered Steam afar
> Drag the slow barge, or drive the rapid car.

Richard Trevithick built a steam-driven model in the 1790s and followed it up with a full-scale vehicle (figure 1.4) which he demonstrated at Camborne on Christmas Eve 1801. Unfortunately the carriage was burnt out a couple of days later while Trevithick and his friends were in a nearby inn, 'comforting their hearts with a roast goose and proper drinks'—they had forgotten to extinguish the boiler fire!

Trevithick built a second engine and had it fitted to a new body in London, where the resulting vehicle was demonstrated in public at Euston Square and near what is now Lords Cricket Ground.

The continuing development of the high-pressure engine in the first years of the nineteenth century resulted in lighter, smaller engines, but even so they were still heavy, and most people thought that vehicles powered by them would be too heavy and cumbersome to run on the roads and not powerful enough for steep gradients. Consequently railway locomotives were developed instead. However, the success of the Rocket in 1825, and of the later railway locomotives, gave engineers and promoters incentive to work on steam road carriages. For example, nineteen year old James Nasmyth built a working model of a steam carriage in 1827, and it so interested the Scottish Society of Arts that he was given funds to build a full-scale carriage, which he did, and it worked very well to everyone's satisfaction. It carried up to eight people and made runs of four or five miles for some months. However, the Society was not interested in its commercial possibilities and Nasmyth wanted to return to his studies, so the carriage was broken up and the bits and pieces sold. A number of attempts were made to develop practical road locomotives in the 1820s and 1830s and Gurney, Hancock, and Scott-Russell made vehicles which actually ran on regular services. Goldsworthy Gurney built several experimental steam coaches and Sir Charles Dance ran three of these coaches between Gloucester and Cheltenham in 1831. In all, 296 journeys were made, 2666 fare-paying passengers carried, and 3644 miles covered in four months before Dance gave up, 'beaten by the ignorant obstructiveness of the provincial horse and turnpike interests'. It is likely, however, that the service was not making enough profit for its proprietor to persevere.

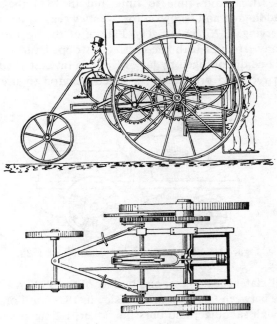

Figure 1.4 Trevithick's steam carriage.

Walter Hancock built a number of steam carriages between 1828 and 1838 including the Infant, Era, Enterprise, Autopsy ('See for yourself') and Automaton (figure 1.5). He developed a reliable boiler operating at a

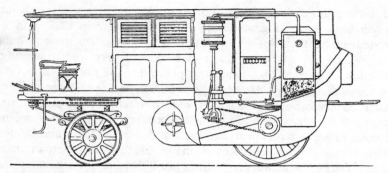

Figure 1.5 Hancock's steam carriage 'Automaton' of 1836. It had 22 seats and weighed about four tons unladen.

pressure of 100 lb in^{-2}, it worked two vertical cylinders and the engine crankshaft was connected to the driving axle by a chain drive. The comfort of the passengers was considered, the carriages were well sprung, and the engine and stoker were in a separate compartment. The carriages were used on regular services from time to time and in 1836 they ran between Stratford, Paddington and Islington for twenty weeks, travelling 4200 miles in all and carrying 12 761 passengers. Failure of the operating companies rather than any troubles with the carriages stopped further growth, and Hancock became discouraged by lack of public interest. In 1838 Hancock built the first recorded private car, a steam phaeton (figure 1.6) carrying three passengers plus the driver.

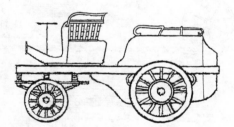

Figure 1.6 Hancock's steam phaeton of 1838.

Scott Russell, later the designer and builder of the SS 'Great Eastern', ran steam coaches between Glasgow and Paisley in 1834 until one overturned, probably through sabotage, and burst its boiler, killing several people.

Many other people—Hill, Maceroni, Squire, Gibbs, Summers and Ogle etc—worked on road locomotion but their vehicles were too heavy and awkward, and the roads were too bad for any to be an unqualified success; people would rather invest in the new railways. However some of the carriages were surprisingly reliable. In addition, a number of inventions were made and a number of ideas tried which were used later in the motor car. Hancock improved artillery wheels, and used chain drive and half-elliptic springing, Hill invented a differential to allow the driven rear wheels to turn at different speeds when cornering. James (1824) used a fore-and-aft propeller shaft with Cardan or universal joint. Gibbs used stub axles for steering instead of a swinging solid front axle. Further progress was prevented, however, by the opposition of the horse-coach companies and the trades associated with the horse. In the 1830s and 1840s turnpike charges for steam carriages were enormously increased. To some extent increases were probably justified as the vehicles must have played havoc with the roads, but charges were multiplied from five to twelve fold. For example, the 4s coach toll between Prescot and Liverpool became £2 8s 0d; on the Bathgate road the 5s toll became £1 7s 1d, between Ashburnham and Totnes the 3s toll became £2 1s 0d, and these increases were despite efforts by an enlightened government to reduce tolls in 1833.

Small steam engines were useful to farmers and were even more so if they could be moved from place to place. At first 'portable' engines were moved by horses but the obvious thing was for them to move under their own power and so traction engines appeared about 1840. Traction engines did not threaten the railways or other interests, though the turnpike trusts were suspicious of them, and so they were able to develop; the military authorities were also interested and this helped as well.

The increasing use of traction engines brought about agitation to make tolls more equitable, and also to control the weight and speed of the engines.

An Act of 1861 limited speeds to 10 mph in the country and 5 mph in towns, and related tolls to those of horsedrawn vehicles. The 1865 Red Flag Act reduced speeds to 4 mph in the country and 2 mph in towns; it also required that at least three people had to drive or conduct the vehicle, with one of these walking at least 60 yards ahead of the vehicle carrying a red flag, and signalling the driver when to stop. Finally, the vehicle had to stop if the driver of a horse-drawn vehicle put up his hand.

These Acts regulated the speed of the cumbersome traction engines but as a side effect completely killed any further development of steam coaches or private carriages.

A later Act of 1878 did away with the red flag but the vehicle still had to be preceded by a man on foot twenty yards ahead to assist horse traffic. A number of people still made experimental vehicles but these were little more than backyard jobs, though some were more ambitious. Boulton in 1848 and Rickett in 1859 built steam carriages, and the latter offered to build his to order.

In 1860 the Earl of Caithness made a well publicised run from Inverness to Borrowgill Castle (near Wick) a distance of 146 miles in two days on a Rickett steamer (figure 1.7). Its two-cylinder engine was claimed to deliver

9 hp and to give the carriage a top speed on level ground of 19 mph. His lordship remarked that, as far as frightening horses was concerned, people seemed more frightened of the carriage than did the horses.

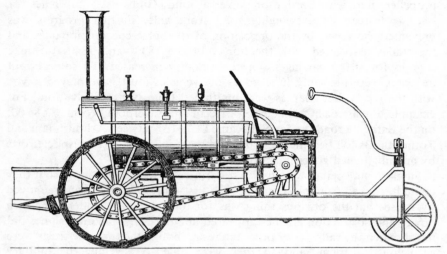

Figure 1.7 Rickett's road steamer bought by the Earl of Caithness.

R Garrett of Marshal and Co. of Leeds, in 1861 built a steam carriage for Mr Salt of Saltaire, but Salt did not use it much, probably because of the law, and gave it to a Mr Frederick Hodges who was much more adventurous and ran it 800 miles at speeds up to 30 mph in various parts of Kent. It received the significant name of 'Fly-by-night' and caused its owner and his friends to have several interviews with local magistrates. Yarrow and Hilditch, Hayball and Tangye all built apparently successful vehicles in the 1860s. There were apparently many other one-off jobs made, but they had no effect on future development and indeed few were ever described in the technical literature.

Not that there was much incentive to make private steam carriages for sale. An excellent railway system was being built and for shorter distances people rich enough to buy a dirty, uncomfortable, steam carriage would stick to their familiar and generally well made horse-drawn carriage. As far as steam coaches were concerned they could have been used as feeders to the railways, and in the big cities.

So though the United Kingdom was the foremost industrial nation until the 1870s, the initiative as far as the self-propelled vehicle was concerned passed to the Continent where there was no restrictive legislation to keep the vehicles off the road. Surprisingly very little progress was made even on the Continent until the 1880s when Bollée, Serpollet, de Dion and others tried again to adapt the steam engine, and Daimler and Benz the internal combustion engine, to road locomotion.

The Bollée family had been bell founders for generations. Amedeé Bollée was inspired, by the sight of a Michaux Velocipede and a steam tractor at the Paris World Exhibition of 1867, to make a fast private carriage and in 1873 his 'L'Obéissante' (The Obedient One) was finished. Bollée started with a clean sheet of paper and the vehicle was in no way an adaption of existing carriages; it had independent front suspension, geometrically accurate steering and each rear wheel was chain driven by a V-shaped twin engine (V2). 'L'Obéissante' was shown in Paris in 1875 and made a sensation but Bollée got no orders for more. In 1878 he made 'Le Mancelle' (The Maid of Le Mans). This had Victorian coachwork in the centre and an engine in front of the front wheels which drove the rear wheels through a shaft, differential and chain drives. In 1880 a gear change was incorporated as a single unit with the engine.

La Mancelle was exhibited, a customer from Alsace placed an order for a similar barouche and a vehicle was delivered which gave satisfactory service for at least twenty years. In 1881 a number of steam vehicles, five Mancelles, five omnibuses, and four tractors, were in the course of construction at a new factory near the bell foundry, but plans to expand further were unsuccessful, a German firm making Bollée vehicles going bankrupt. Bollée, however, continued in a small way making small 1100 kg Rapides some of which could travel at 63 km/h. He also built a steam tram (in 1876), road trains, and traction engines, or rather, road hauliers. In 1896 Bollée turned the car-making side of the business over to his sons, and returned to his first love, the bell foundry. Amadeé fils was only 19 years of age when Benz was successfully making his three wheeler in Germany.

The Internal Combustion Engine

Papin in 1673 and Van Huyghens a few years later tried to make internal combustion engines based on exploding gun powder. The Rev. W Cecil made an engine in the early nineteenth century in which hydrogen gas was ignited, and Professor Farish of Cambridge demonstrated in his lectures a small engine driven by the explosion of gas and air; apparently he had another engine which worked off gunpowder. Isaac de Rivaz built engines after about 1805 in which a hydrogen–oxygen mixture was ignited electrically and the engine used to propel vehicles. Rather earlier, in 1794, Street patented an engine in which droplets of turpentine were injected into the cylinder, then air drawn in through a non-return valve and at half-stroke of the piston, flame was drawn in to ignite the mixture. Samuel Brown in the 1820s made an internal combustion engine but it was more like an atmospheric steam engine than the later internal combustion engines. He forced air out of a closed chamber by igniting gas (hydrogen), isolated the chamber, and then injected water into it to make a partial vacuum, using the pressure differential between the atmosphere and the chamber to drive a piston.

L W Wright designed an engine with offset combustion chambers and double-acting cylinders with ignition at top dead centre (TDC). William Barnett, a little later, compressed the mixture by the piston and ignited it by a gas flame in a hollow, rotating cock. In 1848 a petroleum vapour non-compressive engine was exhibited at Philadelphia. None of the early internal combustion engines was successful commercially, however, and the first to be made in any quantity was the gas engine of Jean Joseph Etienne Lenoir. In Lenoir's engine, gas (town gas) and air were drawn into the cylinder, at half-stroke the inlet ports closed and the piston, driven by the flywheel, compressed the mixture which was ignited electrically at TDC. On the return stroke the burnt gases were discharged. The engine was double acting, mixture being admitted on each side of the piston on alternate strokes. About 500 of these engines were in use in Paris by 1865. Otto and Langen improved on the Lenoir engine and more than halved its specific gas consumption.

Otto was working as a travelling salesman for a wholesale grocer when he read about the Lenoir engine. He had a model engine built, experimented with it and considered that the basic problem of the gas engine was to 'moderate' the explosions so as not to damage it. He tried various ways and one partial solution, the atmospheric engine, which was patented in 1866, became a commercial success, and thousands of these engines were made by the firm of Otto and Langen. In the atmospheric engine the explosion of the gas charge drove a heavy, free, piston up the vertical cylinder; the gas then cooled, when atmospheric pressure on the piston, together with its weight, pushed it down as it engaged and rotated the output shaft.

The output of the engine was limited, it was noisy and bulky, and so Otto attacked again the problem of cushioning the explosion coming up with the idea, in 1876, of a stratified charge: the charge should be rich near the ignition point and thin out and therefore ignite less violently near the piston. If exhaust gases from the previous cycle remained near the piston they would be even more effective in protecting the latter. Earlier work had shown him that compressed gas gave more power than uncompressed.

His new engine took in gas and air, in a suction stroke, compressed the charge, which was then exposed to a flame, the ignited gas expanded doing work, and finally an exhaust valve was opened and the piston forced the spent gas out of the cylinder. A slide valve admitted charge and flame at the appropriate times.

The silent Otto engine worked well, even though the inventor's pet idea of the stratified charge had nothing to do with its operation. Otto was not, however, the first inventor of the four-stroke engine; fourteen years earlier, in 1862, Alphonse Beau de Rochas read and published a paper describing the four-stroke cycle, namely, suction, compression, ignition and exhaust ('suck, squeeze, bang and blow', according to our local driving instructor) and what principles should be followed in designing a four-stroke engine. Beau de Rochas, however, never made such an engine and probably had no intention of doing so.

A gas engine had to be connected to a gas supply, but if it was fuelled by a volatile liquid it could be used anywhere, including on a moving vehicle. Fortunately convenient liquids were available. The paraffin lamp had created such a demand for oil that an oil industry had been established, and wells drilled in Europe, Russia, and, in particular, in the USA. The oil was distilled to obtain the middle fraction, paraffin; the heavy fractions were used for lubricating oil and the lighter fraction, petrol, though it was not called that until 1900, was more or less waste, and it was very volatile, indeed its volatility and low flash point made it a dangerous material to have around.

The first practical petroleum engine was made in 1870 by Julian Hock in Vienna. The piston drew in a charge of air and petroleum spray and this was ignited by a flame jet. It was a two-stroke engine, but it was not long after the introduction of the Otto cycle that four-stroke engines were made which worked on petrol instead of gas. This was done by passing air on its way to the cylinder through petroleum liquid to form an inflammable vapour, and then mixing it with further air to form an explosive mixture which was then compressed and ignited.

The early gas engines were very heavy and cumbersome, generally working at 150 to 200 rpm. Otto and Langen, however, employed a very good engineer G W Daimler who, after playing a large part in making the four-stroke engine a practical proposition, had ideas about running the internal combustion engine at much higher speeds and making it much smaller and lighter. He also wanted the engine to run on liquid fuel so that it would be self-contained and portable. We shall return to him in a moment.

Karl Benz

Karl Benz was born on the 25th November 1844 at Muehldorf, the son of an engine driver. He was educated at the local Gymnasium and at the Karlsruhe Polytechnic where he studied mechanical engineering. Redtenbacher, an outstanding engineer, and Grashof the thermodynamicist were among his teachers and by the time he finished at the polytechnic at the age of 20 he had had an excellent technical education. His first job was at a local engineering works helping to build locomotives. After a couple of years valuable workshop experience he moved to the drawing office of a design firm and then to another engineering works. In 1872 Benz married and in the same year went into a partnership which was not successful, but with the help of Berthe's dowry he bought out his partner and became the owner of a small light engineering works. The turnover of the little firm was, however, small and he had to struggle to survive.

Benz had been interested in the internal combustion engine since his polytechnic days, and considered such an engine would be a useful product for his workshop. Because the four-stroke engine appeared to be covered by the Otto patents, Benz worked on a two-stroke engine, charging the working cylinder by air and gas pumps. The engine was successful, but Benz could not finance further expansion by himself and in 1882 he founded, with two partners and a banking house, the Mannheim Gas Engine Works.

Benz in the 1860s had bought a bicycle and convinced himself that vehicles for personal transport would eventually be power driven and have three or preferably four wheels. He wanted to put his new engine in a road vehicle but his partners not unnaturally were more interested in selling the engines. He consequently left them and on the 1st October 1883 founded Benz and Co. Rhenish Gas Engine Works, again with two partners. His two-stroke sold so well that money was available for experimenting on road vehicles, but as the Otto patent had been invalidated he could now use a four-stroke engine. Accordingly he designed an engine of 90 mm bore, 100 mm stroke working at 400 rpm and developing 0.9 hp. It weighed 95 kg.

In the spring of 1885 the engine was installed in a newly developed three-wheeled vehicle (figure 1.8) and in July 1886 the Benz motor car ran successfully on public roads. By 1888 the car was so reliable that his wife, with his two eldest sons as passengers, drove from Mannheim to Pforzheim 100 km away without incident.

The car attracted little attention at first and sales were small. But in 1888 Benz drove the car two hours daily during an exhibition in Munich and the horseless carriage created quite a sensation. Benz's two partners withdrew in 1890 but Benz was fortunate in finding Frederick von Fischer and Julius Ganz who shared his faith in the future of the automobile and they joined Benz and Co.

In 1892 Benz solved the problem of successfully steering a four-wheeled vehicle and in the following year the twin cylinder Victoria and Vis-a-vis cars appeared. The cheaper single-cylinder 1894 Velo in the 1.5 hp form sold at 2000 Marks and was so popular it was built on quantity production lines; it became known as the 'Doctor's Car'.

Production figures are shown in table 1.1. Benz was in many ways conservative and once his cars were satisfactory he was not concerned to improve them greatly. He did not develop the engines to any extent; for example, the original 0.9 hp engine still developed only 0.96 hp in 1898, and he regarded them only as a means to an end. His perseverance, or rather stubbornness, had been a tremendous asset in developing the car but eventually it became a liability. He did not want to build more powerful engines or faster cars, and though a number of different models were made in the 1890s they were relatively primitive so that the Benz cars became less and less competitive. This was particularly so when Maybach introduced the Mercedes in the early 1900s. Benz's attitude led to trouble on the board and

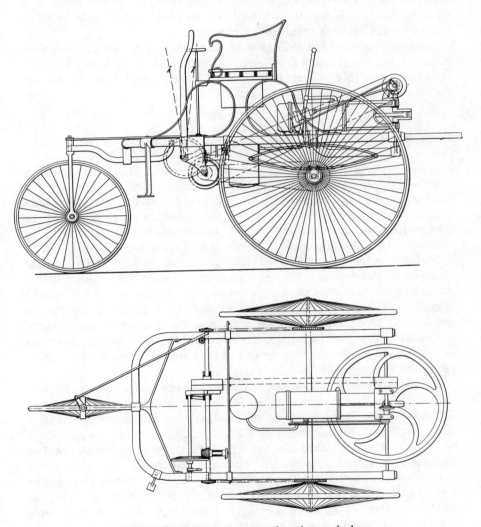

Figure 1.8 Benz's first car, elevation and plan.

Table 1.1 Benz production figures 1893–1900.

Before 1893	1893	1894	1895	1896	1897	1898	1899	1900
24	45	67	135	181	256	434	572	603

he left in 1906 to found with his sons the firm C. Benz Sohne to build cars and the firm was not unsuccessful; it continued until 1926.

The original company adjusted itself to the changed conditions and regained a considerable share of the market. Benz cars captured the world speed record in 1909 and held it from 1910 to 1919. The motor industry in Germany was wrecked in the aftermath of the 1914/18 war and in 1926 the firms founded by Benz and Daimler merged.

Benz died on 4th April 1929 aged 84 years. He was a quiet self-sufficient introvert, but cheerful and not unsociable, and very much a family man. He remained close to the shop floor and held craftsmanship in the highest regard. Frau Benz died in 1944, aged 95.

Gottlieb Daimler

Gottlieb Wilhelm Daimler was born in Schorndorf in Wurttemburg on March 17th 1834, the son of a master baker. He was apprenticed to a gunsmith at an early age; on finishing his time he worked in Stuttgart and then in Grafenstadt with the Werkzeugmaschinenfabrik. He attended the Stuttgart Polytechnic from 1857 to 1859 before going to England where he worked for S J Whitworths in Coventry and Birmingham.

Nicholas August Otto in 1863 established a workshop at Deutz near Cologne to develop gas engines. Privy Councillor Langen supplied the financial backing and, as mentioned previously, Otto invented a four-stroke engine in 1876. In 1872 Daimler was appointed the Technical Director to the firm of Otto and Langen; he soon obtained the service of Wilhelm Maybach, a foreman in a neighbouring works, as Chief Designer. Maybach collaborated with Daimler for the rest of the latter's life and after his death continued to work with Daimler's son Paul.

Daimler remained with Otto and Langen for ten years, and was responsible for considerable improvements in the internal combustion engine. He fell out with the Board, however, resigned, and in 1882 started a workshop in the garden of a large house he bought in Cannstatt (now a suburb of Stuttgart). Maybach joined him and they carried out the research work they had not been able to do when with Otto and Langen.

Daimler considered that the contemporary internal combustion engine ran too slowly and that speeds could not be sufficiently increased using the existing method of ignition. The latter was fairly simple—a slide valve exposed the explosive mixture to a flame at the appropriate instant. In 1879 Funk improved things somewhat and instead of a flame used a hot, hollow tube. Daimler simplified things further by doing away with the slide valve; the explosive mixture was forced into the hot tube and ignited. He had a great deal of trouble finding a suitable metal for the tube and only platinum was found satisfactory.

Daimler was successful in increasing the speed of the engine from the usual 100 to 900 rpm and he obtained a patent for a fast running engine in 1883.

Daimler was convinced that a high-speed internal combustion engine would have an enormous range of uses, but he first demonstrated his engines on a motor bicycle and not on the more obvious choice of a carriage. At the time the bicycle had hardly advanced beyond the bone-shaker and the penny farthing.

Daimler's motorcycle (figure 1.9) had two equal-sized wheels with iron tyres and the rear wheel was driven by a 0.5 hp engine via a belt and pulley; there were two speeds and the drive could be disengaged. The engine was cooled by a fan and had an exhaust silencer. The bicycle was put through its paces at night and on the less frequented roads near Cannstatt, but on November 10th 1885 Daimler's eldest son Paul drove it from Cannstatt to Ventertürkheim and back, a total of 6 km. Apparently the bicycle did not impress its inventor much.

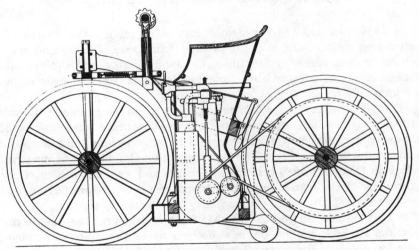

Figure 1.9 Daimler motor cycle of 1885.

In 1886 Daimler put one of his engines on a carriage. He bought an American horse-drawn carriage from Wimpf and Sons, carriage builders of Stuttgart, saying that it was to be a birthday present for his wife. He specified a number of alterations mainly to strengthen the vehicle. The carriage was duly delivered after dark to Daimler's workshop. He had the shafts removed and arranged that the whole front axle swung on a central pivot. A 1.5 hp engine was fitted between the seats and through a system of belts and pulleys drove small pinions which in turn drove internally-toothed sprocket wheels attached to each rear driving wheel. Two different speeds could be engaged by tightening the appropriate belt and pulley. There was no differential, a type of clutch letting one wheel rotate faster than the other.

The car first ran in 1886 and it was the first four-wheeled motor car (figure 1.10). It was successful in demonstrating that the engine was satisfactory and that the horseless carriage was a practical proposition, but Daimler soon realised that a satisfactory car could not be made by adapting a carriage; the car had to be designed as a whole if it was to run at reasonable speeds.

Figure 1.10 Daimler four-wheeled car of 1886.

In 1889 two more experimental cars were made. The first was a belt-driven model with a centre-pivoted front axle. The second was a tubular framed quadricycle with Ackermann steering and sliding-pinion change-gear mechanism. Both were fitted with a V2 engine developing 3.5 hp at 700 rpm.

In 1890 Daimler turned his business into a public company, Daimler Motoren Gesellschaft, one of whose aims was to make road vehicles. He almost immediately quarrelled with his directors, resigning and setting up as a consultant with Maybach. The first production Daimlers were apparently based on Daimler's 1889 belt-driven vehicles and were designed by Georg Schrödter. These rear-engined belt-driven cars were in production from 1893 until 1900; they were simple, fairly quiet and very reliable.

In 1895 Daimler and his faithful Maybach returned to Daimler Motoren Gesellschaft and worked on new designs using the Système Panhard. Daimler, however, had worn himself out with hard work; he was a sick man from 1895 and died on 6th March 1900. Daimler had a weak heart and incidentally did not drive his cars himself.

However, despite these vehicles and some later efforts, Daimler's vehicles had far less impact on the development of the motor car than did his engines. His prime interest was the development of compact high-speed engines and his work on cars and boats was done in the first place to show the potential of the engines, not to make cars and boats.

The 1896 Cannstatt Daimler was redesigned by Maybach and Paul Daimler, which led to the 1901 Mercedes which was a phenomenal success.

Maybach was Technical Director of Daimler from 1895 to 1907 when he was forced out of the company. He joined Count Zeppelin and founded a company for making aircraft engines which became Maybach Motorenbau. He died in 1929.

The automobile stems directly from the efforts of Benz, Daimler and Maybach. They had to create a new technology out of nothing. Benz, for example, had to design and make practically everything himself. In addition to being inventors and engineers Benz and Daimler had to act as their own entrepreneurs and had to be successful managers and business men. They had to be, and were, very remarkable men.

Other Early Inventors

A number of other inventors had built motor cars about the same time as Benz and Daimler, and indeed some earlier. The cars may have been individually more or less successful but none ever went into production, possibly for no other reason than that financial backing could not be obtained. Even if apparently technically successful they would probably have needed much more development to be successful commercially. Consequently they had little influence, if any, on the development of the motor car. Some of them deserve mention however.

Following Cugnot's demonstration of his vehicle the Swiss François Isaac de Rivaz built some successful steam engine carriages. In the early years of the nineteenth century he made an engine in which a mixture of hydrogen and oxygen was admitted to the space below a piston in a cylinder, the mixture was ignited by a spark and as the piston travelled upwards a system of chains and cords turned a pulley which in turn drove the wheels. This was probably the first vehicle to be powered by an internal combustion engine. By 1813 de Rivaz was demonstrating an 18 ft monster with 2 m wheels weighing 950 kg which moved at 3 km/h.

Brown put his engine into a chassis and demonstrated it about 1824. Lenoir put his very heavy and inefficient engine into a vehicle of sorts in 1862 but it was so slow that it would have been much quicker to walk, though in 1863 Lenoir made an 18 km round trip from Paris to Joinville-le-Pont and back in three hours travelling time. The vehicle was bought by the Czar but its subsequent history is not known.

Edouard Delamere-Debouteville fitted a small portable gas engine to a tricycle in 1883 and later installed a two-cylinder gas engine on a four-wheeled brake. At first the engine ran off gas stored in two reservoirs at 140 lb in^{-2}, but later it was converted to operate on light volatile hydrocarbon spirit, a wick-type carburettor being used. Neither vehicle appears to have ever run on a public road.

Hans Johansen of Denmark built the Hammal car in, probably, the late 1880s; it had a two-cylinder horizontal engine of 2720 cc capacity

(104.5×160 mm^2), automatic inlet valve and hot-tube ignition. Transmission was by twin-cone clutch, constant-mesh forward and reverse gears, and a central chain. The car never went into production. Top speed was, or rather is, 6 mph, for the car is not only still in existence but was brought over from Denmark and actually completed the London to Brighton Run on 14th November 1954.

Siegfried Marcus worked on powered vehicles as early as 1864 and in 1870 had mounted an engine, presumably of the Otto type, in a simple wagon and he exhibited a vehicle in an exhibition at Vienna in 1873. In the late 1880s he built a motor car that is also still in existence; (figure 1.11). Indeed it was overhauled and given a run in 1950. Its top speed was 5 mph and the engine developed 0.75 hp. The engine had a capacity of 1570 cc (100×200 mm^2), mechanically-operated slide inlet valve and poppet exhaust valve, low-tension magneto ignition and surface carburettor. The vehicle was simple—it had an oak frame, half-elliptic springs at the front, rubber buffers for the rear wooden wheels with iron tyres. It was steered by a wheel (centre pivot steering), there was no gear box, a cone clutch was used and the rear axle was driven by a belt with a slipping clutch differential on one wheel.

Figure 1.11 The car built by Marcus. By permission of Unwin and Hyman Limited (from A Bird 1967 *Early Motor Cars*).

Although the car was first developed in Germany initiative passed to the French in the early 1890s. They had fine roads and no restrictive legislation to prevent or hamper progress. There were also a number of enterprising and very good engineers in France, men like Levassor, Bouton, the Bollées and men like de Dion and Peugeot to finance and encourage them. The contributions that Levassor, de Dion, Bouton and Peugeot made were so important that they will be described in some detail.

Emile Levassor

The next important name in the history of the motor car after Benz and Daimler is that of Emile Levassor of the firm of Panhard et Levassor et Cie.

EMILE LEVASSOR

Panhard and Levassor were among the leading constructors of woodworking machinery in France.

A solicitor in Paris named Edouard Sazarin had acquired the patent rights of the Daimler engine in France. Sazarin had to have some engines made in France to hold the patents, and as Levassor and Sazarin had once worked together in the same firm Sazarin asked Levassor to make the engines.

Unfortunately Sazarin died, but before his death he impressed upon his wife the importance of the patents. Madame Sazarin visited Daimler in Cannstatt, finding that Daimler was willing that she should act as his representative in France. Incidentally she brought back one of Daimler's engines and had great trouble getting it through the Belgian customs. Levassor continued to act for Madame Sazarin as he had for her late husband, and indeed carried cooperation to the point of marrying her on 17th May 1890.

Levassor was impressed by Daimler's engine, but apparently not at all by his motor car (the 1889 quadricycle vehicle) and set out to design and build his own motor car. His first effort had the engine amidships and after many trials and tribulations he eventually, in the spring of 1891, achieved his first target and drove non-stop from the factory in the Avenue d'Ivry to the Point de Jour and back. The factory was decorated with flags and everybody celebrated.

He experimented with another vehicle with an engine at the rear and then moved it to the front and in 1891/2 the Panhard–Levassor car (figure 1.12)

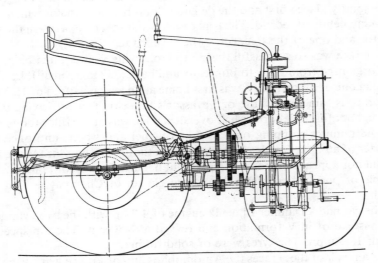

Figure 1.12 The Panhard–Levassor car of 1891.

was launched commercially. It had a Daimler engine at the front connected through a friction clutch and change-speed gearbox to a counter shaft which drove the live rear axle through a central sprocket and chain. The axle carried a differential. This car was the direct ancestor of the modern car.

A number of factors contributed to Levassor's success. The firm was successful and long established, so skilled workmen were available as well as a variety of machine tools and all that goes to make a large works. Money for prolonged development was available. Levassor furthermore, did not have to spend time and effort developing an engine, although he had to do some work on the Daimler engine; he had, for example, vibration troubles to overcome. Indeed, in the Système Panhard it was the Système that was new, and not the component parts, for most of the latter had been used before in one context or another.

On the other hand Benz was the owner of a small struggling factory and started from scratch. Daimler and Maybach had a very small organisation behind them.

One of the major reasons for Panhard–Levassor's commercial success was the publicity gained through a long series of successes in the infant sport of motor racing, although even before the first races the engines and cars were selling well, for by July 1894 Panhard et Levassor had made nearly 350 Daimler engines and about 90 cars. They had thus by 1894 accumulated considerable experience in building and running motor cars.

The first classic race (nominally a trial) was run in 1894 between Paris and Rouen (about 79.4 miles) and was organised by Pierre Giffard for the *Petit Journal*. Fourteen vehicles, including a steam example, out of twenty five satisfactorily completed the course, a Panhard–Levassor and a Peugeot car were jointly placed first and the de Dion Bouton steam tractor which pulled a landau behind it second. The trial developed into a race between the steam tractor and one of the Peugeots—the steam tractor won.

The race was so successful that in the following year a much bigger race was organised from Paris to Bordeaux and back (732 miles in all). Levassor, driving one of his own cars, was first home at an average speed of 15.2 mph. The race is famous because of Levassor's fantastic feat of driving the car throughout (his relief driver had overslept). He was at the tiller for 48 hours and 48 minutes stopping only to take on fuel and water, and once for a quarter of an hour for cleaning the ignition timing mechanism. At the finish he did not appear to be overfatigued, but he 'drank a cup of bouillon, ate a couple of poached eggs and drank two glasses of Champagne with great relish'.

The car had a Daimler Phoenix engine of 4.2 hp with the two cylinders in line instead of in V formation and revved at 800 rpm. The top speed was about 18.5 mph. The tyres were of solid rubber.

As a result of these races the Automobile Club of France was formed and a major road race was held every year. In 1896 the race was from Paris to

Marseilles and was won by the Panhard et Levassor works manager driving an 8 hp Panhard. Levassor also entered the race but lost control of the tiller steering when he collided with a dog near Orange and crashed. He did not appear to be badly hurt at the time but nearly a year later he suddenly collapsed and died from internal injuries received during the accident.

Panhard cars, however, continued to win most of the major races for a number of years. Races and distances are listed in table 1.2. The way in which the average speed increased speaks for itself, but it was not just a question of using more and more powerful engines. Other parts of the vehicles had to be developed simultaneously.

Table 1.2 Some classic road races.

Year	Race	Distance (miles)	Average speed (mph)	Make	Number of entrants
1895	Paris to Bordeaux and back	732	12.2 15.2†	Peugeot Panhard	22 (9 finished)
1896	Paris to Marseilles	1061	15.7	Panhard	32 (16 finished)
1897	Paris to Dieppe	106	25.2	Bollée 3 hp	
1898	Paris to Amsterdam and back	889	26.9	Panhard 8 hp	22
1899	Paris to Bordeaux	351	29.9	Panhard	
1899	Tour de France	1350	30.2	Panhard 16 hp	

† Disqualified as it was only a two seater.

For completeness the results of the other classic road races are given in table 1.3. The Paris to Madrid race caused tremendous public interest. The crowds along the route were so enormous, literally millions, and the cars so numerous and fast (one car was averaging over 65 mph) that there were a number of fatal accidents and the authorities panicked and cancelled the race at Bordeaux. No more major road races were held and future races were held on closed roads.

The speeds say a lot about the skill and courage of the drivers—they had to contend with bad roads (and crude suspensions) and clouds of dust; gear changing was an art and the brakes were very poor.

To return to Levassor, or rather to the firm Panhard et Levassor, the firm produced thirty or so different models up to 1914. They gradually reduced their racing activities and made touring cars with a very sporting performance which were popular with rich young men.

Table 1.3 Some classic road races.

Year	Race	Distance (miles)	Average speed (mph)	Make	Number of entrants
1900	Paris to Toulouse and back	837	40.2	Mors 24 hp	Over 70
1901	Paris to Berlin	687	44.1	Mors 60 hp	110
1902	Paris to Vienna	616	38.9	Renault 16 hp	137
1903	Paris to Madrid		65.3 mph for 342 miles	Mors 70 hp	over 200

de Dion and Bouton

Albert de Dion was born in 1856 and was a member of the old aristocracy of France. Despite his background he was very interested in things mechanical, and one day in 1881 a miniature steam engine in a shop window caught his eye. He found that the engine had been made by Messrs Trépardoux et Bouton and he got them to work for him. A very successful light boiler was produced and it found a number of applications. An experimental steam quadricycle was built and this was followed by heavier steam carriages which could carry enough fuel and water to be practical. Some of the vehicles were articulated—the steam-driven tractor carried a pivot and the wheel plate of a conventional carriage could be mounted on this pivot.

In the late 1880s quadricycles and tricycles were being produced as well as tractors and commercial vehicles but de Dion was aware of the disadvantages of the steam engine and decided to examine the possibilities of the new petrol engine. He got an engineer called Delalande to make a couple of rotary engines for him with the results being sufficiently encouraging for things to continue and from 1892 to 1895 Bouton worked on developing a small engine. Trépardoux disapproved of this, for successful petrol engines would cut into the sale of their steam engines, and he withdrew from the firm which then became de Dion Bouton et Cie. The 1895 engine was aircooled and ran at the very high speed of 1500 rpm; it was quite different from the Daimler and Benz engines and much more efficient than either (particularly the Benz). It was tried out on racing pedal-driven tricycles and worked so well that de Dion Bouton (DDB) made a special tricycle to carry the engine. This tricycle was very successful indeed, by June 1899 no less than 21 800 having been sold. In 1898 a 2.25 hp quadricycle was produced, which carried the passenger's seat in front of the handlebars and between the front wheels, and a small car or 'voiturette' followed in 1899. An improved version of this car was brought out within a few months. This had a 3.5 hp watercooled engine at the rear driving through a 2-speed gear and utilised the de Dion rear axle suspension in which the differential was attached to the body. Passengers and driver sat *vis-à-vis*. The hp of the engine was increased to 4.5 and then to 6 before production ceased in 1903. An 8 hp front-engined car appeared in 1902, which was on Panhard–Levassor lines. From 1904 onwards four-cylinder models were made and in 1910 V8 engines were made for their larger cars.

Though DDB were one of the pioneer manufacturers and did much for the smaller car they had indirectly an even bigger effect on the infant industry because from the beginning DDB sold their engines 'loose' (as separate units) and these engines were used by scores of manufacturers, and the small cheap voiturettes the latter made helped make motoring popular. The engine helped to establish a number of manufacturers on a firm basis. Many firms, once they had built a successful DDB-engined car, were tempted to build their own engines, often at first a copy of the DDB engine, and so branch out on their own. Incidentally in so doing they entered into competition with DDB so that the success of the latter firm contributed to its own decline. However, DDB made over 150 000 engines up to 1911, the overwhelming majority being sold loose.

Peugeot

The Peugeot firm became involved in motor cars at the same time as Levassor. Peugeot Frères was a very old firm located at Valentigny who were manufacturing ironmongery in the 1880s. Armand Peugeot spent some time training in England and when he returned to France brought back the idea of making bicycles and tricycles. Armand got Serpollet to try out one of Serpollet's small steam engines on a three-wheeled vehicle, which managed to cover 300 miles from Paris to Lyons in three days, but was otherwise hardly satisfactory. Armand knew Levassor, and the latter suggested that it would be better to make a quadricycle and power it with a Daimler internal combustion engine, and he visited Valentigny in 1889. The quadricycle was duly designed by Rigoulot and appeared in 1891. It resembled the Panhard et Levassor vehicle in that it used a clutch and sliding gear transmission, the Daimler engine with surface vaporiser, and Ackermann steering, but the engine was at the rear which meant that engine and passenger overloaded the rear wheels and there was not enough load on the steering wheels. The front axle was carried on a transverse spring and kept in place fore and aft by quadrant guides. The driver sat on the back seat and the two front seats faced backwards. Peugeot continued to make the same basic model (figure 1.13) for six years.

Production figures are instructive (table 1.4).

In 1897 Armand broke with the family firm and established S A Automobiles Peugeot et Ardincourt. In 1898 about 500 vehicles were made.

In 1897 Peugeot developed their own engine and placed it horizontally and transversely so that bevel gears would not be needed in the transmission. They changed over to forward engines in 1902.

Peugeot made nearly 50 models between 1901 and 1915 including the 5 hp Bébé in 1902.

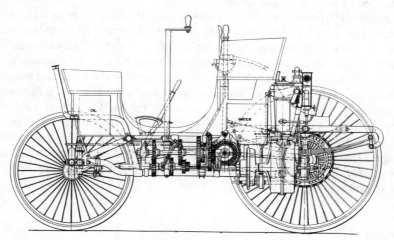

Figure 1.13 Peugeot car of 1896.

Table 1.4 Peugeot production figures 1891–1896.

1891	1892	1893	1894	1895	1896
5 made	18 sold	35	40	72	92 made

Other Continental Manufacturers in the Nineties

By the end of the century there were a considerable number of firms making motor cars. Many manufacturers started by making Benz cars under licence or by copying them. After acquiring sufficient experience they could then branch out on new designs, generally adaptations of the Système Panhard.

Benz made Emile Roger his French concessionaire in Paris and Roget assembled Benz cars probably as early as 1888 possibly making parts himself. Also in France Delahaye, Hurtu, Georges Richard, Rochert-Schneider, Parisienne Rossel and others copied and sometimes improved considerably on the Benz. Lutzmann in Germany built Benz-type vehicles.

In France Amadeé Bollée's son Amadeé fils had a practicable petrol-engined car about 1896. It had a number of original features such as chainless transmission and Amadeé fils followed it by a series of well designed but expensive models until he gave up car production after the First World War. His younger brother Leon designed a car for Darracq in 1897 which had five gears. Leon became much better known for his very successful 2 hp tricycle of 1896. Later he built conventional, but luxurious, cars.

Mors built a vehicle with a V4 engine in 1895; the engine had low-tension ignition from dynamo and battery—the Mors concern was interested in electrical instruments. Louis Renault put a de Dion Bouton engine into his little car and drove the rear wheel by a propeller shaft fitted with universal joints. According to taxation schedules, 1438 private cars were in use in France in 1899 and 2354 in 1900.

Further afield Metallurgique in Belgium produced a Panhard-type vehicle in 1898, and M Vivinus of Brussels produced in 1899 the successful voiturette, the New Orleans, that was to be built under licence in three countries. The firm of Fiat was founded in 1899 and took over Ceirano who made bicycles and had started making voiturettes.

Situation in the UK to 1900

The first petrol-engined vehicle in the UK appears to have been the tricycle of Edward Butler. He exhibited this in 1885 and it was apparently still running in 1895. Butler worked on a four-wheeled vehicle but 'abandoned further development because the authorities would not countenance its use'.

Frederick William Bremner of Walthamstow built a four wheeler based on the Benz, completing it in January 1895, but he apparently had no thought of its commercial development. It is still in existence and successfully completed the Brighton run in the middle 1960s. J H Knight built a three wheeler (figure 1.14) in 1895 and converted it to a four wheeler in 1896; he was fined 2s 6d with costs 'for permitting a locomotive (his tricycle) to be at work'.

The first commercially produced motor car to be driven in this country seems to have been the 3 hp Benz owned by Mr Harry Hewetson who purchased it in Mannheim for £80 in 1894. At first he found the local police friendly but then they got orders from Scotland Yard that he must conform to the law; so he sent a scout ahead on a bicycle and when the scout reported a policeman a small boy got down from the car and marched ahead holding a lead pencil upright with a tiny scrap of red ribbon tied to its end. Hewetson later became the leading agent for Benz cars in this country.

The Hon. Evelyn Ellis in June and Mr J A Koosen in November 1895 imported vehicles into the UK and Koosen was duly booked by the police. People in the UK, however, were becoming increasingly restless when they saw the developments across the Channel. Pressure groups were formed like the Self-Propelled Traffic Association and the Motor Car Club to lobby parliament, company promoters and financiers saw pickings and helped give the motor car publicity, and a powerful group ably led by Sir David Solomon Bart had the Light Locomotive Act passed on 16th November 1895. The speed limit was raised from 4 to 12 mph and a man no longer had to proceed in front of the vehicle to give warning of its approach. The infant journal *Autocar* celebrated by printing its next issue in red, and motorists in general celebrated by taking part on the day in the famous Emancipation run from London to Brighton. This was really organised by the crafty H J Lawson to give Daimler cars free publicity, for he apparently intended that only Daimler cars, in which he had a very great financial interest, should take part.

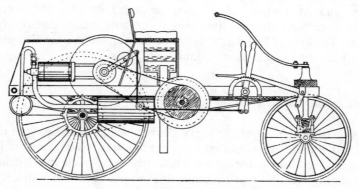

Figure 1.14 Knight three-wheeler of 1895.

Fifty eight vehicles entered and some thirty odd reached Brighton. It was a wet and miserable Saturday but the run was watched by enormous crowds; indeed the vehicles could scarcely thread their way through the packed streets of the larger towns. Lawson gave one of the after-dinner speeches that night at the Hotel Metropole and he made some very accurate predictions about how the motor car would eventually change the face of Brighton.

The first commercially produced cars in the UK were made by the Daimler Motor Co. F R Simms, a young engineer and business man, met Daimler at an International Exhibition at Bremen in 1890, where Daimler showed a passenger trolley car powered by one of his engines. Simms was impressed by the engine and acquired Daimler's patent rights for the UK, and also became a personal friend of Daimler. Back in the UK Simms sold a number of motor boats fitted with Daimler engines, and found other applications for the engines and in 1893 formed the Daimler Motor Syndicate to make motor cars when conditions were more auspicious.

Simms incidentally was the founder of the RAC and the SMMT. H J Lawson, mentioned above, was an engineer of sorts who had made some of the first safety bicycles but he was also a financier and a pretty successful one at that—he probably made half a million just out of floating the Dunlop Pneumatic Tyre Co. He foresaw the future of the motor industry and decided to cash in on it, and as a result of his machinations the first years of the infant motor industry were patchy. Lawson tried to control the industry by buying patents wholesale so that any firm outside his empire would have to pay him royalties, and he even demanded royalties on cars and components imported into the country. He was also responsible for a lot of doubtful practices with his various companies, so that eventually the financial world had no confidence in the motor industry. Fortunately his empire collapsed in the early 1900s.

Lawson bought Daimler's patent rights for the UK from the Syndicate in 1895, floated the Daimler Motor Co. in 1896, and in 1897 English Daimler cars were produced in Coventry. They were in reality a modified Panhard and Levassor design, and though the cars were well made the changes were ill-conceived and the cars (figure 1.15) were heavier than the originals. Their two-cylinder engines developed 4 hp. In 1898 an 8 hp four-cylinder car with wheel instead of tiller steering was introduced, and in the following year a 12 hp four-cylinder car.

Figure 1.15 1898 English Daimler two-cylinder double phaeton ('Siamese' body). By permission of Unwin Hyman Ltd.

A number of much less publicised concerns produced cars in the UK. Some were hybrids and had English bodywork on a continental chassis, indeed on some cars only the nameplate was British! Others were copies of vehicles like the Benz or the de Dion Bouton or the Bollée tricars, and in some cases these cars were an improvement on the originals. For example, Richard Stevens built an improved model of the Benz in 1898, his engine developing 10 hp instead of the 1 hp of the Benz itself. Walter Arnold and Son Ltd of East Peckham also made an improved version of the Benz.

A number of engineers such as Frederick Lanchester, Herbert Austin, George Johnston and Blackwood Murray were working on more original designs but did not get into production until 1900 or later. The various cars that were produced in the 1890s were made in very small batches and the total number of cars produced in the UK probably amounted to a few hundred at most, and there were possibly a couple of thousand cars in the UK at the turn of the century.

Situation in the USA until 1900

Development of the car in the USA lagged nearly a decade behind Europe, possibly because of the poor roads. The first American car appeared in Springfield, Massachusetts in September 1893 and was the creation of two bicycle mechanics, Charles and Frank Duryea. They had read a description of Benz's car in the *Scientific American* in 1889 and set out to build their own.

They built a second car, this time with a two-cylinder engine, and won the Chicago Times Herald race in 1895 (taking 9 hours to travel the 55 miles). The Duryea Motor Wagon Co. built 13 similar cars. These cars had horizontal engines, three-speed belt and jockey wheel transmission, and dynamo-powered low-tension ignition. In 1896 two of these took part in the London to Brighton Emancipation run. Shortly afterwards the brothers parted company and neither played much further part in the industry, the firm being taken over by the Stevens Arm and Tool Company.

Elwood Haynes and the Apperson brothers of Kokoma, Indiana produced a car which ran in July 1894. The three remained in partnership for the next ten years making the Haynes–Apperson car; they then separated and Haynes and Apperson cars were produced until the 1920s. Hiram Maxim put an engine on a tricycle in 1895; as a result he was invited to become chief engineer of a motor carriage department of the Pope Manufacturing Co. of Hartford, Connecticut, the country's largest producers of bicycles. Pope Manufacturing Company became in 1897 the first large scale producer of motor vehicles in the USA. Most of the vehicles, however, were electrically powered and only a small proportion had internal combustion engines (500 electric to 40 internal combustion vehicles in the first two years).

On January 4th 1896 Henry Ford's first vehicle, a quadricycle, made its appearance. Ford had to knock down the wall of his landlord's barn to get the contraption out on to the street. The landlord arrived to protest but was talked into giving a helping push.

Alexander Winton a Scottish engineer and bicycle manufacturer built a satisfactory motor car in 1897 and in 1898 the Winton Motor Carriage Co. was formed and sold its first phaeton in April of that year and indeed sold 22 vehicles in the first eight months of its existence.

Winton car No 12 was sold in 1898 to J W Packard and according to legend something went wrong with it. Packard complained to Winton, they had an argument, and Winton told Packard to go and build a better car himself, and Packard went off and proceeded to do so. Packard's firm, the Ohio Automobile Co., was taken over by a group of financiers and its name changed to Packard Motor Co. with production being shifted to Detroit.

It appears that about six hundred cars were made by thirty or more manufacturers in 1899, and about four thousand, including electric- and steam-powered vehicles, were built in the US in 1900.

Production

For many years a great deal of hand work and skill went into the making of the car. The components of the engine, for example, were machined on relatively simple tools like the centre lathe and the shaper and sent to the fitting bench in a semifinished state. Here the fitters working with hand tools alone—chisels, files, scrapers—spent hours finishing the parts and assembling the engine by fitting one part to another. Each engine was a unique piece of machinery, and its parts could not be interchanged with the same parts of another engine. The time involved was fantastic, for example, on a six-cylinder engine the bottom half of the bearings had to be fitted in their housings, then the crankshaft bedded in by laboriously scraping and checking, next the top half of the bearings had to be fitted and bedded in. After the fitter had eventually finished the engine it was passed to the engine test section for running-in. After running-in it was stripped and had a further scraping and adjustment, then reassembled and eventually installed in a chassis. It was then subjected to further tests on the road after which parts could again be stripped, checked and adjusted. When the chassis had passed all this it went to the coach builder and had an individual body tailored to it. The car was now ready for its final testing and if everything passed muster the car was ready for its owner!

It was no wonder that cars were expensive. But before 1914 many changes were taking place and these were accelerated by the demand for engineering products of all sorts during the First War. Traditionally the marker-off worked off drawings, and scribed lines and centre-punched dots on the work piece for the machinist to follow, but soon jigs, tools and gauges were used and the skilled fitter was moved to the tool room to make the tools with which unskilled workers could turn out accurate parts. At the same time the work was planned to use resources efficiently and shorten production times.

These ideas were not of course new for simpler production methods had been used for years in a number of industries. Isambard Brunel, father of the Mighty Atom, mass-produced blocks for the rigging of the Royal Navy ships at the end of the eighteenth century, had done this by making special machinery and driving it with a steam engine.

Eli Whitney in America was given a contract in 1798 to make 10 000 muskets and he proposed to do this by using special machinery to make all the parts interchangeable so that the muskets could be assembled without any further fitting or matching; he found setting up the machinery took him much longer than he expected! A little later Simon North made pistols on the same principle, and Samuel Colt and others followed. The idea of assembly from interchangeable parts was largely limited to the United States for many years and was in fact known as the American system. The 1851 Exhibition and the Crimean War changed attitudes in Britain to some extent but the infant automobile industry employed craftsmen who used relatively simple tools and built up complicated subassemblies; there was one significant exception—Lanchester used interchangeable parts and after designing his car spent two years, to his backers' disgust, designing specialised equipment to make it. It is not surprising that most people stuck to the traditional methods at first for production runs were not long enough to warrant the cost of special tools.

The other factor involved in volume production, which has a long history, is the assembly line fed by subsidiary lines. In the early fifteenth century the Venetians fitted out their galleys by towing them along a long canal lined with storehouses and as the galley passed the storehouses, cordage, oars, arms, food and even the crew were placed on board and the fully equipped galley passed out off the canal into the sea ready for action.

Olds used an assembly line of sorts and some division of labour in making the Oldsmobile. Leland introduced the interchangeable parts on the Cadillac and in 1908 gave a telling exhibition in this country. Three cars were taken to bits, the components mixed with a number of parts replaced by spares to make things more difficult, then the pieces put together and the three cars driven off and they returned perfect scores in a 500 mile test run. The Cadillac was not intended for the mass market.

Ford had to use interchangeable parts and in 1913 began working on the moving assembly line; the idea of the moving line came from the overhead trolley used by the Chicago meat packers when dressing beef.

Mass production methods were first used in the Old World by Citroen in France and Morris in the UK in the twenties, but for a number of years in many firms the man still brought the bits and pieces to the spot where the vehicle was being assembled. The moving assembly line was further improved when transfer machinery was introduced in the twenties—a component would be drilled or machined automatically and then passed on automatically to another station for further operations to be carried out. Progress has continued right up to the present day when robots carry out many of the tasks and microprocessors and computers control the production lines.

Chapter 2

Developments 1900–1917

Technical Developments

Emile Jellinek, the wealthy Austro-Hungarian Consul at Nice, was interested in automobiles and, in particular, in the 1899 Daimler 23 hp Phoenix. He criticised it and got Maybach at Cannstatt to build a car with various improvements he had suggested, and this was delivered to him in 1900. He was very impressed by his new car and undertook to sell a batch, but for complicated commercial and legal reasons the car was given the name of his daughter Mercedes. He had previously raced under the pseudonym Mercedes—he was a big burly man.

The new Mercedes set a new fashion in the design of the larger motor cars. It incorporated a number of improved mechanical details, such as a four-cylinder engine, mechanically operated inlet valves, and a carburettor with a butterfly valve connected to the accelerator pedal and, like the Daimler, such novelties as a honeycomb radiator, and a gated gear change. It also had a distinctive appearance—it was not a horseless carriage (figure 2.1).

The 1901 Mercedes was followed by a number of other Mercedes cars of similar design but different ratings. The cars had a successful racing record, and, until almost the end of the decade, racing had a tremendous effect on the development of the ordinary touring car. New ideas were tried out on racing cars and if they were successful were soon used on ordinary cars. However, racing cars did go out on a path of their own when to obtain more power the engines were made larger and larger. The 1901 Mercedes had a 35 hp engine of 5.9 litres which gave the car a top speed of about 55 mph. The 60 hp car of a couple of years later had a 9.2 litre engine and a top speed of about 75 mph, and the 90 hp Mercedes had an engine of nearly 12 litres capacity. The 1902 Panhard-Levassor 70 had a 14 litre engine, but one of the biggest engines of all was put into a 1906 Metallurgique rebuilt in 1910 with a 21 litre engine.

More powerful engines were wanted not only to improve the performance of the car but also, on touring cars, to minimise the amount of gear-changing needed. Changing gear was a difficult and skilled operation, so there was a demand for powerful, flexible engines that would develop a good torque over a wide range of speeds so that once he had got going a driver could stay in top gear for all but the steepest hills. The RR Silver Ghost showed its paces in 1907 by going from London to Edinburgh and back in top gear.

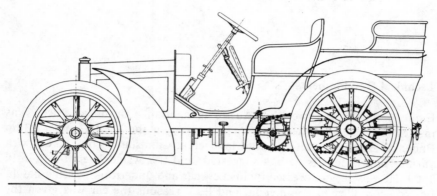

Figure 2.1 1901 Mercedes touring car.

Although the easiest way to increase engine power on private cars was to increase the capacity of the engine, this increased its weight, and the weight of the chassis had to be increased correspondingly, so before long, engine capacity levelled out and increased power was obtained by increasing the engine revs and improving breathing. Engine speeds were greatly increased by reducing the weight of the reciprocating parts and by improving lubrication. The crude splash lubrication systems used in the first cars were replaced by pressure systems, the oil being pumped to bearings and this too permitted higher engine revs. It also made the exhaust cleaner as excess oil was not thrown on the cylinder walls to be burnt away.

So far as breathing was concerned the atmospheric inlet valve did not open and close at the optimum timings and very rapidly gave way to the mechanically operated valve.

At first the inlet valve was on one side of the cylinder and the exhaust valve on the other. This arrangement was displaced by side valve and overhead valve configurations which improved the shape of the combustion space, and required less operating gear, and by 1914 or so side valves were usual. Attempts were made to replace the poppet valve by rotary and other valves, but only the sleeve valve had any degree of success; its main advantage was that it was quieter than contemporary poppet valves.

Hot-tube ignition became progressively less efficient as engine revs increased and in any case it was inherently dangerous and was replaced by electric ignition. The spark could be advanced to ignite the charge at the optimum instant. The Simms–Bosch magneto was the most popular form of ignition—movement of an armature in a magnetic field produced a low-voltage current which induced a high-tension voltage in a secondary coil. The magneto largely displaced the accumulator system as used by Benz, though some cars fitted both systems to obtain easy starting.

Carburation was also improved to make the mixture strength better suit requirements which changed with speed and load. As a result of all these improvements four-cylinder engines of 1910 could deliver three times as much power as 1899 four-cylinder engines of the same capacity.

Another major problem was noise and vibration. Early engines had only one or two cylinders and were governed on the 'hit and miss' system: when less power was needed the exhaust valve was kept closed preventing the engine from working for the next cycle. This made the engine vibrate, and vibrate particularly badly when idling. Controlling the engine output by controlling the airflow and mixture through the carburettor overcame this. Again lightening the reciprocating parts helped reduce vibration but, in addition, torque variations could be reduced and out-of-balance forces minimised by increasing the number of cylinders. Increasing the number of cylinders also increased the power/weight ratio of the engine. The dynamics and advantages of multicylinder engines were well understood; the difficulty was making the engines, and ensuring that all cylinders were contributing equally. Four-cylinder engines had generally been confined to racing cars before 1900, but within a couple of years the problems had been sorted out and bigger cars had four-cylinder engines. Napier made a practical six-cylinder engine in 1904 and sixes became popular because of their smoothness and flexibility. Straight eights were made but were not successful until much later, in the late twenties. de Dion Bouton produced a V8 engine in 1910 and this was the basis of the successful Cadillac 8 of 1914, the first of the hundreds of millions of American V8s.

Other parts of the car were also improved. The gear wheels in the gearbox could either be slid in and out of mesh, or be kept in constant mesh, and the drive taken up by dog clutches. By the end of the first decade it was usual to have direct drive through a dog clutch in top gear, with a consequent reduction of noise in top. Gearboxes were enclosed, shafts made shorter, and bearings bigger; these modifications and improved mountings, together with improved tooth geometry, gave an all-round reduction in noise. Besides the now conventional gearbox, epicyclic boxes were used and a number of ingenious arrangements tried.

At the beginning of the century many cars had chain transmission on to dead axles, but lighter cars followed the lead of Renault, driving through a propeller shaft, with universal joints at one or both ends, and bevel or worm gears to 'live' or revolving axles. Chains picked up muck and were messy to clean, they were noisy and they could break. This resulted in chain transmission largely dying out through the decade, though it was used on heavy racing cars until 1914 or so, presumably because of doubts about what loads the live axle could withstand. Alloy steels improved matters. Brakes acted on the wheels or on the transmission, and the internal expanding brake made headway against band brakes. Towards the end of the first decade premature attempts were made to introduce front-wheel brakes. Wooden, spoked artillery wheels were used on all except a handful of cars, for wire wheels had a reputation for collapsing when cornering at speed. Pneumatic tyres completely displaced solid tyres but punctures took a lot of fun out of motoring though detachable rims and detachable wheels saved the situation to some extent. Tyres, however, became stronger and it was realised that suitable treads could reduce the risk of skidding. Worm and sector steering gear was generally used and the geometry was Ackermann.

Progress in the UK

Long ago a small horse was sometimes called a hobby horse and the name stuck to the horse head on a stick used by Morris dancers. Later it was applied to a toy which consisted of a stick with a horse's head at one end and wheels at the other, the child straddled it and pretended he was riding a horse. About 1810 a contraption called a hobby horse or 'dandy horse' or, on the Continent a 'velocipede' was in fashion briefly, this was a sort of bicycle without pedals and the adult driver propelled the thing with his feet. It was more or less killed by ridicule. A number of people tried to fit the dandy horse with a system of pedals, and the simple method of attaching pedals to the hub of the front wheel was used successfully by Pierre Lallamant of Paris in 1864. Machines were made on this principle by Michaux in France and in about 1868 Rowley Turner of the Coventry Sewing Machine Company introduced the Michaux Velocipede in Coventry; it was promptly called the Boneshaker.

Turner was the nephew of the General Manager of the Coventry Sewing Machine Company and while studying in Paris he had become enthusiastic about the Michaux velocipede to such an extent that he brought back an order to his uncle for four hundred velocipedes for the French market. Unfortunately the Franco–Prussian War broke out before the first consignment was ready; indeed Turner was in France at the time and he missed the last train out of Paris before the siege began so he had to escape on his trusty velocipede! The order was cancelled so the firm, now the Coventry Machinists Company Ltd, had no choice but to develop the home market.

Coventry had been noted for silk ribbon weaving, its factories having attracted skilled weavers from France, Flanders and Italy. But in 1860 the Government removed the ban on the importation of foreign silks. Coventry could not compete on price with the resulting flood of imports and the industry rapidly collapsed. Similarly competition from Switzerland caused its watchmaking industry to close. Consequently, manufacturers were looking out for other outlets and some seized upon the infant cycle industry.

The Boneshaker was slow; one way to speed it up was to make the driving wheel very large so that one revolution of the pedal took the rider a much longer distance and so the pennyfarthing or 'ordinary' cycle came about. It was fast and just the thing for racing, but could be very dangerous—if the rider took a header over the front wheel he could break his neck. A number of people including J H Lawson went on to make bicycles with wheels of more or less equal size with one of these driven by a chain and sprocket wheels, one sprocket wheel being bigger than the other, but the first commercially successful safety bicycle was the Rover of J K Starley which appeared in 1885. By the end of the decade the bicycle had acquired the design which it has retained to the present day though with, of course, improvements in detail. The bicycle was popular but became even more popular after Dunlop introduced the pneumatic tyre. Indeed in the nineties there was a bicycle boom and enormous numbers of people rode bicycles. The bicycle caused a minor social revolution.

In the early years of the century the bicycle market was becoming saturated so that manufacturers looked for ways of diversifying, and the motor car trade was an obvious thing to move into. Some naturally tried making motor bicycles and then became more ambitious and made four-wheeled vehicles. Others built Continental cars under licence or brazenly copied them. Proprietary engines such as the de Dion Bouton and various components could be bought with firms constructing cars largely from such bits and pieces. When they had obtained sufficient experience they designed and built engines and chassis of their own. Firms like Humber, Hillman, Riley, Rover, Singer, Sunbeam and Swift were at first bicycle manufacturers. Allday and Onions, Crossley, James and Brown, Napier, Rolls-Royce, Vauxhall & Wolseley were engineering firms who diversified into cars and only a minority of firms such as Albion, Arrol-Johnston, Austin, Iris, Lanchester and Standard were formed to build cars which were of their own design.

It was not a big job for an engineering firm with some resources to build cars in small numbers. The capital investment was small, for no specialised equipment was needed and most of the cash went in wages and materials. Consequently, by 1900 59 firms had been founded and by 1905 279, though of these 80 had failed by 1905. A number of firms used foreign engines and the only British thing about some of the cars was their names. Edge was rather indignant about this and in 1904 listed the following as bona fide British: Albion, Arrol-Johnston, Ariel, Albany, Achilles, Alldays, Bellsize, Brook, Dent, Duryea (*sic*), Humber, James & Brown, Lanchester, Langdon Davis, Lea & Francis, Ludgate, Marston, Maudsley, MMC, Napier, Pick, Rex, Ryde, Ryknield, Star, Siddeley, Standard, Swift, Thorneycroft, Vauxhall, Wilson & Pilcher and Wolseley. In 1907 11 700 cars were made in the UK of which 2441 were exported. There were 6530 imports.

It was one thing to make cars, it was another to get the public to accept them. Though the man in the street was most interested in the car, he could not afford one. Rich people in general preferred their horse and carriage, and though on the Continent wealthy sportsmen bought cars, the English sportsman preferred his horses. Indeed for some time the owner of a car was generally an enthusiast.

Cars were unpopular on several accounts. All the trades associated with the horse naturally disapproved of them, the army thought that they might reduce the supply of remounts, and of course, the poor horse itself objected to cars until it got used to them.

A lot of people did not take the car seriously. Many others positively disliked it; it was noisy and smelly, 'it barked like a dog and stank like a cat', it raised the dust and damaged the roads. It seemed to be a complicated thing to drive and did not appear to be very reliable, though this was partly because many owners and even chauffeurs did not have much idea of how their vehicles worked and could not correct trivial faults.

The most successful attempt to show that the car was a practical and reliable means of transport was the Thousand Mile Trial organised by the Automobile Club in 1900. The ideas of the Club were ambitious to the point of being foolhardy for the route was: Hyde Park Corner—Bristol—Birmingham—Manchester—Kendal—Carlisle—Edinburgh—Newcastle-on-Tyne—Leeds—Sheffield—Nottingham—London, stopping overnight in each city so that 1107 miles had to be covered in eleven days. This was when a 20 mile trip took some organising and if completed without incident was considered quite a feat. However, 65 vehicles started and 46 finished. Great numbers of people saw the cars *en route* or when they were exhibited in the cities, and followed their progress in the papers. It was realised that the car had to be taken seriously.

Cars, however, were expensive to buy and expensive to run; the tyres were the chief culprits here. Even so motor cars were cheaper than the corresponding horses and carriages, which were very expensive. A two-wheeled vehicle plus cob or pony cost about 10d per mile whereas a two-seater car cost less than 6d per mile, and could do 100 miles a day instead of the 30 miles a horse could manage. Professional men, particularly doctors, soon realised that the car was cheaper, and often the first owner of a car in a village was the local doctor. He generally chose a relatively small two-cylinder vehicle. The larger, faster cars were bought by sporting young men, who had to be fairly rich. Indeed the British industry owed quite a lot to well-to-do owners who were prepared to pay for good design and high quality finish; they in fact paid for the development of the vehicle. As a result British vehicles were much better, for example, than the majority of contemporary, often very crude, American cars but the latter were effective, did the job they were meant to do, and were much cheaper.

As mentioned above racing also kept the motor car in the news. There were long distance events like the Peking–Paris race, though this was more of a rally than a race, but most of the leading countries banned straightforward racing on public roads, so that the cars had to perform on closed circuits, though these could be lengths of ordinary roads reserved for the occasion. Many of these races attracted enormous interest—special trains brought tens of thousands of spectators to Le Mans for the first French Grand Prix of 1906.

Until recently cordonning off of public roads for racing has never been permitted in the UK except in the Isle of Man, Jersey and Ulster, and only a few classic races were ever run in these places. Brooklands was built by Locke-King in 1906/7 as a test track for British motor manufacturers; S F Edge marked the opening by driving a Napier single handed round the track for 24 hours at an average speed of 65.9 mph despite stops for 24 tyre changes. He then drove back to his hotel! The Brooklands Automobile Racing Club was soon formed with the first races being held in 1907, and the Club went on from strength to strength. Hill climbs and sprints were also very popular activities but, like racing at Brooklands, were generally confined to a relatively small number of enthusiasts.

The legislator was busy in this decade as well as the engineer. An act of 1903 set a general speed limit of 20 mph; in addition, local authorities could reduce the limit in towns and villages, and where high speeds would be dangerous. Cars also had to be licensed and carry a licence plate so that they could be identified, and from 1904 onwards motorists were more or less persecuted by the police. Police set up traps and the local magistrate would generally accept the policeman's often crude estimate of the driver's speed rather than the driver's. Things reached such a state that the *Autocar* published weekly maps showing where the police were particularly active, and AA scouts would warn motorists of nearby traps by not saluting as a direct verbal warning would have been breaking the law. In contemporary journals and magazines one of course reads the motorist's side only, the other side did have a case for cars were a nuisance, they sent up the dust in summer, or made muddy ruts in the road in winter, they were noisy and the rules of the road were loose. Most drivers liked to use the crown of the road which was less rutted than the sides, and pulled over to the nearside to allow approaching or overtaking vehicles to pass. If the driver did not do so he 'was guilty of an impoliteness that was difficult to excuse'. Also drivers no doubt often did exceed the speed limit; indeed if you were on a country road and your car could do 40 mph it would be surprising if you did not on a straight and clear road.

The railways had taken most of the traffic off the country roads, and trams and trains had also helped in the suburbs of the big cities. The country roads must have been quiet and pleasant places (though cyclists complained about their surfaces) but they were hardly suitable for the motor car. The roads were the responsibility of the local authorities. In 1909 the motorist accepted increased taxation and a tax on petrol provided the money was spent on improving the roads, and so the Road Fund was born.

A Road Board was set up and came in for a lot of criticism but the dust nuisance was gradually overcome, the macadamed main roads being sealed with asphalt. Dust coats, goggles and protective clothing were then no longer necessary for the motorist and his passengers. The general use of windscreens from about 1910 also helped things. My mother (RTS), however, wore a long linen duster coat, big hat, and sunglasses on Australian roads even in the late twenties.

Driving was still an art. Gear changing was not easy, braking was poor because only the rear wheels were braked, with too much braking on these causing a side slip and the driver had to make adjustments, like advancing or retarding the spark, which were later to be done automatically. As brakes were poor the driver relied on his steering wheel to get him out of trouble, but as the gearing was very high and tyre pressures high the steering was very responsive. He did not, however, normally have the traffic to contend with that his great-grandchildren have today, at least not outside the big cities. It is difficult to estimate the number of cars in this country in the early 1900s but in 1904, when the first national registrations were collected, 8465 cars were in use, or at least that was the number registered. Registration was 15 895 in the following year and 53 196 in 1910.

The most popular type of car was the tourer, still large, heavy and well finished. The closed car like the landaulette or limousine was even heavier and considerably more expensive.

There was still a considerable market for voiturettes, now called 'coupés', or scaled-down cars which usually had one- or two-cylinder engines. A newcomer to the scene was the cycle car which was cheaper than the cheapest of the voiturettes. They were generally powered by a proprietary motorcycle engine and assembled from parts bought out and could be very crude. This type of vehicle is exemplified by Robert Bourbeau's Bedelia of 1911. Cycle cars could cost as little as £60 whereas the 1910 one-cylinder Rover complete with hood and screen cost £155, and the 1912 Morris Oxford cost £175. Though most of the cycle cars were primitive contraptions some like the Morgan and the GKN were quite successful. The cycle car appealed to the sporting character who could not afford a proper motor car.

Before the First World War the car did not have much social impact largely because only the well-to-do could afford to buy and run a car of any size, and in many cases their car simply replaced a horse and carriage.

A year or so before the war a 15 hp chassis cost £300–£450, a first class four-cylinder chassis £450, a six-cylinder £650. The cost of the body ranged from £80 for a small open car to £3–400 for a large luxurious closed carriage. Tax started at £2 2s 0d per annum and insurance on a 15 hp car valued at £500 was about £12 12s 0d. A chauffeur was paid £2 per week and some wily owners gave their chauffeur a bonus which depended upon the mileage covered without a breakdown. Tyres could cost £50 and petrol £10 per annum for a 15 hp car, so for an annual mileage of 10 000 miles the cost per mile was a little over twopence, which was less than the cost of a first-class rail ticket, but first-class travel was for the well-to-do.

There were about 132 000 cars in use in 1914. In 1983 about 126 000 Vauxhall cars were sold so there were about as many cars of all sorts on the road in 1914 as there were B-registered Vauxhall cars in 1984. For every car on the road in 1914 there were a hundred in 1973.

Private motoring continued through the War, the number of cars on the road reaching a peak of 141 000 in 1916 and falling to 78 000 in 1918. Production of cars continued until 1916 when manufacturers went over to making munitions, aircraft engines and tanks, etc.

Reginald McKenna was made Chancellor of the Exchequer in 1915. He raised incredible amounts of money in various ways, and one thing he did was to impose an *ad valorem* duty of $33\frac{1}{3}$ % on imported luxury goods including clocks and motor cars. The intention was to raise money, not to protect the British motor car industry, but it had that effect. The duty was repealed briefly in 1923, but there was such a howl from the industry that it was very quickly restored.

During the First World War the need of the Army was for motor transport and not cars, though car chassis were used for armoured cars and the like. The emphasis was on making vehicles in large numbers and making them reliable, not on technical innovation. A large proportion of the vehicles required was imported from the USA.

The petrol engine, however, was enormously improved because of the demand for high-performance aircraft engines. Rotary and radial engines branched out in completely new directions, but in-line engines were developed from automobile engines. Some idea of the development in aircraft engines that took place is shown in table 2.1. The figures represent averages in service each year.

Table 2.1 Aircraft engine development in the Great War.

Year	Power (hp)	Weight (lb)	(lb/hp)
1914	112	437	39
1915	133	512	37
1916	185	570	31
1917	234	605	28
1918	267	693	26

The improvements were brought about by the extensive use of high tensile steels and aluminium alloys which decreased the weight per hp; and attention to detail design, including improved bearings and balance, which permitted, with the better materials, higher revs. Overhead valves, better combustion spaces, and better breathing together with the higher revs gave the higher outputs.

Progress in the USA

As in the UK many of the early US manufacturers were originally makers of bicycles, or connected with the bicycle trade. Pope, Pierce, Jeffrey, for example, made bicycles, the Duryeas repaired them and Willys was a bicycle salesman. Studebaker, Overland and Moon made horse-drawn vehicles and a number of other firms were involved in various branches of engineering, for example, Franklin and Marmon. The Stanley twins of the Stanley Steamer had a rather more exotic background; at one time they made violins.

Consequently making cars was an adjunct to their main concerns for many of the pioneering manufacturers, but they had the capital, plant facilities and background needed to produce cars in large numbers. Entry into the industry, however, was relatively easy for by subcontracting out for engines and components, and assembling the bits and pieces, the capital and facilities required could be kept low. There was little trouble selling cars as the market was expanding rapidly.

The most successful car in the first years of the century was the Oldsmobile buggy which had a curved dashboard. Ransom Eli Olds took over his father's machine shop in 1891, and made industrial (gas) engines. He built a three-wheeled steam-engined vehicle in 1887 and his first petrol engine in the nineties. He made several prototype cars and found a backer, S L Smith. A factory was set up in Detroit but before production got going the factory was burnt down and only a little single-cylinder buggy with a curved dash was saved. The company concentrated on this one model and Leland and Faulconer supplied the engines, John and Horace Dodge the transmissions. The little car was very popular, and 600 cars were produced in 1901, 2500 in 1902, 4000 in 1903 and 5000 in 1904 making it the first car to be produced in large quantities. Smith wanted to make larger cars and Olds pulled out and set up the Reo Motor Car Company: Reo had some success but disappeared during the Depression. The Oldsmobile Company prospered and was joined with Buick in 1908 to form the nucleus of the General Motors Group.

The Olds Company must have been a good training ground for Maxwell (Maxwell-Briscoe), Hupp (Hupmobile), Chapin and Coffin (Hudson) all worked with the company at one time or another.

Henry Ford continued working on cars but was dropped by his backers because they thought he spent too much time developing racing cars and not enough on making cars that would sell. Henry M Leland of Leland and

Faulconer took over, got production going and changed the name of the company from the Henry Ford Company to Cadillac. Antoine de la Molte Cadillac was the French explorer who founded Detroit in 1701. The first Cadillacs were single-cylinder 6.5 hp vehicles using engines, incidentally, which had previously been offered to Oldsmobile. Leland was a perfectionist and the Cadillac was distinguished by the high quality of its construction and finish.

Ford found himself another group of partners and at first made gas buggies. Four-cylinder vertical engined cars followed in 1904–5 and, starting from the Model A, design after design, eight in all, followed until the Model T which appeared in October 1908. At first practically all the components were bought out and Ford assembled them.

Ford realised that there was a tremendous market for a cheap car but instead of designing a cheap car as such, he designed a car suitable for the market—something simple, reliable and easy to drive, which was made of high-grade materials. He operated from a sound financial basis, employed good salesmen and had a flair for publicity. As he was doing very well when the first Model T was introduced in 1908 the car could be sold at a very competitive price. It did so well that in 1909 Ford decided to make only the Model T.

Ford had very soon found that he could not get enough skilled mechanics to use traditional methods and so had to have special machinery designed so that unskilled labour could turn out interchangeable parts; once production began to soar costs came down and then Ford was able to cut prices and to erect a new and even more up-to-date factory. For some years the chassis was stationary and parts were brought to it; in 1913 the moving assembly line was introduced and once it was sorted out in 1914 costs again came down.

Many of the American light cars in the early 1900s followed the same general plan. The Americans travelled by rail between the big towns so that outside the town boundaries the majority of roads were mere tracks. To cope with these roads light four-wheeled carriages were developed. The frames, etc, were very light and so designed that destructive stresses were not set up as would be the case with more solid vehicles. The American light car or runabout was based on these light carriages. Examples were the Oldsmobile, Cadillac, Elmore, Ford, Rambler, Pope-Hartford, etc.

The runabouts generally had a single-cylinder engine amidships, some form of epicyclic planetary gear, flywheel and chain drive. The wheel base and wheel gauge were considerable and the ground clearance very great indeed, whereas the body was high and narrow so that the vehicle had a spidery, rather frail, appearance. Finish was generally poor but the runabouts were cheap. The very stark Orient Buckboard in 1904 cost $425 and weighed 500 lbs; the Pope Hartford runabout $1050, weight 1400 lb; the Oldsmobile $675, weight 1100 lbs; the Cadillac B $800, weight

1300 lb. On the other hand, the four-cylinder Packard with tonneau body cost $3000 and weighed 1900 lbs, the Winton Quad cost the same but weighed considerably more (2300 lb) and imported cars cost the earth; the 24 hp Panhard Levassor $7950 and the 28/32 hp Mercedes no less than $12 450.

Life in the States was complicated somewhat by the Selden patent. Selden applied for a patent for a liquid-hydrocarbon-engined vehicle in 1879 and kept the patent pending for sixteen years by judicially amending it. He could not obtain financial support to market his vehicle, and apparently was long-sighted enough to realise he could forestall everyone if he hung on to the master patent. The patent was eventually granted in 1895 and he sold it to a syndicate who tried unsuccessfully to make, rather oddly, electric cabs, but who did sue the Winton Motor Carriage Co. and other firms for patent infringement. Things were settled out of court and an association, the Association of Licensed Automobile Manufacturers, set up and member firms could use the patent for a royalty of 1.25 % of the list price of the car with Selden himself receiving a fifth of the royalties. Most manufacturers were quite happy with the arrangement, for doubtful firms could be kept out of the industry and indeed the members could establish a cosy little closed shop. Ford, however, would have nothing to do with ALAM and after a number of court cases finally demolished it, though in fact the patent had little time left to run when Ford won the last case.

Many other famous firms were formed in the first decade of the century.

Studebakers were a well-established wagon and carriage maker, who in the nineties were the largest manufacturers of horse-drawn vehicles in the world. They experimented first with electrically powered vehicles and built such vehicles from 1902 to 1912. In 1904 they assembled petrol engine cars buying engines and chassis out and putting on their own coachwork. The Studebaker Corporation itself was formed in 1911 when a number of small car manufacturers were taken over and the one organisation set up.

David Dunbar Buick was, not suprisingly, born in Scotland. He had a successful marine engine company in Detroit and in 1902 started to work on cars. Two of his engineers Eugene Richard and Walter Marr invented the first overhead valve (OHV) engine and half a dozen two-cylinder 22 hp cars were built in 1903. 37 model B cars were built in 1904 and 750 in 1905 when William Durant bought the firm. Buick left in 1906 and died in poverty.

William C Durant was a leading carriage and wagon producer in the United States at the turn of the century. He bought the shaky Buick Motor Company in 1905 and by 1908 Buick were the leading producers in the country; 8487 Buicks being made in 1908 compared with 6181 Fords. Then in 1908 he formed the General Motors Company (GM) bringing into it Buick and Olds, and in 1909 Oakland and Cadillac. The companies, however, continued to operate independently within the holding company. Though Buick and Cadillac did very well the Company got into financial difficulties within a few years because of Durant's acquisitiveness—he collected companies like other people collect postage stamps—but was saved by a syndicate of bankers. Durant had to resign and Charles W Nash and Walter P Chrysler moved into top positions. Nash was first a farm labourer then got a job as trimmer in a carriage factory. He eventually became General Manager of the firm, then moved to Buick and in the reorganisation became President of GM. Chrysler was a Kansas farm boy who started as a mechanic in railway workshops and eventually rose to be Superintendent of the American Locomotive Works in Pittsburg. He became very interested in cars, joined Buick and when Nash became President of GM Chrysler took his job at Buick. As a result of the extra money and the reorganisation GM recovered and prospered.

Durant left GM to work with a Swiss mechanic, Louis Chevrolet, introducing the very successful Chevrolet car onto the market. By some shrewd manoeuvering Durant got back control of GM in 1916, bringing the Chevrolet Corporation into the fold with him. Nash left GM on Durant's return and bought the Thomas B Jeffrey Co., who made the Rambler, to form the Nash Motor Car Co.

Even in the period 1910–19 country roads in the USA were dreadful, so cars had to be well sprung with very strong axles and high ground clearance. The engines were rather primitive by Continental standards and American design lagged some years behind French practice. The Old World was very critical of the lack of refinement of the average American car, finish was often rough unless absolutely necessary and there was little attempt to clean up the design. These apparent faults were to some extent advantageous; the heavy, slow and inefficient American engine did not stress parts unduly and the lack of finish of course reduced costs. The cheaper cars were cheap because the manufacturer had to use automatic and semi-automatic plant that could be operated by unskilled labourers, but the cost of the plant was more than covered, provided production was on a big enough scale. Many smaller manufacturers also bought out components including engines and assembled, rather than made, the vehicles.

The big home market also allowed the American industry to export on a large scale. Exports were also helped by the roads in most parts of the world which were as bad or worse than in the USA, and in these conditions the American cars performed better than the Old World cars. The average UK engineer was rather shocked when he saw the crudity of the first imported Model T Fords. More far-sighted people realised that the cheap American cars could be as good as the cheap UK cars and probably eventually cost less to run, but the man who could afford a well designed and well finished car continued to buy UK or Continental cars.

The American industry produced quality cars as well as cheaper cars. Large expensive six-cylinder vehicles such as the Pierce Arrow, Peerless, Locomobile were built regardless of expense. Others like the four-cylinder

Packard were first-class vehicles in which quality was more important than price, and on many more such as the Cadillac, Chalmers, Hudson, etc, price was not the only factor that determined the design.

Although the cars were not remarkable the rate of growth of the industry and the scale of production was quite fantastic. By 1901 about 7000 cars had been built in the USA, by 1908 the annual production had increased to 65 000, then reached a quarter of a million a year by about 1911, half a million in 1913, a million in 1915, and a million and a half in 1917. Production was then cut back a little because of American entry into the First World War. The Model T accounted for a larger and larger share of the production; 5986 vehicles were made in 1908, 40 402 in 1911, 168 220 in 1913, 308 213 in 1915 and 785 432 in 1917. The growth rate was not far from following the classic exponential growth curve.

To put these figures in a British perspective, about 11 700 cars were manufactured in the UK in 1908 and by 1913 the figure was about 34 000 (including trucks and lorries), Wolseley was the biggest UK manufacturer and made about 2500 vehicles in 1913. Morris, Austin, Singer and Rover all turned out about 1000 vehicles per year. But Ford, who started production in the UK at Old Trafford in Manchester in 1911, were making 6000 cars a year in this country by 1913. In the same year Ford production in the USA was of the order of 600 or so per day.

The American manufacturers had a seller's market and there was little incentive for research and development. For example, if the designer in the UK wanted more power he would develop and refine an existing engine. An American designer would find it much simpler and cheaper to put another couple of cylinders on to his engine making a four into a six, or a six into an eight. It also impressed the customer more.

Successful electric starters appeared on the 1912 Cadillac and in the following year Bendix introduced the starter drive, and within a year or so it was practically impossible to sell a car without a self-starter. This was a milestone in the development of the car for now anyone, including women, could start a vehicle, not only muscular and skilled men. Sales were said to have increased by 60 % in a single year because of the self-starter.

The proportion of models with six-cylinder engines increased and the six soon became the most popular as it was smoother and more flexible than the four as well as approximately 50 % more powerful. In 1914 Cadillac introduced the first successful eight-cylinder engine. The demand for cars was very great so that anything that ran after a fashion would sell and so the emphasis was on production and not on design. Indeed one contemporary writer described the American car as a peg on which to hang a highly developed system of producing dollars plentifully and quickly, but he admitted that the cars were cheap and, as such, very good value for money. Business got better and better, and production increased reaching a record in 1916 and prices came down until the USA entered the War in 1917. The seller's market still persisted but basic design was stabilised, though much attention was given to the appearance of the body. Prices also went up. At this time half the cars bought in the USA were bought for utilitarian purposes, primarily as a means of transport. The majority of the cars bought in the UK before the War were used for recreation and sport so that the owner was much more particular about detail and finish; he ordered what he wanted and then waited for his car to be made. The American owner bought a car off the shelf. Many rich Americans wanted expensive well made vehicles and these were made, but numerically they were only a drop in the ocean. Few, however, could have been more expensive than the Studebaker exhibited at the New York Show in 1916 which was displayed on a dais covered with purple velvet, and which was said to be valued at £5000—this was quite likely, for its chassis was gold plated!

Chapter 3

Developments Between the Wars

Technical Developments

Ricardo investigated engine knock in the early twenties and showed its dependence on the chemical nature of the fuel and on combustion chamber design. The work was financed by Shell; Ricardo did the engineering using a variable-compression single-cylinder engine and Tizard the chemistry. Tizard later became famous for his organisation of the radar defences of the UK in the years before the War. Parallel work was done in the USA and T Midgley Jnr, after trying a great number of additives, discovered the anti-knock properties of tetraethyl lead. The octane rating was developed (Ricardo originally used cetane as his standard) and all this greatly helped refineries for they could process crude fuels to give fuels with higher and more consistent anti-knock properties and these could be improved further with tetraethyl lead.

The improvements in fuel and in combustion chamber design resulting from this work permitted higher compression ratios to be used. Aluminium pistons and improvements in lubrication and materials allowed higher piston speeds. Higher compression ratios and higher speeds meant greater power. Overhead valves were used on more expensive cars but side valves were used on most cars. Electrical starting and lighting had become standard in the USA by 1915 or so and appeared on post-war cars in the UK. Magneto ignition was almost universal in the UK in the early twenties but was gradually replaced by coil ignition during the course of the decade.

A major advance during the twenties was the general adoption of four-wheel braking systems during 1924 and 1925. Maximum deceleration was almost doubled and, equally important, skidding and side-slips were much reduced so that driving, and particularly driving in traffic, was much safer.

Vehicle ride improved, not because of any great improvement in suspension but because low-pressure tyres came into general use in the second half of the twenties. These were not an unmixed blessing as they made the steering less positive and, together with the extra weight of the brakes, could cause wheel shimmy. The driver and passengers also now had more protection, for by the end of the decade the saloon car had almost completely displaced the open tourer, which had been the standard body in 1920. Dodge had all-steel bodies in 1916 and these eventually displaced composite bodies of steel panelling on wooden frames. Another important advance was the introduction of synthetic lacquers in 1925 to replace the traditional paint and varnish; drying times were enormously reduced and cars did not have to be stored for long periods while the paint dried.

Some technical improvements, for example, improvements in carburation and ignition, took place slowly over the years, but others—like four-wheel brakes—were implemented very rapidly. Synchromesh appeared in the USA first on the Cadillacs in 1928, and on the Buick in 1929. It was first introduced on a British-made car, the Vauxhall Cadet, in 1932 and it made gear changing so much simpler that it was in general use within a couple of years. On the other hand independent front suspension (IFS), although it had been used on occasional vehicles in this country for many years, was not given the enthusiastic reception that it had in the USA in 1934, and elsewhere; it was not used on many UK cars before 1939. Vauxhall were the first to use it on quantity-produced cars.

Mechanical braking systems were much improved, but these were progressively displaced by hydraulic brakes during the thirties.

Steel bodies were now normal and the body was attached to a chassis frame. Integral construction, in which the body carried the wheels, engine, etc, without a separate frame, had been used on the Lancia since 1922 and was used on the Lincoln Zephyr in 1932. The Vauxhall 10 of 1937 was one of the first British cars without a conventional frame. Vauxhall were one of the few firms in this country that were prepared to innovate, though this was partly because much of the development work on IFS, for example, had been done by the parent GM in the USA. Conservatism paid off in body styles, however, for British cars avoided the worst extravagances of streamlining which became popular in the early thirties in the USA and on the Continent, and many of the more expensive cars had a very attractive appearance.

Progress in the UK

When the First World War ended there was a very great demand for cars, particularly for cheap cars. Many people had learned to drive in the forces and a lot were prepared to spend their gratuities on cars. The pre-war car firms had had to go into armament production during the War which resulted in greatly expanded production facilities which were suitable for car production. A number of other firms such as Alvis, Aston Martin, Bean, Bentley, Clyno and Cubitt also started making cars.

Most of the pre-war manufacturers misjudged the market; they naturally built at first 1914/15 cars, and big powerful cars which only well-to-do people could afford, whereas the demand was for cheaper cars. Engine designers had learned a great deal from their work on wartime aircraft engines, and they naturally wanted to use this experience on automobile engines, but this too meant emphasis on more powerful and more expensive engines. Again, cars were built using skilled labour and few manufacturers were interested in using quantity production methods.

A number of small firms, however, realised the market for very cheap cars. There was a boom in cycle cars for these could be put together quickly and cheaply, particularly if proprietary parts were used. However, by about 1925 these had had their day; people much preferred to buy the Austin 7, and the small cars displaced the cycle cars almost entirely, only a few like the Morgan persisting.

Some manufacturers realised that the industry could only survive if they adopted the mass production methods of the USA. Citroën were one of the first to do this on the Continent. Andre Citroën had, before the war, managed Mors and had studied American production methods. He was an artillery officer during the War and was so perturbed by the shortage of shells he offered to build within a year a factory to turn out 20 000 shells a day, given the finance. His offer was taken up and the Citroën works eventually produced 45 000 shells a day. After the war he turned his organising ability to making cars and, by using mass production methods, was able to sell the 10 hp Type A Torpedo for about a quarter of the price of a comparable car in 1914. In 1922 he produced the very successful little two-seater 5 CV 'Citron' which did for the French public much the same as the Austin 7 hp of 1922 did for the British. The Citron was a fine little car—one of the authors (RTS) owned one some 25 years later.

In January 1921 the Treasury introduced the infamous tax based on the RAC rating. Even when it was introduced the rating was out of date. It encouraged the design of inflexible, high revving, long-stroke engines, and as no-one else would tolerate these things there was practically no overseas market for the mass-produced bread and butter English cars. This situation persisted until 1950. The UK could have received enormous sums over this period—successive governments were told of the situation in no uncertain terms by the industry but were too short-sighted and inept to do anything until after the war when they were forced to. The rating did, however, help keep out American cars and cut into the sales of the Trafford Park Model T Ford.

The big firms had to face a difficult transition period between the termination of war contracts and the sales of their first post-war cars; during this time no cash was coming in and a great deal going out. Later many firms were in trouble when the seller's market collapsed, and there was a slump in 1921. A number of firms went out of business but recovery was rapid. In 1922 there were nearly ninety car manufacturers and production was about 73 000 vehicles.

The structure of the industry was soon to change drastically, largely because of the adoption of elementary volume production methods by the bigger, more successful firms. The other firms went out of business, were taken over, gave up making cars, or in many cases found their niche in making luxury vehicles or high-performance sports cars in small numbers.

This process had reached such a stage by 1929, the last year before the Depression affected things, that only 31 firms were producing cars, making 182 000 cars altogether, but Morris and Austin were making about 60 % of all cars between them and Singer, the third big producer, about 15 %.

Austin themselves were in a serious position financially in the early twenties but at the critical time Austin himself designed the Austin 7 on his billiard table at home with the assistance of one very junior draughtsman. It was a true motor car, but small enough to go anywhere a motor cycle and sidecar would go. It sold very well indeed and after a reorganisation which involved Engelbach taking over the production and Payton the financial side, the firm prospered and production increased from 2600 cars in 1922 to 25 000 in 1926. By 1927 Austin was strong enough to bid for his old firm Wolseley when Wolseley passed into the Official Receiver's hands, but Morris was determined to outbid Austin whatever the latter offered and Austin had to withdraw.

Morris had brought out the two-seater Oxford in 1913 and a four-seater version, the Cowley, in 1915. After the war production was resumed but sales were poor and in 1921 Morris, like Ford many years earlier, decided to cut prices. The results were dramatic—1932 cars were sold in 1920 at £525 each, 3077 in 1921 (at £425 in February, £341 before the Motor Show), 6937 in 1922 and 20 024 in 1923 £255 making Morris Motors the biggest producers of cars in the UK and in 1925 they produced 41 % of all UK cars made. Morris bought out his suppliers when he could and as mentioned above he outbid Herbert Austin for Wolseley—and so for the first time acquired a design office! Morris had of course a very efficient production organisation which did some of the early work on transfer lines.

The bigger UK firms were very conservative, Austin was an excellent engineer, but if something was doing a job well he saw no point in altering it; he aimed at designs that were sound and very reliable. Possibly his only major innovations were the first Wolseley of 1899 and the Austin 7. Morris still bought many of his components out and at first was a car assembler rather than manufacturer and so was more interested in the production side than the design of the car as such. He was a business man rather than an engineer.

Innovation was therefore confined to the smaller firms but even here many of the best cars were well tried designs highly developed by skilled engineers and drivers. One unkind person stated that the contemporary Rolls-Royce was a triumph of craftsmanship over design, and Bugatti considered Bentleys to be the fastest lorries in Europe. A lot of the impetus for development came from racing with Brooklands being an enormous boon in this respect. However, it became derelict during the Second World War and was very short-sightedly sold after that; it is now largely covered by a housing estate.

The industry was handicapped at first by the lack of large-scale component manufacturers. There were some specialist manufacturers like Dunlop (tyres), Sankey (wheels), Lucas, CAV (electrical equipment), Ferodo (brake linings and clutch facings) who had been involved in the motor industry almost from its inception. A number like Solex and Zenith (carburettors), Triplex (windscreens) started just before the Second World War. Many others like Automotive Products (brakes and clutches) began from small beginnings in the twenties. Few began in a big way, but most grew with the industry. There was a lot of competition and a great deal of research and development work was carried out, which is still being done by the component manufacturers—indeed until recently more than that done by the motor manufacturers themselves.

The total number of cars in use climbed fairly regularly from 109 715 in 1919 and reached the million mark in 1930. The car was now making an impression on society. Roads were built and improved, though the Depression cut this work drastically. Measures were begun at the end of the decade to control traffic, and trams started to disappear. It was not of course cars alone that caused all this. In 1919 there were about twice as many vehicles of all sorts registered as cars, but by 1930 the ratio had dropped to be about the same number of other vehicles as cars. Increased mobility by both bus and car led to disorganised urban sprawl and, worse still, to ribbon development. Petrol stations and garages became part of the scene and the motorist did not have to set out with a supply of two-gallon cans as well as a full petrol tank.

The Depression had surprisingly little effect on the British motor industry compared with that of the American and other industries. Even in the worst year the production of cars fell by only about 15 % compared with 75 % in the USA and after 1932 production was above the 1929 figure and by 1937 twice that of 1929. This was partly due to people buying smaller and cheaper cars, for example, the proportion of 10 hp and under cars of the total number of new cars registered increased from about 35 % to nearly 60 % between 1929 and 1933. Also the real income per head of the people who could afford cars was not much affected by the Depression.

More firms fell by the wayside and there were further regroupings. Throughout the thirties Morris and Austin were the biggest producers, next came Ford and then Vauxhall, Rootes and Standard, and in 1938 these firms between them produced 90 % of the output in this country. Morris and Austin, however, progressively lost a lot of the market to the other firms, particularly to Ford who had begun production in a new works at Dagenham in 1932. The two bigger firms lost out by spreading their resources too thinly, for example, in 1933 Morris offered 8, 10, 12, 16, 18 and 25 hp cars all in two or three styles, and Austin in 1934 made 44 separate models based on nine alternative chassis. Ford, on the other hand, concentrated on their very successful 8 and later 10, and a V8 engined car. Rootes and Standard had the very successful Minx and Nine, though there were other models as well.

Manufacturers copied one another. If one firm produced a successful model others tended to bring out a similar car of the same size and price, and to compete on quality and other grounds. One aim was to bring out a small family car to sell for £100; the first £100 car was the Morris Minor of 1931 but this was too basic. Ford's £100 8 hp car of 1935, on the other hand, was remarkably successful. The attempts to reduce costs in a very competitive market meant that design and finish often suffered. Cars could be underpowered, noisy and badly sprung so that after two years' use, by which time steering and braking as well as bodywork could deteriorate badly, they could be positively dangerous. More people were killed on the roads in 1934 than in 1960, though there were nearly four times as many vehicles registered in 1960 as in 1934. Road improvements, etc, of course also played a part in reducing accident rates.

Before the 1939–45 war the UK was second to the USA in production. There were 33 British manufacturers but only about twenty of these were really independent and as mentioned above the Big Six—Morris, Austin, Ford, Vauxhall, Singer and Rootes produced 90 % of the cars made in this country.

By August 1939 registration had exceeded the two million mark. Road building programmes had recovered from the Depression but traffic jams and traffic congestion could be very bad. Queues stretched for miles on Bank holidays. Weekend motoring was very popular, people flocked to the seaside and to the country, generally to the latter's detriment. Ribbon development continued despite belated efforts to control it, and towns miles apart could be joined by a ribbon of houses one house wide on either side of the road. Suburban sprawl was made possible by cars and buses; towns spread and engulfed villages and merged with other towns to form conurbations. Six or seven thousand people were killed on the roads and more than two hundred thousand injured each year.

Traffic caused more congestion and complications in the cities of the Old World than the New. The Old World cities were relatively stable in size and were built in the first place for pedestrians and horse-drawn vehicles.

Progress in the USA

Production of private cars in the United States recovered rapidly after the war and by 1919, at two million, was greater than in 1916. Production was down a little during the slump year of 1921 but from then on production of cars romped away to four million in 1923. The seller's market finished about halfway through the decade and the rate of increase slowed somewhat, but even so production exceeded five and a half million in 1929. By 1925 more than half the families in the country had their own car and in California one car was registered for every 2.8 people. The cost of cars steadily decreased until about the middle of the decade, and in any case time payment schemes had been introduced, making payment comparatively painless. The more prosperous owner replaced his car every couple of years so that second-hand cars were remarkably cheap. To counter this General Motors started to change their models each year in order to date the second-hand car as rapidly as possible. The current UK practice of incorporating a letter in the car's registration number does the same job rather more subtly—your neighbour knows that, though your particular model of car has been in production for many years, your car is only one year old. To return to the twenties the other companies soon followed suit.

The economic ramifications of the automobile boom were very wide spread. Not only did the car manufacturer benefit but so did the steel, rubber, glass and oil industries. The notoriously bad country roads were replaced by properly surfaced ones and well engineered highways were built. A great number of people were involved in selling cars and servicing them.

At the beginning of the twenties the major companies were Ford, General Motors and then, a long way behind, Hudson (and Essex), Studebaker, Dodge, Maxwell, Willys/Overland, Nash, Packard and Durant. There were also a number of much smaller firms, many producing expensive and specialised vehicles such as Reo, Peerless, Franklin, Hupmobile, Marmion, Stutz, Andrea, Duesenberg and Rickenbacker. Production was on such a large scale that no really new organisation successfully broke into the mass market, and as in the Old World there were a series of regroupings and takeovers, and firms went out of the automobile business, so that though there were 108 firms in 1923 there were only 44 in 1927 and that two years before the Depression.

General Motors replaced Ford as the leading firm. Ford's top management worked in a bear garden and Ford, like Benz more than 20 years earlier, continued obstinately to turn out the Model T (15 million in all) long after they had become obsolete. Ford stopped manufacturing in May 1927, went into hibernation for a year and put the Model A on the market in 1928. It was a very ordinary car and within a year it was being outsold by the Chevrolet.

Durant was tipped out of GM—now a corporation—yet again in 1921 and this time for good. Apparently he considered the corporation to be rather a sideline and was more interested in the stock market. The 1921 slump caught him badly and to avoid bankruptcy, which would have shaken confidence in GM, he had to sell out his holdings and resign.

Pierre DuPont took over with Alfred P Sloan as his second in charge.

Sloan became President in 1923. He was an excellent administrator, and he had W S Knudsen who had been largely responsible for Ford's success, on the production side. Knudsen became President in 1937. Chrysler stayed with GM but got fed up with Durant and had left to reorganise two firms, Willys and Maxwell, which had suffered in the slump. John Willys eventually regained control of the Willys-Overland Co. but Chrysler remained with Maxwell and when the car he designed was successfully on the market the Chrysler Corporation came into being. In 1928 Chrysler bought out Dodge Brothers and with their extra manufacturing capacity and sales organisation Chrysler got into the mass market with the Plymouth car in 1928.

In 1929 US production reached the record figure of 5 337 087 cars, a figure that was not matched again until 1949, but plunged to 1 331 860 in the depths of the Depression in 1932. People, of course, did not give up using motors cars, they just did not have the cash to buy new ones. Registrations fell by only about 10 %.

Some of the medium-sized firms like Packard, Nash and Hudson kept going. Studebaker, who rashly bought Pierce-Arrow in 1928, was at one time in the hands of the receiver but reorganisation got them going again. Willys-Knight underwent the same traumatic experience. The Depression, however, was disastrous for most of the smaller companies though some managed to move into other industries, Reo and Marmon made trucks and Peerless made, of all things, beer. Many firms went out of business altogether.

GM not only came through the Depression intact but also expanded its interests. Work on diesel engines was undertaken, largely at Kettering's instigation, which revolutionised railroad transportation and led to diesels beginning to appear on trucks and buses. GM also went into the manufacture of aircraft engines. In 1937 GM comprised Cadillac, Buick, Oldsmobile, Pontiac (replacing Oakland) and Chevrolet, and had 40 % of the US market. They were the biggest privately owned manufacturing enterprise in the world.

Chrysler (Chrysler, Dodge, de Soto, Plymouth) saw their production drop drastically but by careful management not only recovered, but paid off the debts they had acquired when they bought Dodge.

Ford survived, despite its mismanagement, because of its size and reputation, but dropped to number three position after Chrysler, despite the Ford V8 engine of 1932. Henry Ford made no attempt to develop a reasonable top management structure, he had his favourites and stirred up ill-feeling as a matter of policy.

In 1942 production of private cars stopped in the USA. The automobile industry changed over to making munitions and its contribution constituted one fifth of the nation's entire output of war *matériel*.

Chapter 4

Developments Since 1945

Technical Developments

There were few technical innovations in the first twenty five years after the Second World War that had not been foreshadowed earlier. Most changes involved modifications due to improved materials, changed production methods, or deeper understanding of the principles involved.

Engine designers took advantage of fuel of progressively higher octane number to increase compression ratios, and improved bearing materials and lubricants permitted higher engine speeds. The higher compression ratios and higher revs gave much greater power for a given engine capacity. Better breathing and combustion chamber design also helped. The way in which the power output of the average family car increased over the years is shown in table 4.1. Output has therefore increased fairly uniformly by about 6 hp/litre per decade since 1910.

Tooling for the production of a completely new engine became more and more, and eventually, almost prohibitively expensive so that the basic block often remained in production for many years. However, production methods gradually improved to reduce costs, and the engine performance was stretched. In time all engines used overhead valves which were operated by pushrods on cheaper cars and by overhead camshafts on more expensive ones. Four cylinders were usual in the UK and on the Continent, six cylinders in line being used on bigger, more expensive, cars though some V8 engines were made. V8s became popular in the USA and through the fifties these engines became more and more powerful with some rated at 250 bhp and more. By the end of the sixties 350 to 400 bhp was not unusual. The rest of the world could not afford the great thirsty brutes powered with these engines.

Though the period to about 1970 was one of steady development, the last fifteen years or so have seen some remarkable changes which are still taking place. These have largely been brought about by US legislation to control noxious emissions. Quite simple modifications to the engine brought about some reduction in emissions but to meet increasingly stringent requirements catalytic converters were necessary. These reduced the oxides of nitrogen and oxidised the hydrocarbons in the exhaust but to allow them to operate efficiently, petrol injection and transistorised ignition systems were needed; the older systems did not control the mixture and timing precisely enough.

Another way of reducing emissions was to use a 'lean burn' engine. A weak mixture burns more slowly and may cause the engine to misfire; to overcome this turbulence had to be increased further still, which was achieved by using specially shaped combustion chambers. Another approach was to use a small prechamber which was filled with rich mixture, the latter was ignited and the flame spread to the main region of the combustion space which contained a weaker mixture. Lean burn could encourage NO_x emissions, and as the latter tended to form at surfaces lean burn was not suitable for bigger engines (two litres and more). It was therefore of interest to European and Japanese manufacturers.

Table 4.1

Year	1914	1940	1950	1960	1970	1980
Out-put (hp/litre)	10	25	30	36	47	51

A remarkable development in the last decade or so has been the use of microprocessors to control mixture, ignition and timing. Sensors measure the relevant engine variables such as the airflow into the engine, engine revs and temperature, exhaust gas composition, etc, and microprocessors continuously adjust timing and mixture to give the least emissions and minimum fuel consumption over a wide range of operating conditions. The use of sophisticated electronics and microprocessors is not only revolutionising engine design and management but it is also having a considerable effect on other components of the car, for example, improving automatic gear shifting (with lock-up clutch), antilock braking, active suspension systems (in which the suspension parameters are continuously adjusted), power steering, and the diagnosis of faults. Eventually one processor will control all these functions, and multiplexing will reduce the wiring required to simple ring mains.

Returning to the engine, diesels were made quiet and smooth enough for private cars though it is still possible to identify a diesel engine when it is idling. Their extra costs tend to outweigh their advantages and so they have only made headway very slowly in the UK.

The Wankel engine, in which a three-lobed rotor rotates in a specially shaped chamber, was installed in the sixties on the German NSU and the Japanese Mazda. It bankrupted the former but has remained in production in Japan. It has a high fuel consumption and difficulty in meeting emission requirements.

A number of firms including Rover in the UK experimented with gas turbine engines but their high fuel consumption, even when used in conjunction with a heavy heat exchanger, made them an impracticable proposition for powering private cars. The heat exchanger recovered some of the energy that otherwise would have been lost in the hot exhaust gases.

Manual gearboxes changed little but by the sixties even the cheapest cars had three forward gears with synchromesh on all three. For quite a time it was fashionable to have the gear change lever on the steering column. The automatic gearbox made tremendous strides in the US where it almost completely displaced the manual box in the fifties, whereas even now only a small proportion of cars sold in the UK have an automatic box.

Independent front suspension soon became universal; most of the systems used were based on pre-war systems except perhaps for the McPherson strut. Few manufacturers found independent rear suspension (IRS) worth its weight and cost on conventional cars. Detailed studies and much development work, however, improved vehicle ride considerably. Similarly the steering was improved and eventually rack and pinion steering became the most popular system. Powered steering became standard on big US cars but the overwhelming majority of UK cars still use manual steering.

Radial tyres were available in the early fifties and gave much improved road holding but they cost more and it was a considerable time before they displaced crossply tyres.

In the fifties disc brakes were used on racing cars because of their stable output and their better heat dissipation. They were very successful and began to appear on sports cars and high-performance cars, and by 1970 their use had become quite general, at least on the front wheels. Legislation caused the introduction of split and dual systems so that if one subsystem failed the other would still give some braking. Vacuum servos were fitted to reduce pedal efforts on larger cars and slowly became more generally used. Antilock systems have been available on some expensive cars for many years but are now appearing on other cars.

Integral construction replaced composite bodies in the UK; one of the last firms to go over to such construction was Rolls-Royce in 1965. The appearance of the car changed but the changes followed pre-war trends; the body got wider and lower, headlamps and wings were swallowed up by the body and the boot became an integral part of the body. In the late forties the bodies tended to be bulbous but the designs were refined and by the sixties many vehicles had clean attractive lines. American styles deviated more and more from European ideas. To European tastes they became worse and worse, the nadir being the juke box style of the fifties but things improved in the following decades.

Returning to the UK and Europe the stylist, often Italian, varied the appearance of the car so that fashions have come and gone, but extremes have generally been avoided and many cars have had quite attractive lines.

The inside of the car has changed considerably. More glass was used and so the interior of the car became much lighter and airier. More and more switches and lights have appeared on the fascia, which is a happy hunting ground for the stylist. Light switches and trafficators, etc, were gradually moved to the steering column, a much more convenient arrangement, as was realised nearly a century ago.

Citroën persisted with front-wheel drive (FWD) and a number of manufacturers, particularly in France, followed their lead but it was not until the seventies that it became fashionable. It was notably used on the Mini in 1959 and later on the BMC 1100 and 1800 models, with Ford of Europe bringing out the Fiesta in 1976 and Opel the Cadet in 1979. The GM Tornado of the early sixties had FWD but the car was not a success and so FWD was not used again on the big domestic US car until the eighties.

The rear-engined VW had some imitators. Renault, Simca and Fiat all produced small rear-engined cars on the Continent, Hillman made the Imp in Scotland and GM the Corvair in the USA. It is difficult to design a rear-engined car that will handle well and predictably under all conditions of motoring, so it is unlikely that these cars will have successors. Front-wheel drive is a much better proposition.

Progress in the UK

Production of cars for export continued for some time after the war began as foreign currency was desperately required. When France fell the whole industry had to turn over to the production of munitions, and overseas debts were paid off by selling British assets.

By 1943 it seemed reasonable to assume that Germany would eventually be defeated but it was also obvious that when the war was over Britain would be in a bad state economically. Manufacturers were therefore allowed to give some consideration to their post-war plans, but the staff and time were not available to do much in the way of new designs. Consequently when post-war production began practically everyone took over from where they left off in 1940. Getting production going was quite enough of a headache without introducing new models.

The Big Six naturally had the lion's share of the market. The remaining 10 % was shared by rather more than a dozen firms, including Rover, Singer, Jaguar, Rolls-Royce, BSA, Jowett, Armstrong Siddeley, Allard, Healey, Morgan, Lea Francis, Aston Martin, Jensen and AC. A couple of newcomers appeared—the Bristol Car Co. in 1947 and Lotus considerably later.

At first there was a shortage of just about everything, but the most serious and persistent was the shortage of sheet steel which was particularly bad about 1951/2. This was tremendously frustrating for a seller's market existed until about 1956. The industry had other troubles including government interference. Naturally the government wanted everything exported; indeed steel was rationed and doled out to the various companies according to their performance in the export market. This made things unstable because countries could and did impose restrictions on imports without warning and there was not a big home market to act as a cushion. Indirectly the reputation of British cars was damaged, for most of these were designed for British conditions; they did not take kindly to bush tracks and hard work. Smaller firms could not set up adequate dealer and servicing networks and provide sufficient spares in remote areas, and so again reputations suffered. On the other hand, some cars like the Jaguar XK120 and the MGB were excellent advertisements for British industry.

Other troubles were the penal taxation which made investment difficult, and the government soon found that they could help manipulate the economy by changing the rate of purchase tax on cars. And all the time the government was taking their money and throttling their market the vehicle manufacturers had to listen to exhortations telling them to increase production. Pressure was also exerted to reduce the number of independent manufacturers, and for each to produce only one or two models.

The motorist too had his frustrations. First, he might wait, literally for years, after ordering his car for it to be actually delivered. It might be much quicker to buy a second-hand car but these could cost more than a new one which was the case for many makes until 1953. Purchase tax, introduced in 1940, was $33\frac{1}{3}$ % to 1947, when it rose to $66\frac{2}{3}$ % on expensive cars and in 1951 was $66\frac{2}{3}$ % on all cars. The tax on petrol was increased but even worse petrol was rationed until the end of 1952. The petrol was called pool petrol and had an octane rating of little over 74 so compression ratios were limited to about 6.8 to 1.

In the same way as cycle cars enjoyed a period of popularity after the First War, imported 'bubble cars' such as the Messerschmidt, Heinkel, Isetta, Goggomobil cars, helped in the fifties to fill the gap until bigger, 'proper' cars became available. These little cars had the advantage that they did remarkable numbers of miles to the gallon of rationed petrol. There was one change for the better and that was the abolition of the RAC hp rating as the basis for tax. The government of the day was willing to consider alternative ways provided total revenue was not reduced and after some indecision by the industry, which could not make up its collective mind, a tax based on capacity was introduced and then in 1949 a flat £10 rate.

Australia was the biggest market for UK cars until 1956 or so, when sales tax and a quota system began to squeeze UK manufacturers out. GM managed to obtain extraordinarily favourable terms to manufacture in Australia and began building the Holden car based on an elderly Detroit design, so that competition from the Holden was very strong indeed. Consequently, Austin built factories in Australia and switched exports to the USA, where sales rose from 74 000 in 1956 to 310 000 in 1959, but by then competition from the compacts was strong and sales fell back to 30 000 in 1961. Competition was also coming from other sources. The German motor car industry had been devastated by 1945—for some years after the war GM put the nominal value of the Opel works at one dollar—but Opel was back in business in the fifties. The VW works had been offered for sale to both European and American manufacturers without a buyer being found, but was built up by Dr Nordhoff into one of the largest car firms outside the USA. Similarly Mercedes-Benz were got going again. Though in 1950 the UK was the biggest exporter of cars in the world, by 1955 West Germany had displaced it both as the biggest exporter, and as the biggest producer (other than the USA). In 1960 the UK had 24, France 21, and West Germany 36 % of the world market.

To return to the UK, production in the late forties increased rapidly. The steel shortage was a limiting factor, particularly in 1951 and 1952, but the situation eased when new steel mills opened in South Wales.

The most important event in the fifties was the merger of Austin and Nuffield in 1952 to form the British Motor Co. Merging had been discussed by Austin and Morris themselves as early as 1924 but they did not like one another, Morris wanted complete control, and nothing happened. There was more talk in 1948, and in 1952 negotiations were successful. Both Nuffield and Leonard Lord of Austin realised that the two organisations would have to join forces to compete with the American giants, if not with Ford (UK). Besides Nuffield wanted to retire, which he did in 1954. In that year BMC had 38 %, Ford 27 %, Standard and Rootes both 11 % and Vauxhall 9 % of the UK market. The merger did not do a great deal of good. Morris, in particular, had factories all over the place for historical reasons and Nuffield had preferred decentralisation. It proved an impossible job to bring such a heterogeneous collection into a single well integrated and efficient organisation.

The Austin-Morris merger was the biggest of the many takeovers and mergers that took place until BMC itself merged with Leyland in 1968.

The Leyland Motor Co. made commercial and special vehicles, were doing very well and decided to enter the car industry. They acquired Standard-Triumph in 1961 and Rover in 1967 and in 1968, after long and complicated negotiations in which even the then Prime Minister had his say, Leyland Motors and BMC merged to form the British Leyland Motor Co. (BLMC). The idea was that the combined resources of the two organisations could finance the development of new cars and produce them in such volumes that they would be competitive on a world scale. The new board had the formidable task of running an industry made up of a number of firms that employed directly nearly 200 000 people. It badly needed pulling together. Though some of the smaller firms were doing quite well BMH were currently running at a loss; there were too many ageing models in production—no new models were in the pipeline, except the unhappy Maxi—and the whole concern was grossly overmanned. Consequently, by 1970, there were only two major producers in the UK, namely BLMC and Ford with 48 and 27 % of the total market respectively, and two smaller producers Vauxhall and Chrysler with 11 and 13 % respectively, and of these only BLMC was not American owned. The way the various post-war amalgamations took place is shown in the following figure 4.1.

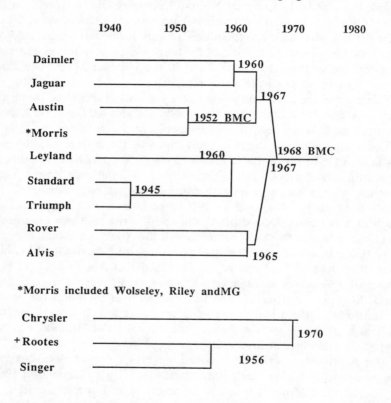

Figure 4.1 Post-war amalgamations in the British motor industry.

One of the firms to sell out was Rootes. Rootes did very well in the fifties, taking over Singer in 1956, but not so well in the following years and so they were prepared in 1964 to sell a minority of their equity to Chrysler who were looking for a foothold in Europe. Chrysler acquired a controlling interest in 1970 and changed the name of the firm to Chrysler (UK). Despite investment the firm was too small and too troubled to flourish, and by 1976 so much money had been lost and the parent firm Chrysler (USA) was in such a bad financial state itself, that it threatened to liquidate its UK subsidiary. The UK government gave financial help but things did not improve and Chrysler (UK) was sold to Peugeot in 1978 and there was a change in name to Talbot.

To return to the industry generally, the Suez crisis in the mid-fifties and competition in the USA by the compacts in 1961 caused problems but otherwise production had increased fairly steadily from 1946 to 1964, but 1964 was a turning point in the history of the industry. Except in 1972 production never again exceeded the 1964 figure and after 1972 production decreased year by year. Successive governments continued to use the motor industry to help regulate the economy, chopping and changing purchase tax rates and hire purchase restrictions. The economy tended to 'stop and go' and the motor industry followed suit, for example, production in 1964 increased partly as a result of a reduction in purchase tax, and fell in 1966 and 1967 because of the introduction of more stringent credit restrictions. Sterling was devalued in 1967 and further means were taken to restrain home demand and encourage exports; production was very high in 1968 and exports increased by over two fifths. Output in 1970 was restricted by industrial troubles but there was a rapid rise in the standard of living in the next few years and production was very high, nearly 1.92 million in 1972. About half a million people were employed in making cars or parts for them. However the industry was not as prosperous as the production figures suggest; profits were poor—profits on exports were low even after devaluation—investment had lagged, and BL in particular was slow in developing new models, productivity was very low and industrial unrest endemic. Production lost in 1973 and 1974 as a result of industrial action (including a miners' strike) approached a million cars.

These weaknesses became very apparent in the recession that followed the oil crisis of late 1973. Measures to conserve energy reduced industrial activity generally, and so car production was reduced. Demand also fell in 1974 and 1975, but then recovered and increased until the 1980 recession. The motor industry, however, did not recover. There was a 25 % overcapacity in the Western European car industry in 1974, and Japan was exporting very reliable and cheap cars, so competition was intense. A manufacturer had to compete on price, quality, design and delivery, and the UK industry in general could not cope with the competition. To make its task more difficult reductions in tariffs and entry into the Common Market made foreign cars cheaper. Imports rose from 0.15 million cars in 1970 to 0.96 million in 1979, between the same years UK production fell from 1.64 to 1.07 million.

Because of the enormous amounts of money needed to tool up for volume production, losses become correspondingly large when production falls below the breakeven point. In 1972 BLMC produced 916 000 cars, in 1975 only 604 000, so the company was nationalised to save it from bankruptcy. Incidentally, BLMC became British Leyland in 1975, and Austin Rover Group in 1986. The recession precipitated the collapse of

BL but BL was fundamentally unsound. As mentioned above, too many models were made in too small numbers and some competed with each other; new models like a new Mini, a new Jaguar, were not pressed hard enough, even worse new models were put on sale before all the faults had been removed. Quality control was often poor. The major cause of BL's misfortunes, however, was labour troubles; it appeared at times to have a virtually uncontrollable work force. Restrictive practices abounded, there was a lack of discipline and productivity was abysmally low. Management was weak, particularly at the top, and appeared to have lost confidence; it feared a confrontation that might lead to a strike which would destroy the firm, and the unions took advantage of the management's weakness. Things reached such a stage that employees were taking out 92 % of the added value in wages and social services, leaving precious little for investment, tax collector or shareholder.

Strikes were not confined to BL, in 1978 BL and Ford between them had had to contend with 350 strikes in six months and the cumulative disruptive effect was enormous. In the period 1975 to 1980 eight times as many days were lost per employee in the motor industry as a result of industrial action as in industry generally. Ford extricated themselves by integrating with Ford of Europe so that they could use the higher productivity and more reasonable working practices of sister organisations, though as a result less than half of what goes into Ford cars assembled in the UK is actually made in the UK. Later GM integrated Vauxhall and Opel production in a rather similar way. Less than half the GM cars sold in this country in 1984 were made in the UK.

Returning to BL, it was reorganised after nationalisation and money invested to enable production to increase. It did not, and in 1977 a new chairman was appointed and a strong board eventually got together. The unwieldly organisation was divided into manageable autonomous sectors which were given specific targets to meet, money was provided and political backing given. Factories were closed that were inefficient and which had been plagued by labour troubles, so that the total work force reduced. Production, however, still decreased, reaching a minimum of 395 820 in 1980 and increased slightly in the following years to 440 000 in 1984. By 1984 the tax payer had given BL more than £2 bn, and yet the long-term viability of the Austin-Rover sector was still doubtful. The Jaguar sector was profitable and was privatised in 1983.

About one million cars were made in the UK in 1984 and the total work force, including that in the components industry, was about 282 000 people. The components industry also suffered badly though a number of firms with foresight had set up factories abroad. The fall in UK vehicle production drastically reduced the home market which was halved between 1972 and 1980 with EEC imports being very competitive. A number of the biggest component firms made losses in the early eighties and had to shed large numbers of employees.

Progress in the USA

Post-war experience in the United States was similar in many ways to that in the UK. In addition to the Big Three, six independents (Studebaker, Packard, Nash, Hudson, Kaiser-Frazer and Crosley) all tried desperately to catch up on the backlog of orders caused by the war. Shortage of materials and strikes slowed things up and most companies did not introduce new models until 1949. The Korean War brought about price and production controls and when that ended there was a rapid expansion, but there followed a recession which put the smaller firms in trouble. Crosley was bought out in 1952, Kaiser-Frazer merged with Willys-Overland in 1953 and car production was given up in 1955 so that they could concentrate on the Jeep. Hudson and Nash merged in 1954 to form the American Motor Co. (AMC) and in the same year Studebaker and Packard merged and kept going until 1966 when they stopped car production.

One remarkable effort was the attempt of Henry Kaiser to break into the market. Kaiser had done very well in the war years making ships, including Liberty ships and he started making cars. At first things did not go too badly but the cars were very ordinary, production methods pedestrian, and Kaiser had trouble building up a sales organisation so that when the market contracted he merged with Willys-Overland.

General Motors went on from strength to strength.

Henry Ford retired in September 1945 and Henry Ford II, his grandson, took over. The company had been wretchedly managed but Henry II obtained some very able people and within a few years the company was pulled round and in 1950 displaced Chrysler, which made them second to GM.

The major companies could not go in for wholesale price competition—GM could easily have undercut the others but if GM had forced them into bankruptcy the US Government would have accused GM of establishing a monopoly and GM probably would have had to pay to put them back on their feet again.

So instead of competing on price, competition was in respect of engine horsepower and appearance, and the stylist tried to make the car look as prestigious as he could. Manufacturers also tried to give their cars some individuality by offering them in a variety of colours, trim and accessories. To English eyes the resulting cars were ghastly—they were far too big, smothered with chromium plating, and looked like mobile juke boxes. When driven the big cars wallowed and rolled on corners, they were underbraked and their petrol consumption was enormous; they guzzled gas. On the other hand, the big cars were comfortable, easy to drive and as the engines did not have to work hard they were reliable.

Styling became more and more outlandish and eventually a reaction set in and Ford was caught with the Edsel; it appeared in 1957, withdrawn in 1959 and is said to have cost Ford a quarter of a billion dollars. There is now a flourishing club for the proud owners of Edsels. The typical car of the sixties was about 18 ft in length, weighed about 4000 lb and was powered by a 7 litre V6 engine developing 350 bhp or more, however, there were many people who not only wanted a less ornate vehicle but wanted something smaller, more manoeuverable, and easier to park. European cars, particularly the Volkswagen, fitted the bill and imports increased from 1 % of domestic production in the mid-fifties to 12 % in 1960. The industry eventually reacted and about 1960 introduced the compacts. GM had the Tempest, Corsair and Chevy, Ford the Falcon and Comet, Chrysler the Valiant and Dart, and Studebaker the Lark. AMC made the Rambler as early as 1950 and its production shot up from 80 000 in 1955 to half a million in 1960. Not that the Big Three wanted to build small cars, small cars were not proportionately cheaper to make and profits were smaller. People who wanted cheaper motoring were advised to buy good second-hand cars. Consequently, the big firms tended to import cars made by their European subsidiaries. The compacts incidentally were not only bigger than the European cars they were intended to supplant but tended to grow year by year and to have bigger and bigger engines as optional extras. The 'subcompacts' which appeared at the end of the sixties were smaller and by definition less than 100" in length.

Imports, however, again had 10 % of the market by 1968 but the Big Three now reacted much more strongly for a new generation of compacts would now be competing with imported cars for a defined market and not with their own big brothers. But 90 % of US production facilities were still geared to making the big cars.

In 1965 a young lawyer Ralph Nader published a book *Unsafe At Any Speed* criticising the way the industry, government and other organisations dealt with road safety generally, and criticising the GM Corvair in particular. The Corvair had a rear engine and originally had swing axles so it could handle viciously. The publicity killed the Corvair even though by the time that Nader's book appeared the swing axles had been replaced by a much safer suspension system.

The US Government became very safety conscious and eventually the US safety regulations directly and indirectly influenced the industry worldwide. Hitherto most work had been on accident avoidance, on improving the braking and handling of the car, the driver's field of view, lighting, etc, but the emphasis was now on primary safety, on protecting the occupants of the car in the event of a crash. Anti-burst door locks were fitted, projections, both internal and external, removed, possible impact areas between occupant and vehicle padded, high-impact resistance windscreens and safety belt anchorages fitted, etc.

Some extraordinary ideas were seriously considered such as the airbag which was to be explosively inflated during impact. Also cars in the USA were to be fitted with bumpers that would withstand a 5 mph impact; this was not to protect the occupants but to reduce insurance claims. These great changes did nothing to improve the appearance of the car.

The climatic conditions and topographical situation of Los Angeles, together with its very high car population (vehicles in the State of California consumed more petrol per year than in the UK and West Germany combined) helped produce the notorious Los Angeles smog which became so bad that legislation was passed in 1966 in California to reduce vehicle emissions. Later the Federal Government harmonised standards throughout the USA and set industry a set of future targets to meet.

Considerable improvements were made by quite simple modifications to the engine such as recycling the crankcase gases, but to reduce emissions to the targets set, catalytic converters to reduce the oxides of nitrogen, and to oxidise the hydrocarbons and carbon monoxide were needed. Complicated electronic control systems were required to minimize the emissions under all conditions including starting with the engine cold. The cost of the car was considerably increased. As the expensive catalysts were poisoned by lead, the engine had to run on lead-free petrol which reduced its efficiency. From 1967 to 1976 the average US car increased 800 lb in weight in meeting safety and emissions requirements.

In the years following 1973 there was an energy crisis and in 1975 the US Government brought in mandatory legislation to reduce petrol consumption. The CAFE (Corporate Average Fuel Economy) regulations laid down that the consumption averaged over all cars of a given producer, from mini-compacts to large cars, should be reduced in stages from 18 mpg (US) in 1978 to 27.5 mpg (US) in 1985. There was room for improvement—the champion gas guzzler of 1974, the Olds Toronado, averaged 6.8 mpg (US). The industry met the targets from 1978 to 1982 and were not far below in the following years but could carry over credits from previous years when they had exceeded the targets. The targets were met primarily by reducing vehicle weight (average down 24 % between 1974 and 1984) and engine displacement (down 34 %). Front wheel drive (on half the 1984 cars) was more space efficient than conventional rear wheel drive and gave some reduction in weight, as did the greater use of weight saving materials. Reductions in tyre rolling resistance and in body drag coefficients also helped. Incidentally, legislation like the CAFE regulations does not cause a proportionate decrease in fuel consumption, for if his car is cheaper to run, a driver tends to use it more.

Meeting CAFE requirements did not involve any particular engine developments in the period to 1985 but much work went into developing more inherently fuel-efficient engines such as the lean-burn engine, and into developing electronic control of the engine so that the latter operated under near optimum conditions under most circumstances.

The quadrupling of oil prices and the recession following the Yom Kippur War had a great impact on the industry and production fell from 9.6 million in 1973 to 5.9 million in 1975. People postponed buying new cars but when they did they still preferred the conventional big V8s—in 1977 no less than 76 % of the cars made had big (generally 7.2 litre) V8 engines. The US driver very sensibly preferred his big powerful car to a smaller one even if the latter did more miles to the gallon. His roomy air-conditioned car was comfortable, was fitted with power steering and automatic gear change, and was easy and relaxing to drive. He could overtake a truck or haul a caravan or boat without any trouble, and roads and parking lots were tailored to big cars.

The recession following the beginning of the Iraqi–Iranian War in 1979, however, was even more disastrous than the earlier recession for not only was the total market reduced but people were now prepared to buy smaller, lighter cars which the industry was hardly in a position to supply and foreign cars came flooding in, imports increasing from 15 % in 1973 to 23 % in 1979. GM showed a loss for the first time since 1921, and GM and Ford between them lost $15 bn in the period 1979 to 1982. Chrysler almost went under; they had only 9 % of the depressed market, were making the wrong sort of car and at the end of 1979 were losing money at an appalling rate. They were only weeks away from bankruptcy when they were bailed out by the Treasury which would not lend them money but guaranteed loans they could raise up to $1.5 bn. Losses continued in 1981 and 1982 but by 1982 Chrysler had turned the corner and later did so well that they could pay off their loans. The Treasury incidentally profited for they had insisted on a share of the equity as a return for their guarantee and Chrysler bought it back. The other companies too had to borrow money to increase their output of compacts and subcompacts at the expense of the traditional car. However, the industry rallied and recovered, and helped by the work forces who forewent pay rises for several years, made a profit of $18 bn in 1984—which was more than what most European manufacturers could do.

Japanese Competition

Before the Second World War the Japanese motor industry concentrated on trucks and buses, and automobile plants were fostered for military purposes. There were only 180 900 motor vehicles in the country in 1938. After the war progress was retarded by damage and the Japanese automobile industry and technology was far behind that of the US and Europe. To modernise the automobile industry some large companies concluded agreements in 1953 with European manufacturers to introduce European technology into Japan over a seven-year-period. Toyota, on the other hand, developed its own technology and in the fifties made the first Japanese car (previous cars had been based on Western cars). Other manufacturers also started making motor cars, Honda for example, after becoming the biggest makers of motor cycles in the world.

Many of the Japanese firms assembled cars from bought-out components rather than manufacturing *ab initio* and reduced costs by carrying minimum stocks, the components arriving 'just in time'. The work force was hard working and cooperative, and quality control excellent. For many years Japanese cars were not as advanced technically as many European and US cars, but though conservative in design they were well built, competitively priced, and very reliable. Marketing was shrewd with sales and servicing organisations being built up. Production was only 32 000 vehicles in 1950, but about 100 000 in 1955, 480 000 in 1960, 5.3 million in 1970 and 7.26 million in 1980 when US production was 6.37 million. This rapid rate of growth was based on an enormous potential home market. Only 534 000 vehicles were in use in 1956, but nearly 8 million in 1966 and 28 million in 1975. Production was later sustained by exports. In 1974 about 40 % of national production was exported (a smaller percentage than the corresponding figure for Germany, France, Sweden and Italy but a little greater than that for the UK). Exports were insignificant in 1965 but were one million by 1970, 2.5 million in 1975 and 3.9 million in 1980. By 1970 Japan had captured 13.75 % of the world export market, by 1975 21 % and by 1980 46 %. To overcome quota restrictions in some export markets Japanese firms collaborated with local firms or set up local factories.

Part II
The Technical Development of the Motor Car

Chapter 5
The Engine

Developments to 1900

Not only had the gas engine to be turned into a petrol engine, with all the complications of carburation and ignition, it also had to be adapted to power a moving vehicle. That meant it had to be made much lighter, much more powerful for its weight and size, it had to function properly while being badly jolted and shaken, and it had to develop power to match the speed of the vehicle, and so deliver power over a range of speeds, not at just a constant speed like a gas engine.

The new engine had to be reliable, which was difficult to ensure because it had so many novel features, and its cost as a prime mover had to be competitive with that of the horse it was displacing.

There were many problems, and generally each problem had a number of solutions so that some curious engines appeared. The more successful solutions, unless they were very well protected by patents, were soon copied so that at any one time not only did the ordinary cars look alike, they were also very similar under the bonnet. Small firms did not normally have the staff or facilities to work out the proper dimensions of the many components that made up a car and so shamelessly copied successful models. Big firms copied too; one manufacturer in the thirties boasted that he never innovated, he let the other companies take the risks and if the new idea worked he would eventually find ways of copying it!

Some solutions were technically very good, but could not be adopted, or at least not without undue expense, using existing materials and manufacturing methods, and so had to wait for years before they were finally put into practice. Patents could also delay the implementation of a good idea for many years but when the patent expired it would be taken up immediately.

Benz engines

The first Benz engine was based on contemporary gas engine practice, and the gas engine in turn owed a lot to the steam engine. The engine had a single horizontal cylinder, a vertical crankshaft and a large horizontal flywheel which was placed over the rear driving wheels. The cylinder bore was 91.4 mm and the stroke about 100 mm (656 cc capacity); these dimensions resulting from the limitations of the casting art at the time. At the upper end of the crankshaft a bevel wheel gearing drove a horizontal shaft at half crankshaft speed and valve cam and ignition timer were mounted on the shaft. The inlet valve was of the sliding type fitted to the side of the cylinder, and driven by a rod attached to a cam. The exhaust valve was of the mushroom type and operated by a crank pin at one end of a lever, the other end of which carried a roller which engaged a cam on the horizontal shaft. A spring on the exhaust valve kept the lever roller in contact with the cam. The mixture was supplied by a surface vaporiser and an air mixing valve; ignition was by accumulator, coil and sparking plug, and the cylinder was watercooled.

When run on industrial gas the engine developed about 0.67 hp at a crankshaft speed of between 250 and 300 rpm. Later tests showed that 0.88 hp was developed at 400 rpm using petrol as fuel. This engine was fitted to a three-wheeled vehicle which was capable of a maximum speed of about 7 mph.

Later another single-cylinder engine of bore 116 mm and stroke 160 mm (1690 cc capacity) was designed, it developed 1.5 hp when running between 250 and 300 rpm and this engine was fitted to the three-wheeled vehicle which was the first production car to be sold (by Emile Roger, Benz's Paris agent, in 1888).

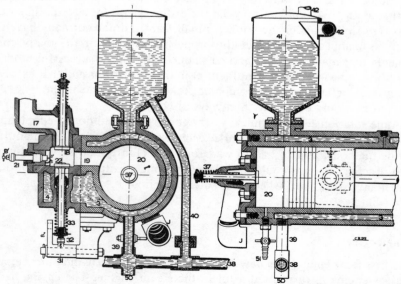

Figure 5.1 Two sectional views of the Benz 3 hp single-cylinder engine.

The Benz engines were heavy and underpowered. A more powerful engine (figure 5.1) was fitted to a four-wheeled vehicle in 1893 which became very popular, but otherwise there was little improvement in the basic design for the next ten years. The 1893 engine had both bore and stroke of 110 mm and it developed about 3 hp at 600 rpm. The valve arrangement was inlet over exhaust at the side of the horizontal cylinder, and the inlet valve was now atmospherically operated. The flywheel rotated in a vertical plane and was driven by spur gearing at the end of the horizontal crankshaft. A double cam was used to operate the exhaust valve, the large cam was used for normal running and a small cam made starting easier by lifting the exhaust valve to allow a small quantity of compressed mixture to escape just before ignition. The electrical ignition contact disc was mounted on the same shaft that carried the double cam. A second inlet valve was located at the end of the cylinder and this valve automatically admitted air into the cylinder to prevent the formation of a vacuum when the throttle was almost closed. No governor was needed, for the throttle valve enabled the driver to vary the engine speed over the range 250 to 900 rpm.

Followers of Benz. These early Benz engines were made under licence by E Roger in Paris and later by such firms as Delmer in Belgium, Arnold in England and Mueller in the USA. Roger used high-tension ignition and placed the engine above the rear wheel, but the crankshaft was horizontal and not vertical as in the original Benz engine. Other imitators of Benz, but who mainly made twin-cylinder engines, were Audibert-Lavirotte, Rochet-Schneider, Delahaye, Hurtu-Diligeon, and G Richard. Most of these firms made few changes to the basic model. The Audibert-Lavirotte engines were made with either one or two cylinders and to make starting easier the compression was reduced by opening a discharge valve by a lever and cam. Rochet-Schneider concentrated on balancing the moving parts. The Delahaye engine had two horizontal cylinders driving two cranks 180° apart to improve balancing; two cams were used to operate the exhaust valve, one functioning during normal running and the other advanced the exhaust timing during compression when starting the engine. A centrifugal pump circulated the cooling water and the engine could develop 6 to 8 hp at a speed of 700 rpm. These changes were typical of the small modifications made to Benz's design though most people also made improvements to the electrical ignition system.

Daimler engines

Daimler concentrated on making a small light petrol engine capable of running at higher speeds than the Benz engines. The first engine was aircooled, mounted vertically and consisted of two parts, the cylinder and the crankcase. The inlet and exhaust valves were located above and below one another at the side of the cylinder, the inlet valve functioning automatically, while the exhaust valve was operated mechanically by a rod arrangement worked by a spindle driven by a wheel at the end of the crankshaft. The engine speed was 750 rpm and the speed was controlled by a governor which put the exhaust valve out of action when the speed became excessive. Ignition was by hot tube, the petrol being supplied by a surface carburettor. This engine was the first to be fitted with a starting handle. The Benz engines were started by swinging the flywheel.

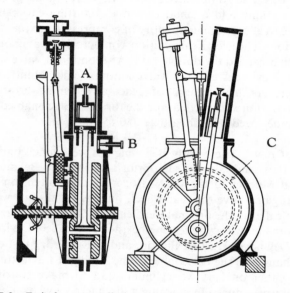

Figure 5.2 Daimler two-cylinder V-engine. Valve A in the piston is opened at bottom dead centre to allow air to be admitted through valve B to moderate the over-rich mixture supplied by the carburettor. By permission of Seeley Services and Co (from E C H Davies 1965 *Memories of Men and Motor Cars*).

Daimler next worked on a two-cylinder 'V' engine (figure 5.2), he wanted an engine that was no larger, but which had a greater power-to-weight ratio, and ran more smoothly than the single-cylinder engine. The new engine had a bore of 60 mm and a stroke of 100 mm, the cylinders were inclined at 15° to one another, and as in the earlier engine the inlet valves were placed immediately over the exhaust valves. Initially the pistons were provided with valves to improve scavenging but this arrangement was soon abandoned owing to overheating of the valves. The crankshaft was made in two parts, each part consisted of a disc flywheel enclosed in the crankcase, and a common pin between the two discs carried the two connecting rods. The exhaust valves were operated by pushrods actuated by sliding pieces which carried a roller and ran in two cam grooves cut in one face of a flywheel. The grooves were concentric but on one side they were interconnected by curves and the sliding pieces passed into the inner and outer grooves so that each exhaust valve was raised every other revolution. Overspeeding of the engine was prevented by a centrifugal governor which at a preset speed moved outwards and caused the tip of a rod to deflect the pushrod away from the exhaust valve on its upward movement; this action continued until the engine returned to its governed speed. This medium-speed engine made an important contribution to the development of the automobile in France. Messrs Panhard and Levassor acquired a licence in 1889 to build the engine and they used it on their first and highly successful cars of 1891. The earlier Peugeot motor car was also fitted with a V-type Daimler (supplied by Panhard and Levassor) but without the scavenging valve in the pistons.

The engine continued to be improved by Daimler and Maybach and in 1894 the larger of two versions made had a bore of 75 mm and a stroke of 140 mm developing 4 hp at 800 rpm. These engines were complicated and heavy and were replaced by a new engine known as the Phoenix engine. This was a two-cylinder vertical engine which represented a considerable advance on the earlier models. Panhard and Levassor fitted it to their cars in 1895 and it was the engine first built by the Daimler Company at their factory in Coventry.

English Daimler engines. Daimler made some improvements to the Phoenix engine so that the improved English Phoenix engine developed 4.5 hp (figure 5.3). It had a bore of 90 mm and a stroke of 120 mm, and a capacity of 1526 cc. The engine was constructed from a number of castings; the cylinder heads, the cylinder bodies, and the upper and lower halves of the crankcase. The region around the cylinder above the bottom of the stroke was watercooled. Again the inlet valve was atmospheric and the exhaust valve was operated by a cam follower and vertical rod. The bottom of the rod engaged a notch above the cam follower and when the speed of the engine exceeded the governed limit a governor on the crankshaft moved a cam along the shaft which through a system of levers lifted the vertical rod out of its notch and the exhaust valve remained closed until the engine speed became normal. The camshaft was driven from the crankshaft at half engine speed by means of bronze spur gears. A boss on the large gear worked an eccentric which drove a rotary water pump and assisted water circulation round the cylinders. Petrol was supplied by a float-feed Daimler Maybach carburettor. The ignition was hot tube but later models had electrical ignition.

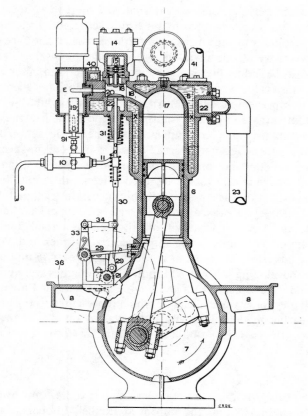

Figure 5.3 English Daimler Phoenix two-cylinder engine of 1899.

Other engines

In the middle nineties there was a proliferation of engines particularly in France which had become the leading automotive nation. Many people began by building Benz or Daimler engines and then with the experience they had obtained went on to build their own.

The single-cylinder engines did not differ greatly from one another, atmospheric (automatic) inlet valves were used, and the exhaust valves were mechanically operated. Engines were generally governed on the hit and miss principle with quite a variety of mechanisms being used. Ignition was electrical by trembler coil, or by hot tube.

All these single-cylinder engines suffered from excessive noise and vibration because of the great difficulty in designing a quiet well balanced, single-cylinder engine. In fact most of them would have shaken to bits if they had been run at high speeds for any length of time (and high speeds could be obtained by over-riding the governor).

de Dion Bouton, however, by reducing the weights of the reciprocating parts, better detail design and improvements to the electrical ignition system, were able to run their engines at 1500 rpm, about twice that of the Daimler and four times that of the Benz engines.

A de Dion Bouton engine is shown in figure 5.4. The piston was very light in weight and the H-section connecting rod passed through a slot in the top of the aluminium crankcase which was made in two halves joined in the vertical central section; the two flywheel discs acted as crankwebs. Oil was fed into the crankcase by a hand pump and distributed by splash. Engine speeds were controlled by a hand lever or pedal which regulated the exhaust valve lift.

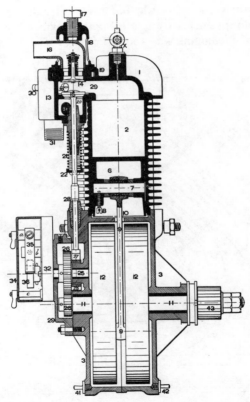

Figure 5.4 The 1.75 hp de Dion Bouton engine of 1899.

Other attempts were made to make the engine run more smoothly but efforts were concentrated on reducing the out-of-balance reciprocating forces and not the impulsive second-order forces (see p 83). The simplest arrangement was to use two cylinders and have one piston ascending while the other was descending as in the Daimler engines.

Two cylinders also gave much more power so that a number of two-cylinder engines appeared and even some four-cylinder engines. The additional cylinders were not different from the original cylinder but the designer had more scope for originality in the disposition of the cylinders, and how their outputs were coupled, as well as in the carburation and ignition arrangements. For example, in the V4 Mors engine of 1896 (figure 5.5) the cylinders (70 mm bore, 80 mm stroke) were inclined in pairs at 90° to each other. The cylinder barrels were aircooled and the heads watercooled. The connecting rods of each pair of pistons were connected to cranks 180° apart on a common crankshaft and each set of cylinders fired alternately to give two impulses to the crankshaft every revolution. Engine speed was varied by throttling the mixture and each cylinder had in effect a separate carburettor and throttle (but automatic inlet valves were still used). Four small pumps enclosed in an oil tank provided automatic lubrication. The output of the engine was about 7.5 hp.

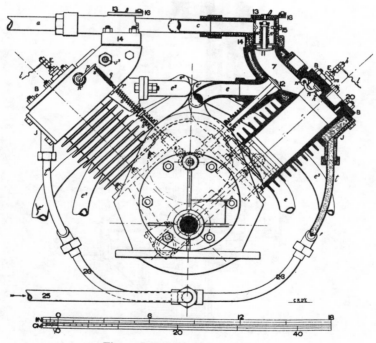

Figure 5.5 Mors V4 of 1896.

There was no agreement where the engine should be located. It could be placed at the rear, amidships, or at the front of the vehicle, and the cylinders could be vertical or horizontal or in V formation with the crankshaft transverse or parallel to the chassis frame.

Developments 1900–1917

The automobile engine at the beginning of the century worked, but it was a primitive piece of machinery, it was very heavy for its power output, and it was not very reliable. People objected to the noise and vibration it made, and it stank.

Most cars had only one cylinder and vibrations occurred because there was only one firing pulse every two revolutions of the slow running engine, and because the reciprocating piston and connecting rod gave rise to out-of-balance forces. Besides the primary forces caused by the piston thumping back and forth, there were also smaller secondary out-of-balance forces that arose because of the angularity of the connecting rod; the piston was not quite at its mid-point in the cylinder when the crankshaft had rotated through 90° from top dead centre, and as a result a secondary unbalanced force was set up at twice the crankshaft frequency. Vibration too was made much worse by the way most engines were governed, which in effect reduced the number of effective firing impulses per revolution in order to reduce speed when the engine was not fully loaded.

The rotating parts of the engine had to be balanced statically, otherwise vibrations would be set up, but static balancing was only a question of good design and workmanship. Dynamic balance was not important as speeds were low.

The engine of the touring car was lacking in power; more power (and a higher power-to-weight ratio) was needed not only to improve the performance of the car, but also to make driving easier. The driver had to be adept at adjusting the ignition and accelerator to keep the engine going properly when he overtook another vehicle, tackled a hill, or even just turned a corner. He also wanted a flexible engine, that is, one that would pull over a wide range of speeds without him having to attempt to change gear (one of the authors (RTS) had in the forties an elderly friend who always bought large, powerful US cars so that he could take off and go anywhere in top gear!).

Multicylinder engines

One way to increase power output was to increase the capacity of the engine, but the size of a cylinder could not be increased indefinitely because the greater the size and therefore weight of the reciprocating parts, the lower the speed at which the engine could run, so that its output did not increase correspondingly. Consequently it soon became more sensible to increase the number of cylinders than to increase their individual size, even though this added to manufacturing problems and cost. Running was also smoother because there were more firing strokes per revolution. Consequently a number of two-cylinder engines appeared in the nineties. The early Daimler engine had two cylinders in a narrow V configuration but generally the cylinders were in line. Two-cylinder in-line engines were balanced as far as primary forces were concerned provided the crankthrows were at 180° to one another, however two firing strokes occurred in one revolution but none in the next, and also the secondary out-of-balance forces were not balanced. There are no primary or secondary out-of-balance forces when the two cylinders are opposed, and a number of engines were built that had two horizontally opposed cylinders. Lanchester went a stage further coupling each cylinder to a flywheel. These flywheels rotated in opposite directions so minimising torque reactions (figure 5.6). Opposed-cylinder engines take up a lot of space and are expensive to build, but even so, small numbers of opposed fours and sixes have continued to be made to the present day.

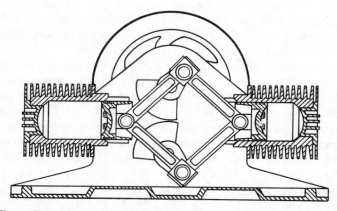

Figure 5.6 Lanchester engine showing two contra-rotating crankshafts geared together to provide primary and secondary balance. By permission of Seeley Services and Co (from E C H Davies 1965 *Memories of Men and Motor Cars*).

It was realised, however, that, except for small engines, two-cylinder engines doubled the disadvantages without doubling the advantages of single-cylinder engines. Three-cylinder engines were tried but had little in their favour and were obsolete by 1914. They were in primary balance provided the crank throws were at 120° to one another, but because the forces were not in the same plane a couple acted on the crankshaft (as it also did of course in a two-cylinder engine).

Four-cylinder engines appeared on racing cars in the nineties but not on tourers until the twentieth century when they were notably used on the 1901 Mercedes. The cylinders were in-line, for a V4 engine was heavier, more complicated and not as well balanced as an in-line engine. Four cylinders in line were in primary balance provided all the crankthrows were in the same plane, and if the two outermost throws were in phase and at 180° to the two inner throws no couple acted on the crankshaft. To prevent the firing strokes from 'winding up' the crankshaft, firing order was a matter of choice between 1, 2, 4, 3 and 1, 3, 4, 2. When speeds increased vibration from the secondary out-of-balance forces became an embarrassment in four-cylinder engines, and to overcome the vibration Lanchester invented his harmonic balancer. This consisted of two balance weights rotating in opposite directions and running at twice engine speed. It was mounted below the centre bearing and driven by spiral gears from the mid-point of the crankshaft. The weights rotated in opposite directions so there was no net horizontal force and they were phased to neutralise the vertical unbalanced force. Though it was inclined to be noisy the balancer made the four-cylinder engine nearly as smooth as a six-cylinder. Mitsubishi used a rather similar balancer in the seventies.

Six-cylinder engines behave like two three-cylinder engines in line, there are no out-of-balance forces and by making the throws of the first three cranks the mirror image of the second three, no net couple appears in the engine. Six-cylinder engines should also give 50 % more power than the corresponding four-cylinder engine as well as running more smoothly, but the six was troublesome to develop. It was difficult to make a long engine rigid, the crankshaft could twist, and it was very difficult to get the same charge in each cylinder and to ensure that each cylinder did its share of the work. On the other hand, some costs did not go up proportionately, for example, only one magneto, one carburettor and one fuel pump were needed.

Spyker made six-cylinder engines in 1903, Napier in 1904 and Sunbeam shortly afterwards. Sixes became established largely because of the persistence of Napier and the salesmanship of Edge who took every opportunity to promote Napier cars, so that Sixes became the mainstay of the Napier Company throughout the latter's existence. Another notable Six was that fitted to the 1906 Rolls-Royce Silver Ghost; it showed what smooth running could be obtained with a Six and there was consequently quite a fashion for Sixes during the next four or five years.

Attempts were made to produce straight-eight engines but with little success. They were too long and the charge distribution, crankshaft and ignition problems of the six were increased.

The V8 configuration gave a more compact and rigid engine than the straight-eight, Ader made one in 1903 and Darracq built a 23 litre 200 hp V8 engine for a racing car which reached 109 mph. de Dion Bouton produced a V8 engine for touring cars in 1910, and though it was not successful the very successful V8 made by Cadillac in 1914 owed a lot to this engine. Even so the Cadillac engine (figure 5.7) was a remarkable achievement.

Increasing the number of cylinders not only increased the output of a well designed engine but also increased its weight, along with the weight of the chassis that had to carry the engine, and a heavier vehicle meant higher cost, greater fuel consumption and more wear on expensive tyres. Consequently the first Fours and then the Sixes were used only on big expensive cars.

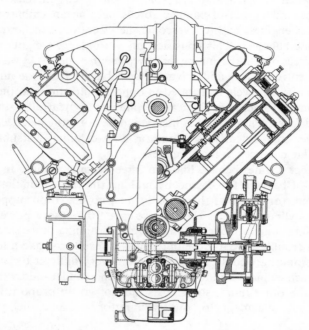

Figure 5.7 End view of the first of a famous line—the 1915 V8 Cadillac engine. By permission of Reed Business Publishing Ltd (from *Automobile Engineer*).

People who could not afford the cost of a Six but who wanted better performance and smoothness than a two-cylinder engine could provide bought Fours which have always been the biggest selling cars in the UK.

Four-cylinder engines were quite big and typically of three litres capacity in 1900 but from about 1914 much smaller engines of about one litre or so were built in considerable numbers for use on small cars. They were scaled-down versions of the bigger engines, and as they were intended to be inexpensive and reliable their design was generally conservative.

In the late nineties de Dion Bouton made a small car or voiturette which was powered by a 3.5 hp 402 cc single-cylinder engine of the type mentioned earlier. It was very successful and was quickly copied by other manufacturers, and the later forms of the voiturette were the forerunners of the light car. The evolution of the voiturette engine followed that of its big brothers, mechanically operated inlet valves, side valves, improved bearings, spray carburettor, etc. The single-cylinder engine was replaced by engines with two or more cylinders; the Rover when it went out of production in 1912 was the last of the voiturettes to be powered by a single-cylinder engine (it was a sleeve valve engine incidentally). Two-cylinder engines continued to be made, the Jowett 8 hp flat twin was in production basically unchanged from 1910 to 1954, but people wanted small cars rather than runabouts and so, as mentioned above, small four-cylinder engines were made.

For people who wanted really cheap transportation there was, after about 1910, the cycle car. Cycle cars were generally powered by motorcycle engines, some manufacturers built their own but many bought proprietary designs like the 8 hp JAP V2 engine.

By 1914 engines were better balanced than in 1900, bearing areas were increased, the crankshaft better supported, and lubrication and oil flow improved. As a result of all this mechanical improvement, and parallel improvements in carburation, ignition and cooling, engine speeds practically doubled between 1900 and 1914, from typically 750 to 1500 rpm, which resulted in engine output increasing from about 2 bhp/litre to about 10 bhp/litre.

Not that everyone was satisfied. Pomeroy in 1914 criticised contemporary engines for having insufficient breathing capacity and inadequate valve gear. He thought cam shape was not properly considered, high-compression ratios were wrongly considered pernicious, and designers showed a hopeless lack of enterprise!

Progress in the USA

The USA lagged behind Europe in engine design until the First World War. Early cars were runabouts powered by single-cylinder engines which were slower and heavier than contemporary European engines like the de Dion Bouton.

The Model A Ford was powered by a two-cylinder engine which was eventually replaced by a four-cylinder engine which was the forerunner of the 1907 Model T engine. The latter (figure 5.8) had a bore of 95.2 mm and stroke of 101.6 mm, a compression ratio of 4 to 1 and developed about 20 hp. Its basic design remained unchanged throughout its production life. Other manufacturers began building four-cylinder engines but these showed no advance over European engines; people wanted cars with powerful, flexible engines to cut down gear changing and the usual way to obtain power was to increase the number of cylinders rather than make the individual cylinders more efficient and work harder. Six-cylinder and later eight-cylinder engines appeared and by 1915 the Sixes were the most popular.

Though for many years most American engines appeared to European eyes to be unsophisticated and crudely finished, they were simple and reliable and did the job they were intended to do very well. Some manufacturers like Cadillac, however, built engines that were as good as the best in the Old World.

So far as manufacturing was concerned the European car industry in 1914 was only a cottage industry compared with that in the USA.

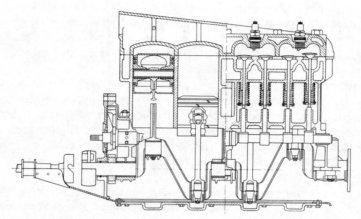

Figure 5.8 The Ford model T engine showing the detachable one-piece cylinder head covering all four cylinders. By permission of Seeley Services and Co (from E C H Davies 1965 *Memories of Men and Motor Cars*).

Valve mechanisms

A camshaft on the Benz engine rotated at half speed and operated a slide-type inlet valve fitted to the side of the cylinder and a poppet exhaust valve on the cylinder head. The inlet valve on the Daimler engine was 'automatic', or 'atmospheric'; the poppet valve was closed against a light spring, opened during the induction stroke and was kept closed during the rest of the cycle by the pressure in the cylinder. The exhaust valve was a mechanically operated poppet valve like that on the Benz engine. The inlet valve was mounted above the exhaust valve in a casting by the top of the cylinder.

Putting inlet and exhaust valves on either side of the cylinder instead of both on the same side, as in the Daimler engine, simplified casting of the cylinder, and this arrangement, the T head (figure 5.9), became the usual arrangement in the early years of the century. It had other advantages, the valves were readily accessible through plugs screwed in the cylinder head above them, and this was very important, because the valves were generally made of carbon steel which could not withstand high temperatures without wear and pitting occurring, so that the valves had to be frequently reground and replaced. The exhaust valve gave most trouble, for the inlet valve (and the sparking plug above it) could be cooled by the incoming charge, but the exhaust valve was heated by the exhaust gases and if the valve did not seat properly the escaping hot gases rapidly eroded valve and seating.

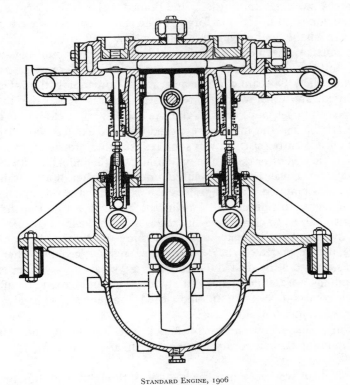

Standard Engine, 1906

Figure 5.9 1906 Standard engine showing T-head layout. By permission of Seeley Services and Co (from E C H Davies 1965 *Memories of Men and Motor Cars*).

The automatic valve was cheap but inefficient, it opened too late and closed too early, which resulted in only an attenuated charge reaching the cylinder and so by 1907, when Daimler at last changed over to them, mechanically operated inlet valves were standard.

The valves themselves were of the mushroom type. At first the valve head was screwed to the stem but later the valve was machined from a forging and this allowed it to be shaped. The cam did not engage the valve stem directly but through a tappet, the valve was then not subjected to a side-thrust and so did not wear unevenly. A small gap was left between tappet and stem to allow for thermal expansion of the latter for otherwise the valve might not seat when hot, and soon provision was made for adjusting the gap to allow for wear; too big a gap and the tappets gave an intolerable clatter. Valve springs closed the valves and kept them closed but these springs could give a lot of trouble.

Early camshafts were built up, the cams being machined into shape separately, mounted on the shaft and pinned or keyed in place. Later they were machined from the one forging. The cams themselves were of simple harmonic shape and engaged a flat on the end of the tappet. More complicated arrangements were also used, and the cam could engage a roller in the bottom of the tappet, or a rocker could be interposed between the cam and roller. The cam contour determined how rapidly the valve opened and closed, and for how long it remained open, as well as the lift.

At first the inlet valve opened and the exhaust valve closed at top dead centre (TDC), and the inlet valve closed and the exhaust valve opened at bottom dead centre (BDC). It was soon realised that pressure was still below atmospheric well after the piston was at BDC on the induction stroke and so the inlet valve was then not closed until well after BDC. Similarly the cylinder was emptied more rapidly if the exhaust valve was opened before BDC and closed just after TDC. Closing the inlet valve just after the exhaust valve also helped to clear the exhaust gases but it was feared that with too much overlap hot exhaust gases might blow back into the carburettor. Timing also depended on the type of valve, engine speed and other factors, and was arrived at by trial and error. For a typical four-cylinder engine in 1914 timings were: inlet opened 10° before TDC and closed 30° after BDC; exhaust opened 40° before BDC and closed 5° after TDC. The valve lift was between 0.25 and 0.375 in.

Timing depends on the performance level of the engine, but for a corresponding contemporary engine the figures could be 15°/45° and 45°/15°.

The T head was very inefficient. It was thought that the combustion chamber should approximate a sphere in order to minimise heat loss to the surrounding metal, whereas the combustion chamber of a T head was a shallow cylinder with side pockets. The T head also needed two camshafts, thus increasing cost and weight. As a result it was obsolescent by 1914 and was displaced by the side valve (or L head) engine and to a much lesser extent by the overhead valve (OHV) and the F head (overhead inlet valve; side exhaust valve) engines. The L head had inlet and exhaust valves side by side in the one pocket, making a more compact combustion chamber and one camshaft served both sets of valves. Access to the valves was obtained by screwed caps in the cylinder head, or, in a few cars, by detaching the cylinder head. The diameter of the valves on average was about half the cylinder bore.

The advantages of mounting the valves over the cylinder in the cylinder head had been realised at an early stage. The combustion chamber could then approximate a hemisphere and breathing was improved, so that an

OHV engine had a better performance and better fuel consumption than a side valve engine. However the valve gear of the OHV engine was more complicated and expensive, it was difficult to lubricate and also it was noisier than on a side valve engine. Valve regrinding was more difficult, as was decarbonising the engine, and if a valve stem should break the consequences were disastrous. Consequently OHV engines made only slow progress and in 1914 only about 10 % of the new cars of that year in the UK had OHVs.

Two OHV arrangements emerged. In one the valves were operated by pushrods which in turn were actuated by a camshaft located in the side of the engine, and in the other arrangement the valves were operated directly by an overhead camshaft (OHC). The hybrid F head was used by a few manufacturers; the overhead inlet valve was operated by a pushrod and the exhaust valve directly through a tappet as on an L-head engine. A large inlet valve could be used in this arrangement.

Racing cars wanted the most powerful engines possible and by 1914 GP and such cars all had OHVs generally operated by two overhead camshafts, one camshaft for the inlet valves and one for the exhaust valves. There could be four valves per cylinder, for the valves could be made lighter, and the breathing of the inlet valves and the cooling of the exhaust valves improved. Some engines on touring cars also had more than two valves per cylinder.

Sleeve valve engines. The Knight sleeve valve engine was used on cars made by Daimler, Minerva, Panhard, Willys-Overland and other firms. In this engine the piston ran in the inner of two concentric sliding sleeves (figure 5.10) which were provided with ports or slots, and the sleeves were moved by two short connecting rods driven by the camshaft in such a way that the ports were in line with a passage to the carburettor, during the induction stroke, and again in line with a passage to the exhaust manifold during the exhaust stroke. The sleeve valve engine was quieter than one with poppet valves, it was also simpler and a higher compression ratio could be used. The trouble was lubricating the sleeves, for although they could be given a reasonable life for the times (about 50 000 miles) the oil on the bores tended to burn and give a very noticeable blue haze to the exhaust gases.

Largely due to the efforts of Lanchester, who was acting as consultant to Daimler at the time, the Silent Knight was successful when it appeared in 1908 and from about 1910 until 1935 Daimler made sleeve valve engines only.

The Burt-McCollum engine had only one sleeve which reciprocated and rotated back and forth in order to line up the ports. It was used on Argyll cars and on some Vauxhalls. The sleeve valves were well protected by patents (litigation over the Knight patents cost Argyll £50 000 at a time they could ill afford it) and were taken up by only a few firms, and so they were not as intensively developed as the poppet valve. Indeed the quietness of the sleeve valve engine caused manufacturers of poppet valve engines to think of ways of quietening their own engines.

Instead of reciprocating a sleeve, a simple slide valve could be moved up and down as on the Benz engine, but it was difficult to seal and lubricate such a valve.

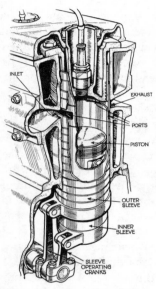

Figure 5.10 A sleeve valve engine. By permission of Haymarket Publishing Co (from *Autocar* and *Autocar Handbook*).

Rotary valves. Rotary valves appeared to offer many advantages, they were quiet, gave good breathing (because the ports could be made large), there were no losses in springs, and they were mechanically simple. Consequently many rotary designs were put forward and in a few instances rotary valve engines were made. Two designs, Darracq (figure 5.11) and Itala, went into production but not many engines were made and significantly Itala's profits took a dive after the engines went into production. All the designs failed because the valves could not be properly sealed and made gas tight, and because of lubrication troubles, though these factors are interrelated. Rotary valve engines, however, continued to be designed and made in the thirties and later, notably by Cross and Aspin, but again nothing came of this work. As Cross pointed out, the poppet valve had been the first in the field and subsequently highly developed so that the rotary valve never got the attention he thought it deserved.

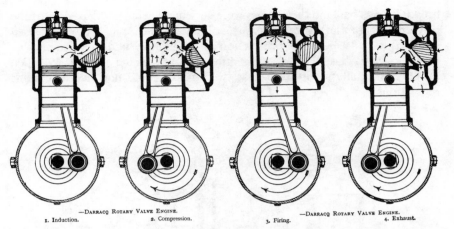

1. Induction. 2. Compression. 3. Firing. 4. Exhaust.

Figure 5.11 Darracq rotary valve engine. By permission of Haymarket Publishing Co (from *Autocar* and *Autocar Handbook*).

Two-stroke engines

Two-stroke engines not only worked but could give quite a respectable output at low speeds, for every second instead of every fourth stroke was a firing stroke. Because of their simplicity they were much used on motorcycles, on which they were quite successful, and also on cycle cars and much later, in the fifties, on bubble cars and minicars. They were also used on some bigger cars such as the Trojan in the UK and the DKW in Germany.

Sir Dugald Clerk invented a two-stroke engine in 1879, and Benz made a two stroke at one time, but most two strokes derive from the Day three-port engine of 1889.

In the latter type of engine (figure 5.12) mixture was drawn into the crankcase during the upward stroke of the piston through a port or a non-return valve, and was compressed in the crankcase when the piston descended until the latter neared the bottom of its stroke, when it uncovered an exhaust port through which the exploded charge escaped and then almost immediately opened a transfer port through which the crankcase gas passed to the cylinder where it was further compressed and fired near TDC. A baffle plate on the top of the piston helped prevent outgoing and incoming gases from mixing. Later the crown of the piston could be shaped for the same purpose.

Another arrangement for the two-stroke engine was for two pistons to share a common combustion chamber and crank pin but with the pistons slightly out of phase. The transfer port was near the bottom of one cylinder and the exhaust port near the bottom of the other.

The great advantage of the two-stroke engine was its mechanical simplicity, there were no valves and very few moving parts. Needless to say

94   THE ENGINE

breathing was bad and the engines were very inefficient. Even so some of the tiny engines of the fifties gave surprisingly high outputs. Trojan ceased production in 1936. DKW production began in 1929 and these engines or rather engines based on them continued to be made until 1969 when production finally stopped because two strokes could not meet emissions regulations.

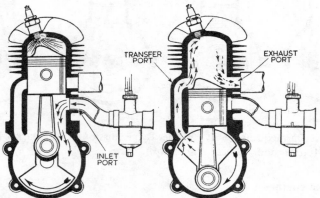

In the two-stroke engine, mixture is sucked into the crankcase as the piston rises and transferred to the cylinder head as it descends. Exhaust coincides with the transfer period and is assisted by the " loop scavenge " effect of the fresh charge

Figure 5.12 Principle of the two-stroke engine. By permission of Haymarket Publishing Co (from *Autocar* and *Autocar Handbook*).

Governing

Early engines ran well at only one speed and that was at approximately full load, mainly because their carburettors supplied a mixture of constant strength, which was too rich or too weak at other speeds and loads. The engines therefore had to be run at more or less constant speed and this was done by fitting centrifugal governors which over-rode the normal valve mechanism. In de Dion Bouton engines, for example, the governor regulated the lift of the exhaust valve so that a full charge could not be admitted through the inlet valve. In the 1901 Mercedes the lift of the inlet valve was limited by the governor, and in other systems the governor, by a complicated system of levers, prevented the exhaust valve from opening. To reduce his speed in traffic the driver operated an accelerator, by means of a hand lever or pedal, which changed the setting at which the centrifugal governor cut in. The engine fired unevenly when the load was reduced which could be very unpleasant, especially when the vehicle was stationary. Consequently it was not long before this 'hit and miss' system was discarded and the governor operated on a throttle in the carburettor. Developments in carburation rendered the governor unnecessary and the driver regulated engine speed directly through the throttle.

The combustion chamber

Simple thermodynamics shows that increasing the compression ratio (CR) of an engine increases its power output, but contemporary fuels limited the CR to about 4 : 1; at higher ratios the engine began to knock. The higher the CR the more difficult it was to make the pistons gas tight and the greater the piston friction. Breathing was poor but this was soon realised so piping and passages were made wider and shorter, and inlet ports and carburettor better matched. As Lanchester said, improving its breathing 'could turn a pig of an engine into a willing horse'.

It was considered that the burning charge should lose as little of its heat as possible to its surroundings and therefore the combustion chamber should ideally be spherical (smallest surface area per unit volume). Some of the early chambers were indeed hemispherical. A 1905 Fiat head was not only hemispherical but had inclined overhead valves and this type of chamber soon became widely used on racing cars. The T head with its two lobes was obviously inefficient. The L head or side valve chamber was more efficient in that it had only one lobe but racing experience showed that the OHV hemispherical head had the most potential.

Most designers, however, thought that they had more important problems to worry about than the precise shape of the combustion chamber.

Cylinder block and crankcase

Development of the cylinder block depended on progress in foundry work and metallurgy. Early engines had detachable heads, but it was found so difficult to make the head gas tight that the heads were cast integrally with the block. It was not long before two-cylinder engines were made with both cylinders cast in one block, and when four-cylinder engines appeared they were at first made by mounting two two-cylinder blocks in line on one crankcase. Similarly six-cylinder engines were assembled from two three-cylinder castings or three blocks of two cylinders. Engines made in this way were not as rigid as when all cylinders were cast in the one block (monobloc), they were longer, and cooling and pipe work were more complicated. Even when sound monobloc castings could be made many manufacturers continued to use bi-cylinder blocks as they were lighter and easier to manhandle during manufacture and when they had to be removed for the engine to be decarbonised or worked on.

However, the advantages of monobloc construction were so great that monobloc engines eventually became standard though they were by no means general before 1914. Improvements in casting not only made the blocks lighter but enabled complicated water passages to be incorporated in the block and so external water jackets of copper, brass or steel to enclose the block were no longer needed. Inlet and exhaust manifolds could be cast in the block so that the appearance of the engine was made much neater though there was not necessarily any improvement in performance.

The engine had to be periodically decarbonised which was a difficult job for it might have to be completely dismantled. A detachable head would have greatly simplified things particularly with side valve engines. This had to wait until reliable copper–asbestos gaskets were available. The Model T Ford, however, had a detachable head on a single cylinder block.

A simple barrel-shaped crankcase sufficed with one- or two-cylinder engines. It was not difficult to assemble an engine with such a crankcase if the crankshaft had only two main (or engine) bearings, and ingenious ways were found of inserting a third bearing. A split crankcase was needed for any bigger engine, with the cylinder blocks bolted to the top part of the crankcase which was of aluminium alloy, and webs cast in the crankcase to carry the intermediate engine bearings. Aluminium is light and therefore thick sections could be used. The bottom part of the crankcase was also generally of aluminium and acted as a sump for the engine oil. It could be ribbed externally for better cooling and carry internal baffles to prevent the oil from surging on hills. The crankcase was provided with hangers or brackets; if these were connected directly to the chassis frame the crankcase could be heavily stressed when the frame twisted and so it became standard practice to mount the engine on a subframe (see Chapter 15).

If the engine was not carried by a subframe it was generally supported by a trunnion bearing at the front and a pair of hangers at the rear of the engine.

By 1915 or so most cars in the USA had engine and gearbox assembled as the one unit which was suspended from the frame at three points and no subframe was used.

Stroke-bore ratio

There was little agreement as to what the ratio of stroke to bore should be in the early years of the century and ratios varied from 0.75 for the Lanchester engine to 2 for the Panhard-Levassor. Many engines were more or less square (equal bore and stroke) and for the others the average was about 1.2.

Changing the stroke-bore ratio changed so many other things that it was difficult to decide on an optimum value or even if there was an optimum. For example, a higher ratio gave a lighter piston but higher piston speeds and a heavier connecting rod. Some long-stroke racing engines did very well and it was eventually thought that the touring car in Europe and the UK should have a relatively small bore and a correspondingly long stroke, and the ratio was generally about between 1.3 and 1.5 by 1914. Short-stroke engines were more popular in the USA. However, another factor, and one that was not technical, influenced the ratio in the case of UK cars, and this was the RAC rating.

RAC rating

There was some argument about how engines should be rated in hill climbing competitions and the Royal Automobile Club (RAC) came up with a rating system that was not only used for competitions, but was approved by the Society of Motor Manufacturers and Traders (SMMT) in 1906, for catalogue purposes, and was adopted by the Treasury as the basis for taxing vehicles under the Finance Act of 1910. The rating put engine horsepower as equal to $nD^2/2.5$ where n is the number of cylinders and D the diameter of the cylinder bore in inches. This formula was correct for a brake mean effective pressure (BMEP) of 67.2 lb in^{-2} and a piston speed of 1000 feet per minute, which were typical pressures and speeds at the time the rating was introduced. Pressures and speeds increased and so the formula got progressively more and more in error, but the real trouble was that the formula considered cylinder bore only and not capacity (or engine speed).

The rating encouraged in the twenties and thirties the development of small engines which had high stroke-bore ratios and high engine speeds. It was not taxation only that was responsible for this situation but also that first cost and insurance charges, etc were related to the RAC rating. The rating did much harm to the export industry, people overseas wanted efficient engines, not the noisy high-speed type that was designed to minimise tax in the UK.

Pistons

Pistons were made of grey cast iron and fitted with cast iron piston rings. These were heavy, but it was difficult to reduce their weight much because if the crown was made too thin it overheated, and the piston could expand and distort, stick in the cylinder, and the lubricating oil carbonise leading to pre-ignition.

Later pistons were made from steel or were composites with steel head and cast iron skirt to reduce weight, but they tended to be noisy. Pistons were flat headed or slightly domed and they were turned to give a high surface finish. Internally they could have quite complicated shapes to help heat flow from head to skirt.

Attempts were made to use aluminium pistons and Rolls-Royce for one had obtained fairly satisfactory results from aluminium pistons in 1903 but aluminium pistons were not adopted on automobile engines in the UK until after the First World War. Aluminium pistons were tried in the USA during the war years, but despite all the work done on them they were found not to last as long as cast iron pistons, and they were more expensive. Consequently many US manufacturers reverted to cast iron pistons about 1918.

Crankshafts

Early crankshafts were built up, but on later engines the crankshaft was forged and machined. On some of the smaller engines only two engine bearings were used but when the number of cylinders increased more bearings were fitted to reduce the tendency of the shaft to spring and to minimise bearing wear. For example, it was considered good practice to fit seven main bearings to a six-cylinder engine. The greater the number of bearings the longer the engine tended to be, and of course the more expensive it was.

The crankshafts of Sixes were long and could be whippy and a serious problem arose if the natural frequency of the torsional vibrations of the crankshaft matched engine speed, or a multiple of engine speed, so that resonance occurred. The vibrations set up were not only objectionable to driver and passengers but could also damage the engine. Royce encountered this problem and overcame it by fitting to the crankshaft an auxiliary flywheel which was allowed a small amount of radial movement on its hub so that it acted as a friction damper. In 1909 Daimler had some six-cylinder engines which vibrated very badly and Lanchester invented his damper in order to release them for sale. The damper consisted of a flywheel mounted at the front end of the crankshaft and coupled to it through a lubricated multi-disc clutch; this assembly acted as an effective viscous damper. There could be trouble with leakage of oil and the change of viscosity of the oil with temperature, so the oil was replaced by rubber. Much later silicone oil was used because its viscosity changes little with temperature, and the damper was welded inside an oil-tight container. Provided the crankshaft was well designed in the first place the Lanchester damper allowed the six-cylinder in-line engine to run free from torsional vibrations and enabled the Six to attain its potential refinement.

Developments 1919–1939

The major UK manufacturers stuck to four-cylinder engines for some years after the First World War, leaving Sixes and Eights to firms making small numbers of high performance and luxury cars.

The situation changed in the mid-twenties and the big firms started making Sixes and some Eights. The six-cylinder engines were of about two litres or more, but soon much smaller six-cylinder engines, some barely a litre in capacity, were produced and there was a craze for small Sixes. A small well designed and well made six-cylinder engine could run as smoothly as a sewing machine, but the smoothness and power were gained at increased complexity and cost. Many of the engines, however, were not good value for money.

At the Motor Show in 1920 only 15 % of the models shown were Sixes, in 1926 35 % were Sixes and at the 1930 show Sixes outnumbered all other sizes. However, by the early thirties a four-cylinder engine could run nearly as smoothly as a six, thanks to resilient rubber mountings and lighter pistons. It could also now deliver a great deal more power, and more power, plus the synchromesh gearbox, made gear changing less frequent and easier. And the four-cylinder engine was cheaper than a Six. Consequently the Four regained much of its previous popularity and in 1939 about three quarters of the cars sold in the UK had four-cylinder engines.

Four-cylinder engines were used on the small mass-produced cars; various ways of making them cheap and easy to produce were used, for example, overhead valves were replaced by side valves, no provision made for adjusting tappets, the central main bearing of the crankshaft omitted, etc.

A number of firms made straight Eights but with varying success. Only a few were technically and commercially successful and some were called Occasional Eights as only occasionally did they fire on all eight cylinders! As to be expected, troubles were experienced with the long crankshaft and in getting more or less the same charge in all cylinders (some contemporary six-cylinder engines could have the same troubles).

The eight-cylinder in-line engine had to be tucked in under a very long bonnet and a good stylist could make a long bonnet attractive, particularly on open cars. There was next a fashion for Eights and most firms tried to include one in their range. Even some family saloons like the 1928 Hillman Vortic had a straight-eight engine and this was despite the Depression.

Development and tooling costs were too great and the potential market too small for most UK firms to risk making a V8 though Daimler built a double-six V12 sleeve valve engine in 1926 which ran so smoothly that it was practically impossible for a passenger to detect whether or not the engine was running.

Things were very different in the USA. In 1920 engines were mostly in-line Fours and Sixes. In 1940 a single-cylinder engine was being made for the Bantam, Crosky and Willys were making four-cylinder engines, and there were 20 six, 25 eight, two twelve and one sixteen-cylinder engines in production. Thirty nine of these were in-line and the rest Vs. Only ten had overhead valves and twenty were more than four litres in size.

The combustion chamber

About 1911 it was realised that another factor besides chamber shape and breathing affected combustion, and that was turbulence. Sir Dugald Clerk showed that the flame front in a gas could be made to propagate 2.5 times faster by making the gas turbulent. Turbulence could therefore cause higher pressures and permit greater engine speeds. A number of people worked on this idea and Ricardo, in particular, investigated turbulence in the early twenties. He also examined the phenomenon of knock and as a result of this work, realised that by shaping the combustion chamber and properly locating valves and sparking plug, compression ratios and thus engine outputs could be considerably increased. After careful experimentation he invented a turbulent head for side valve engines.

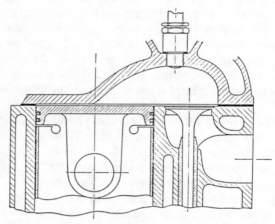

Figure 5.13 Typical Ricardo turbulent head. By permission of The Council of the Institution of Mechanical Engineers (from *Combustion Chamber Design and the Influence of Fuel Quality*).

In the Ricardo head (figure 5.13) the chamber was displaced towards the side valve pocket, so that part of the piston and cylinder head were separated by only about 0.1 in at TDC; this area of close approach was called the squish area. Squish served two purposes, it generated considerable turbulence at the end of the compression stroke, and the gas trapped in the squish area, the end gas, was cooled by the metal surfaces, and therefore was less likely to detonate and cause knock. The sparking plug was placed near the exhaust valve so that the hottest gas was ignited first, and the expanding flame front had the minimum distance to travel. The Ricardo turbulent head enabled the compression ratio to be increased to about 4.8 : 1 and the output of the engine was increased by about 20 %. The octane number of the contemporary fuel was 45 to 50. Manufacturers used combustion chambers based on Ricardo's designs, or combustion chambers of their own but based largely on his ideas, which were often developed for them by specialist firms. The Americans realised the significance of the Ricardo head and used it extensively, often without paying royalties.

Midgley in the USA showed in the early twenties that knock could be delayed and compression ratios increased by doping the petrol with lead compounds, but as the compression ratio increased and the size of the combustion chamber decreased, it became more and more difficult to devise an efficient side valve head, and in any case the side valve engine did not have the potential of the OHV engine. Even so the turbulent type head gave the side valve another twenty or more years life.

Some manufacturers, for example, Austin, Ford and Standard, did not make overhead valve engines until after the Second World War. Firms like Lanchester, Vauxhall, Rover, Riley and Lea Francis were building OHV engines in the twenties, MG and Singer in the thirties and they generally stuck to OHV (or F head for Rover later).

Some manufacturers gave up OHCs and turned to pushrod-operated OHVs.

Overhead valves. The hemispherical chamber with inclined valves was efficient but expensive. Its breathing could be increased by inducing swirl which could be achieved by injecting the incoming charge tangentially. Swirl is exploited in the laboratory trick of emptying a Winchester quart bottle. It is a tedious job emptying the bottle normally but if the bottle is inverted and its contents given a rotary motion so that they are set swirling the bottle is emptied in a remarkably short time.

The compression ratio was limited for a true hemisphere without squish, but this could be overcome by doming the piston crown to bring it closer to the head surface. The hemispherical chamber could be approximated by making the surfaces containing the valves planar, and the resulting pentroof chamber was easier to manufacture, the valve gear simplified, and therefore costs reduced.

The inverted bath tub chamber which appeared in the early twenties was also cheaper to make for the vertical valves were in line on the flat roof of the chamber. The chamber could be made quite small and there was plenty of squish.

Inclining the flat roof of the bath tub chamber generated the wedge-shaped chamber, which was used on some American cars in the late thirties. The sparking plug was mounted on the thin edge of the wedge opposite its apex and the valves carried in line on the top face of the chamber. A number of manufacturers continued to use F heads.

There was a great variety of combustion chamber and head designs, and some of them were very ingenious, but until the thirties and the advent of higher octane petrol the limiting factor as regards overhead valves was generally breathing troubles rather than combustion chamber shape.

Valve mechanisms

Valves at the start of the period were made of 3.5 % nickel steel, but later silchrome and other alloys were used, with a more heat-resistant alloy for the exhaust valve if it was of different size from the inlet valve. Later too materials such as Stellite had to be developed that were resistant to attack by the lead compounds in leaded fuels.

When engine speeds increased the rate of operation of the valve springs could reach their natural frequency, when they would surge and not keep in step with the valves. The forces developed during surging could be high enough to cause fracture of the springs. Various ways were tried to overcome surging, usually by increasing the natural frequency of the springs and making them lighter, or by using other types of spring, such as hair-pin, leaf, or torsion springs, or by improved manufacturing processes, but the most effective way was found to be to use two (or more) concentric springs per valve, each of different frequency.

Another approach to overcoming spring surge and spring fracture was to do away with the spring altogether, and to use desmodromic valve gears (figure 5.14) in which one cam opened and another cam closed the valve. The trouble with such gears was that they had to be very accurately designed and made, and even then it was difficult to ensure gas tightness when the valve was closed without unduly stressing the components. Desmodromic gears have appeared from time to time but have never looked like challenging conventional valve gears.

Figure 5.14 Desmodromic valve gear on 1912 Peugeot. By permission of Reed Business Publishing Ltd (from *Automobile Engineer*).

Tappet rattle was always a problem and palliatives like felt tips on the tappets were not much use. An ingenious way of quietening the tappets had been used by Amedee Bollée Jnr before the First World War, which was developed in the USA in the early thirties. In effect oil at engine pressure took up the tappet clearance and a valve trapped the oil when the valve lifted and released it on the downstroke of the valve. No matter how big or how small the clearance volume it was therefore always filled with oil. The hydraulic tappet worked very well provided the engine speeds were not too high to cause a pumping action on the tappet and prevent the valve from seating properly.

There were improvements in detail design of the valves and in the materials used but otherwise the valve mechanism of the side valve engine changed very little. The situation was different for overhead valve engines for which a number of systems were developed and improved.

In the pushrod OHV engine the camshafts engaged tappets and these in turn reciprocated vertical pushrods mounted in holes drilled in the cylinder block. Rockers were pivoted on the top of the block and the pushrods bore against one end of the rockers and the tops of the valve stems against the other end. Tappet clearance was adjusted in various ways and the simplest and most convenient was for the cup-ended top of the pushrod to engage a ball on the rocker, which could be screwed up and down and locked in place. To reduce reciprocating weights the pushrods were made as thin as possible—they should be 'just like knitting needles' according to Georges Roesch, who designed some very successful pushrod engines for Talbot (figure 5.15). But not all designers were as skilful as Roesch. Valve gear and shafts might not be rigid enough, pushrods could deflect (and even break)

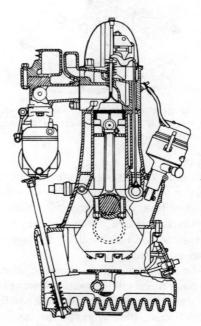

Figure 5.15 Talbot pushrod engine. By permission of Seeley Services and Co (from E C H Davies 1965 *Memories of Men and Motor Cars*).

which caused the timing to go astray and the engine to lose power. The gear could be excessively noisy and need frequent adjustment and maintenance.

It was relatively simple to produce engines with the pushrods vertical, and the valves vertical and in a row, so that system eventually became the most widely used. But other pushrod arrangements were used. Some of these were very ingenious and most were devised so that the valves could be inclined and mounted on either side of the combustion chamber, and still be operated by the one side-mounted camshaft. They were generally complicated and required excessively long pushrods.

Overhead cam systems (figure 5.16) were expensive and the drive could be noisy and troublesome but even so OHCs were used on quality cars like the Lanchester, and the valve system and chain drive carefully designed and very well made.

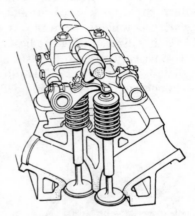

Figure 5.16 Typical overhead camshaft layout of the thirties. By permission of Haymarket Publishing Co (from *Autocar* and *Autocar Handbook*).

The single overhead cam design was quite simple; the cams operated rockers and the rockers operated the valves. The valves could be in line, or inlet and exhaust valves inclined and arranged on either side of the line of rockers. Adjustment was simple and rather similar to that for pushrods. Single OHCs were used on some quite small cars, Morris tried OHCs on the Morris Minor but reverted to side valves.

On twin OHC engines the cams could operate directly on the valves through rockers or through fingers which were rockers pivoted at their ends. Another arrangement was for the cam to engage a small inverted bucket in its own cylinder, and the valve stem engaged the bottom of the bucket in such a way that it was not submitted to any side force. Clearance was adjusted by inserting shims between bucket and valve stem. Twin cams had the same drawbacks as single cams but more so. Chains tended to be noisy and short lived, and maintenance was difficult. Even just adjusting tappet clearance could involve a lot of stripping and reassembling, in addition to the retiming of the valve gear.

Some cars like the Riley 9 of 1926 had twin camshafts but these were on either side of the block as in the old T head and operated the OHVs by pushrods.

Overhead valves fitted in with styling changes. Side valves were readily accessible when the bonnet opened in two halves along the centre-line of the car, but wings were now merging with the body, and the bonnet lid was hinged at one end and opened at the other, and so the OHV mechanism could be made accessible whereas the side valves were no longer so.

The poppet valve was now almost as quiet as the sleeve valve and so the latter was phased out in the thirties. Daimler ceased production of sleeve valve engines in 1935 and took over Lanchester in order to acquire a successful poppet engine.

Though the automobile sleeve valve engine died out in the thirties the sleeve valve was used very successfully on aircraft engines and the most powerful piston engine of the Second World War was a sleeve valve unit. It had 24 cylinders and was reputed to have developed over 5000 bhp. Some sleeve valve aircraft engines required so little attention that they ran for 5000 hours before they were given a major overhaul, and 5000 hours is twice the working life of most cars.

Camshafts and camshaft drives

Camshafts continued to be machined from steel forgings, the cams being formed integral with the shaft which was then hardened and ground. The importance of gas inertia was realised; the valve timing adjusted to take this into account, and much more work now went into optimising the shape of the cams and in particular making the valves close more quietly.

A number of ways of driving the camshaft from the crankshaft were tried. On pre-war side valve engines the drive had been through gears or a train of gears, but later a chain drive was often used. Spur gear drives could be noisy, herring bone or spiral single gears were quieter but they were expensive. Composite metal fabric gears were tried but wore rapidly.

Similar drives could be used with pushrod OHVs, but driving overhead cams was more of a problem. Bevel gears at each end of a vertical shaft were tried, and Bentley used eccentrics. Eventually relatively silent chain drives were developed and these could be of the inverted V type or, later, roller chains. The latter could be double or even triple—the chains being run side by side but with common links on the inner sides. A spring-loaded idler sprocket, or a curved blade spring bearing against a length of chain, took up slack. Chains could be made short on side valve engines and two or more chains with an intermediate sprocket enabled short chains to be used on OHC engines. Chains had to be well lubricated and so were generally enclosed to save mess and make things quieter.

The camshaft was found to be a convenient component from which to drive the oil pump and the distributor.

Cylinder blocks and crankcases

Before long all four-cylinder blocks in the UK were cast monobloc, but some six and eight-cylinder engines continued to be built up from smaller groupings of blocks.

Detachable heads also became standard in the early twenties, though engines with very big bores, and some racing engines, could still be made solid because of the large forces developed with cylinders of large diameter. Aluminium alloy heads were later used on some high-performance engines; the heads ran cooler than cast-iron heads because of the much greater thermal conductivity of the aluminium, and so compression ratios could be increased by about one unit, and the output of the engine increased accordingly.

Once monobloc castings were used the cylinder block could be made even more rigid, though at the cost of extra weight, by casting the top part of the crankcase integral with the cylinder block, and not using a separate aluminium crankcase. This had been done on the Model T Ford but such an integral casting did not become general in the UK until the late twenties. There was then a trend to make the integral crankcase as shallow as possible in order to reduce the weight of the engine, but to do this the stroke had to be decreased and therefore on a four-cylinder engine the out-of-balance secondary forces increased. However, aluminium alloy pistons were now more or less standard and resilient rubber mountings were being introduced.

Aluminium alloy blocks weighed much less than the corresponding cast iron blocks. They had been used very satisfactorily for aircraft engines and were used on some high-performance and luxury cars. The aluminium alloy was too soft for cylinder walls and so cast iron cylinder liners had to be used. The liners could be dry liners when they were inserted into the aluminium bores, or wet liners when they were in contact with the water. Wet liners were thicker than dry liners and had to be sealed or gasketed to prevent loss of water. Hard inserts also had to be developed in which to mount the valves.

Liners were also used on some cast iron blocks, the liners were centrifugally cast from high grade cast iron that had much better wear properties than the free flowing cast iron used to make the block. Liners could also be replaced when they were worn, which was cheaper than reboring the block and fitting outsize rings. A rebore would be expected in the thirties after about 25 to 30 000 miles with an ordinary block. People complained more and more of high cylinder wear in the twenties which at first was attributed to the new aluminium pistons. It was found, however, that wear was much greater if the car was used on short runs than if it covered the same distance in long trips, and it was shown that the wear was accelerated by chemical corrosion of the cylinder surface by acid products of combustion condensing on the cold cylinder walls. Wear could therefore be reduced by bringing the surfaces to operating temperature as quickly as possible, and eventually all cars were fitted with thermostats, so that water was not circulated through the block until temperatures were too high for condensation to occur.

Cylinder wear was further reduced after the Second War when chromium-plated piston rings were introduced, and oil additives used to combat cylinder wear and corrosion. The crankcase, integral or not, was a convenient home for the starter motor, generator, magneto (if fitted), distributor, fuel pump, dip stick, etc.

Stroke-bore ratio. Squarer engines reduced piston speeds for a given engine speed and the larger the bore the bigger the valves could be, particularly on OHV engines (though on some heads the valve openings overlapped the cylinder bore). Consequently when technical interests predominated there was a trend to squarer engines. Curiously piston speeds remained almost constant from 1920 to 1940.

Pistons

W O Bentley sold DFP cars before the First World War and on a visit to the works in France he noticed an aluminium piston on the Chief Engineer's desk. The piston was a sales gimmick from an aluminium firm, but it gave Bentley ideas and he had aluminium pistons made up by DFP which were tried out and got to work successfully in racing engines. The Balloon Factory (later the Royal Aircraft Establishment, Farnborough) were also working on aluminium pistons and during the war these were used very successfully on aircraft engines. Work was therefore carried out to adapt such pistons for automobile engines.

Aluminium has a higher thermal expansion than cast iron so that the aluminium piston had to have some clearance in the cylinder bore when cold, and this was taken up as engine and piston warmed up. This clearance caused objectionable piston slap and oil leakage when the engine was cold which could result in the bore wearing to an oval shape.

Various ways were tried to overcome piston slap. One way was to taper grind the piston so that it was wider at the bottom than at the top, where the expansion was the greatest. The piston could be ground oval so that it was slightly less in diameter across the gudgeon pin and therefore expanded sideways rather than lengthways, for the front part of the piston carried most of the load during the firing stroke. Slots could be cut in parts of the circumference at the top of the skirt to direct heat flow to the less heavily loaded parts of the skirt. The back of the skirt could be slit at a slight angle to the vertical, and the slit closed up slightly to relieve the piston. More than one of these ideas could be used at the same time, and pistons (figure 5.17)

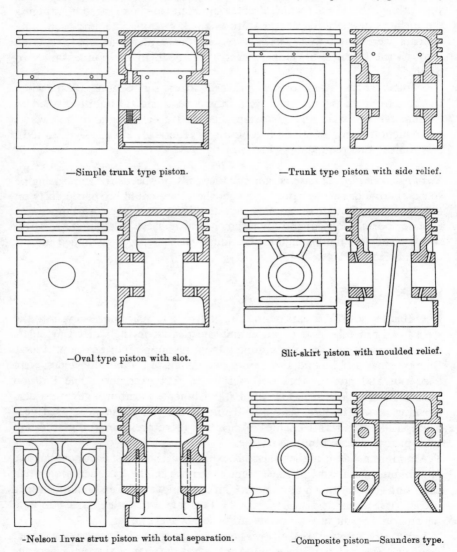

—Simple trunk type piston.

—Trunk type piston with side relief.

—Oval type piston with slot.

Slit-skirt piston with moulded relief.

-Nelson Invar strut piston with total separation.

—Composite piston—Saunders type.

Figure 5.17 Some pistons of the twenties. By permission of The Council of the Institution of Mechanical Engineers (from *Institute Proceedings*).

were cast with various internal struts and ribs to transmit the thermal and mechanical stresses most effectively.

Another approach to piston slap was to cast struts of invar steel integrally in the piston, and as invar expands little with temperature the piston was restrained from expanding in the wrong directions. Steel inserts could be used in the same way. Some manufacturers preferred composite pistons which had aluminium heads for good heat dissipation and cast iron skirts.

Light alloy pistons became almost universal in the Old World by the late twenties, and their light weight enabled engine speeds to be almost doubled. Aluminium pistons also ran much cooler than cast iron ones, easing lubrication problems and permitting slightly higher compression ratios. Deposits were less, reducing the possibility of pre-ignition. In parallel with developments in design improved alloys of greater strength and lower expansion coefficients were developed.

Aluminium pistons were not so extensively used in the USA where steel pistons, made as light as possible, were more popular.

Rings. Two to four rings were fitted together with a scraper ring. The scraper ring was shaped to increase the pressure and on the downstroke to remove oil, rather than float on the oil film on the cylinder, and so control the thickness of the oil film. The ring was often at the base of the skirt and so tended to reduce piston slap.

Connecting rod

The connecting rod could be of steel or light alloy. It was normally of H section but hollow tubular rods were used on some high-speed engines. There were different ways of dealing with the gudgeon pin. The little end could clamp the pin and the pin oscillate in bosses in the piston, or it could oscillate on the pin, or the pin could float and be a sliding fit in both little end and bosses. Circlips or end pads prevented the pin from drifting sideways and damaging the cylinder walls if it oscillated in the piston bosses.

Big-end bearings consisted of detachable bronze half shells lined with white metal, but at the end of the period they were being replaced by thin-wall bearings. These were semi-circular flexible steel strips coated with a thin layer of soft alloy, and bearings, shafts and housing were now so accurately made that the thin-wall half shell could be slipped into place on the assembly line without the need of any further fitting. Engine bearings were also being replaced by thin-wall bearings.

Connecting rods on V8 engines could be mounted side by side on the one crankpin, or one rod could run between the two ends of the other forked rod. On one V8 the second rod pivoted on a bearing near the big end of the first.

Crankshafts

Crankshafts were made from carbon steel or nickel–carbon steel but hardened steel, or case-hardened steel, was used in more expensive cars. Torsional vibrations on Sixes and straight eights were overcome by fitting dampers on the front end of the crankshafts or by reducing their length.

American manufacturers tried to eliminate vibrations on six-cylinder engines by making the crankshaft very massive but such crankshafts were expensive to make and fit; this was one reason why Cadillac had earlier decided to make V8 engines. Other manufacturers used counterweights and Hudson, for example, made a six-cylinder engine with eight balanced counterweights and four main bearings.

Counter-balance weights were too expensive for UK cars and not used until they were really necessary in the thirties, when speeds were high enough to necessitate the crankshafts to be dynamically as well as statically balanced. The counterweights were forged integrally with the shaft by extending the webs on the other sides of the throws, or they were bolted on.

Ford, and later some other manufacturers, cast their crankshafts. This casting was carried out to such fine limits that very little further machining of the casting was needed.

Roller or ball bearings could be used for big end bearings but if used for main bearings the crankshaft had to be of the built-up type rather than forged, for otherwise the bearings could not be fitted, and in any case ball bearings were not so able to cope with unsteady loads.

By the late thirties most four-cylinder engines had three main bearings, even the small cheap engines that had hitherto got away with only two. Six-cylinder engines generally had four main bearings.

Engine mountings

The engine subframe was obsolescent after the First World War and either three or four-point suspensions were used.

Four-cylinder engines were becoming more powerful and their roughness more noticeable, so that attempts were made to insulate the engine from the frame by interposing rubber pads between suspension arms and chassis. A unique mounting was used on the 1926 Riley. At the rear of the gearbox the unit was mounted on the frame by rubber pads and the other support consisted of a circular-section bearer passing across the crankcase between the first and second cylinder. The bearer carried conical plugs which seated in rubber ring bearings in conical holes in bosses on the crankcase. The cross member acted as a rigid interconnection between the frame side members although the engine was supported on rubber.

In 1927 Chrysler mounted the engine of their Plymouth on a rubber pad located at the front end of the engine and rubber brackets under the gearbox, so that the roll axis of the engine passed through its centre of gravity. Pads on either side of the gearbox restrained the engine from rolling. 'Floating power' mounting was very effective and the engine of the Chrysler Plymouth appeared to driver and passengers to be as smooth as that of a V16 or twelve-in-line engine.

Improvements in the bonding of rubber to metal, and in the rubber itself took place and by 1934 resilient rubber mountings were used on about a third of the models at the Motor Show. A variety of brackets and mounts appeared; movement of the engine was controlled by shear of the rubber rather than compression as this allowed greater movement. Mountings were now designed on scientific principles and located so that the inertia of the engine reacted against unbalanced forces and couples. A flexibly mounted Four could now run as smoothly as a solidly mounted Six.

The floating engine could move considerably and so there had to be flexible connections between engine, exhaust and controls, etc. On certain models clutch pedal and gear lever could tremble and the steering wheel oscillate violently, but by better design these faults were eliminated, and flexible mounting of engines became firmly established.

Developments since 1945

When production recommenced after the Second World War most manufacturers simply continued making their pre-war engines with most UK engines being four-cylinder types and there was a considerable proportion that had side valves. The petrol available had a low octane number and there was a seller's market so there was no great incentive for manufacturers to bring out new engines. When true post-war engines eventually appeared they were almost invariably four-cylinder engines with pushrod-operated overhead valves. Taxation after January 1948 was no longer based on the RAC rating and so the stroke-bore ratio decreased and engines were made much squarer. Petrol of higher octane numbers became available and so compression ratios could be increased. Improvements in lubrication, carburation and ignition were made and for twenty years or so after 1945 progress was a matter of steady development along more or less familiar lines. There were parallel developments in metallurgy and in production.

The main spur to development has been competition, and innovation in the competitive components industry, but in the sixties and seventies there were new pressures on the automobile industry and political pressure led to legislation in the USA and, later, in other countries. In respect of the engine noxious emissions and later overall fuel consumption had to be reduced, and targets were set that the industry had to meet. Legislation caused an enormous amount of research and development to be done and a lot of new ideas emerged.

If an engine is selling well the manufacturer has little incentive to improve it, he is much more interested in finding ways to cut production costs. Small changes and improvements over the years, however, can have a cumulative effect.

Some cars and their engines remained in production for many years. The engine of the 1948 Morris Minor was basically the same as that of the Morris Eight of 1934. In 1951 the Minor was given the new Austin OHV engine which remained in production until 1971. The Ford Eight engine was made with little modification from 1932 to 1962, and the Ten engine, which was the Eight slightly bored out, from 1934 to 1959. The VW flat four and the Citroen 2 CV engine from 'the car that time forgot' were two long-lived engines from the Continent.

The same engine or slight modifications of it could be used on different models and even on cars of different nominal manufacturers; this was one manifestation of 'badge engineering' and its absurdities. A car could be sold under one name and then, with possibly slight differences in trim sold under a completely different name and make by a different manufacturer.

When a more powerful engine was needed it was generally much cheaper and quicker to 'stretch an existing engine than to build a new one'. Indeed cars can be overdesigned in the first place to allow for future modification and stretching. In the days when cars were taxed in the UK according to the RAC rating, an engine could be bored with small diameter cylinders for the UK and with larger diameter cylinders to make a more powerful engine for overseas markets. The obvious way of stretching an engine was to increase its capacity either by increasing its bore or stroke, or both, and again if the engine was built with a generous factor of safety in the first place this did not necessitate much redesign. However, one change could lead to another, for example, most modifications increased bearing and big end loads and so heavier duty bearings might be needed. Similarly valve and ignition timings might be altered and so thorough development and testing were required.

If a manufacturer is making a range of cars engine production can be rationalised by using common design and common components for all the engines as much as possible. By making the same basic engine in say two bore sizes, with two different strokes and with either four or six cylinders a range of engine sizes from 1000 to 2500 cc capacity can be covered using only two blocks. The same layout would be used on all engines and the different engines would have many components in common and, indeed, some components would be used on all the engines. Three quarters of the components or more might in any case be obtained from outside component manufacturers, who would do their own research and development. Indeed, as mentioned above, innovation by the competitive components industry often led to engine improvements and developments.

The cost of designing and developing a new engine and the cost of tooling up for the engine are so great that it is rare for a design team to be given a clean sheet of paper and told to make a new engine. Instead they have to make use of existing production facilities and of standard components as much as possible.

Even if the team had full freedom the engine would under normal circumstances probably not be very different from existing engines. Many dimensions and other details are not critical, others are obtained from semi-empirical equations based on enormous amounts of test data, and detail design is based on the experience of many people.

Compromises may have to be made; for example, a large inlet valve may give good high-speed performance, but the torque might fall off at low speeds because a smaller valve is needed to promote turbulence at low speeds. The engine has to be properly matched with other components, particularly with the carburettor.

When the prototype engines appear they are subjected to long and rigorous testing on rigs and cars, deficiencies are shown up and rectified and then the engine is tested, modified and retested until it is found satisfactory and produces its design performance.

The power output of the engine is of course only part of the story. Ultimately the driver wants a powerful engine that is also smooth and flexible, and economical to run and to maintain. The manufacturer wants an engine that is cheap and easy to build and which will compare well with his competitors' products.

Progress in the USA

Post-war six- and eight-cylinder engines in the USA were basically the same as those of 1941, but some minor modifications were made and compression ratios generally increased. Advantage was soon taken of the higher octane fuel that became available to design a new range of short-stroke pushrod-operated overhead valve V8 engines. One of the first was the 1949 Cadillac, it had a capacity of 5.4 litres, a compression ration of 7.5 : 1 and developed 135 bhp. The 1951 Chrysler Firepower engine had hemispherical combustion chambers, a capacity of 5.42 litres and developed 180 bhp at 4000 rpm.

Overhead valve engines were shorter and more rigid, as well as more powerful, than the corresponding side valve types and their attractions were such that by 1955 all the Big Three had OHV V8s; Ford dropping their side valve V8 in 1954. In 1956 American Motors had to design their own V8 as their uprated Six did not meet the popular demand for more power even if this was at the expense of fuel economy.

The engines were generally squarer than engines in the UK and most had wedge-shaped combustion chambers to promote turbulence. They developed maximum torque at about 2500 rpm and maximum power at about 4500 rpm.

The celebrated horsepower race then followed and more and more powerful engines were built as each manufacturer tried to outdo his competitors. These high powers were obtained by increasing the size of the engine and its compression ratio, and by improving breathing by using bigger valves and more lift.

The production of Fours fell away until in 1965 only hundreds were built compared with the millions of V8s. V8 production had its ups and downs like production generally but fell dramatically in 1980 when the numbers of Fours, Sixes and V8s being produced were approximately the same.

Wankel engine

Commercial production of a novel engine, the Wankel rotary engine, began in 1963 and it was installed in the NSU Spyder and later the Ro80. This engine (figure 5.18) consists of a three-lobe rotor mounted freely on roller bearings which operate on an eccentric bearing surface formed integrally with a main shaft. The rotor works in a chamber shaped like a two-lobe epitrochoid—a figure eight broad at its central join. During motion of the rotor its three apices effectively seal against the housing, and three chambers of continuously varying volume are formed in which induction, compression, expansion, and exhaust, occur as in a four-stroke engine. The rotor has recesses in the faces between the lobes and in these recesses combustion is initiated. Single inlet and exhaust ports mounted in the stationary chamber are opened and closed at the appropriate times as the rotor rotates. The engine was not successful in Europe but Japanese companies were interested in the Wankel and after much development work the sealing was improved and the engine made a practicable proposition. Other modifications improved the efficiency and the low-speed torque and gave better fuel economy. Noxious emissions were limited by fitting a thermal reactor system. The third generation of Mazda engines developed 77 bhp per litre and the engine is still in production.

Other manufacturers spent a lot of money on the Wankel engine but gave up because of worries over emissions and doubts about whether the engine could compete with reciprocating engines. The lightness, smoothness and quietness of the rotary engine is, however, well suited to the motorcycle where its fuel consumption is not such a problem so that the Wankel is used on motorcycles and, in particular, on racing machines.

Emissions

From 1960 or so the technical development of the engine has been influenced by American legislation to reduce obnoxious engine emissions and to decrease fuel consumption. The offending emissions were hydrocarbons, carbon monoxide, and the oxides of nitrogen (NO_x). Considerable

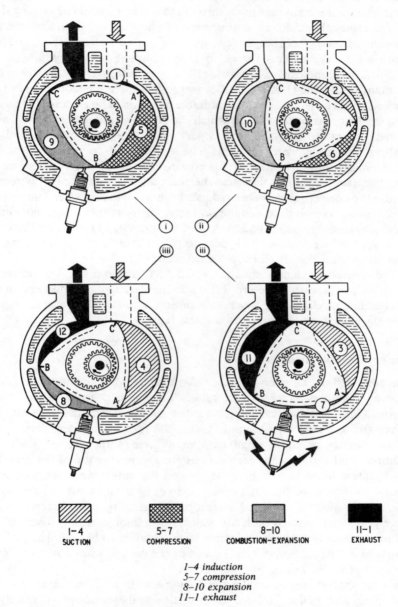

Figure 5.18 Operation of the Wankel engine. By permission of Mazda Cars (UK) Ltd.

reductions in the hydrocarbons and carbon monoxide were obtained by recirculating the crankcase gases through the engine instead of venting them to the atmosphere. Recirculating part of the exhaust gases reduced the amount of NO_x. Another arrangement was to add afterburners and add more air to the exhaust gases to help complete the oxidation of the hydrocarbons and carbon monoxide. The most effective way, and one used on many cars in the USA after 1975, was to fit a three-way converter, a tube packed with carriers coated with suitable catalysts. The exhaust gas passed into the first part of the tube in which the NO_x was reduced and then, with added air, passed into the second part where the hydrocarbons and the carbon monoxide were oxidised. The engine had to operate near the stochiometric air/fuel ratio for the converter to be effective and so petrol injection (see p 152) was used together with an elaborate control system to ensure the correct ratios were used at all times, including when starting the cold engine. As lead poisoned the expensive catalysts (platinum, palladium, rhodium) lead-free petrol had to be used. Compression ratios had therefore to be reduced which tended to increase petrol consumption, but operating near the optimum air/fuel ratio helped matters.

The second problem, reducing overall fuel consumption, was solved in the USA by increasing the proportion of smaller engines (and cars) in the total engine production. Elsewhere lean-burn and stratified charge engines were developed which tended to use less fuel.

The combustion chamber

The combustion chambers of side valve engines were similar in design to pre-war chambers. There was a variety of chambers for overhead valve engines. The simplest was merely an extension of the cylinder bore with the valves and sparking plug mounted in the flat roof above the piston.

The engines of most small cars were fitted with inverted bath tub chambers and vertical valve stems. On some engines the top of the chamber was inclined to form part of a wedge and the valves were again perpendicular to the top of the chamber. These configurations enabled a reasonable compromise to be made between good breathing and relatively simple valve gear. The inverted bath tub was easy to manufacture as four simple milling operations enabled the combustion chamber to be completed with a tolerance of one cc. Sometimes the axis of the bath tub was not in line with the midline of the engine but inclined to it.

Wedge-shaped combustion chambers were often used on bigger engines. The valves were in line, inclined to the vertical and located approximately in the centre of the chamber with the sparking plug placed in the thick end of the wedge. Squish could be introduced by locating the thin edge of the wedge over the piston crown. Rover used an F head on the engine fitted to the Rover P.3 (figure 5.19) and angled the top of the cylinder block and one half of the pentroof-headed piston squished over the inlet valve and the other part of the piston formed one wall of the combustion chamber. The other two walls of the wedge-shaped chamber carried sparking plug and exhaust valve. This layout apparently gave good breathing and turbulence, and good fuel economy, but more conventional overhead valve chambers were adopted in later engines. Rolls-Royce also used F heads for a time. Some OHV engines used angled pistons to give wedge-shaped combustion chambers.

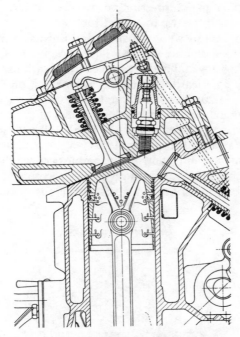

Figure 5.19 The Rover P.3 combustion chamber. By permission of The Council of the Institution of Mechanical Engineers (from *Combustion Chamber Design and the Influence of Fuel Quality*).

Pentroof and hemispherical heads were reserved for high-performance engines. Though most chambers could be classified as bath tub, wedge, and so on, some were intermediates or hybrids between one type and another. For example, one side of a bath tub chamber could be inclined and not vertical and so have something of the shape of a wedge-shaped chamber.

A very different type of head, the Heron or bowl-in-piston head, appeared in automobile engines in the early sixties. A similar type had earlier been used on diesel engines. Inlet and exhaust valves and the sparking plug were located in the flat roof of the cylinder, while the combustion chamber was a bowl in the crown of the piston. Squish and swirl could be introduced, and the flame front propagated hemispherically. The piston, however, tended to be heavy and could be a problem to keep cool.

Legislation to reduce emissions and promote fuel economy caused radical rethinking of combustion chamber design in the seventies. Squish had to be drastically reduced, for mixture in the squish areas might not be completely burnt, and therefore turbulence had to be generated in other ways. To decrease fuel consumption an efficient head with very good breathing and swirl was needed. High-compression ratios were ruled out in some countries because they entailed the use of leaded fuels.

One approach used by some Japanese and European manufacturers was to develop stratified charge engines. The concept of the stratified charge goes back to Otto's day and had been taken up at various times by a number of people including Ricardo when he was working on sleeve valves. The basic idea (figure 5.20) was to initiate combustion in a small prechamber filled with a rich mixture and then jet out a mass of this burning gas at high temperature into the cylinder which was filled with a leaner mixture; excess air in the latter would oxidise completely any remaining fuel and carbon monoxide. Complete and controlled combustion would also assist fuel economy. The straightforward stratified charge engine had a separate inlet valve in the precombustion chamber. Ingenious ways were found to

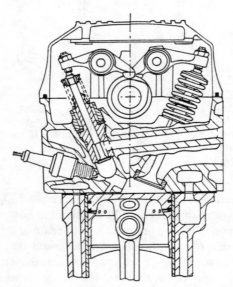

Figure 5.20 Honda stratified charge engine. By permission of Honda UK Ltd.

generate swirl and turbulence, and to project the burning gas rapidly throughout the leaner mixture.

Another way of ensuring good fuel economy and low emissions is to use a weak mixture in a compact combustion chamber and rely on turbulence and a high compression ratio to ensure rapid and complete combustion of the fuel. One of the simpler lean-burn systems was the May swirl combustion chamber. The inlet valve was located near the face of the flat-headed piston with the exhaust valve and sparking plug sited in a bath tub chamber which had a sloping channel leading to the inlet valve and this promoted high swirl. Ignition was initiated during compression and flame propagation induced rapid swirl near the exhaust port. Porsche used a part-spherical combustion chamber offset from the piston crown in order to give high swirl. The compression ratio was very high, 11.5 to 1. In the Ricardo high-ratio chamber the combustion space was below the exhaust valve and there was a shallow-groove transfer passage from the inlet valve to the combustion chamber. Again an intense swirling action was created. Because of the excess oxygen in the charge, hydrocarbon and carbon monoxide emissions were low in lean-burn and stratified engines. NO_x is formed at high temperatures and at metal surfaces, so to limit NO_x emissions the engines had to be kept small, that is under 1.5 litres in capacity.

The complex combustion chambers of stratified and lean-burn engines had to be accurately machined so that these engines were costly compared with conventional types. Driveability and engine response could also be affected.

Valve mechanisms

Most of the true post-war engines built in the UK had overhead valves, and in practically all of them the valves were operated by pushrods and rocker arms. Design was straightforward and followed pre-war practice.

Pushrods were generally formed from solid steel rods with the ends induction hardened. The lower end of a rod seated in the bottom of a barrel or bucket-shaped tappet which moved up and down in its own little bore in the cylinder block. The tappet could be slotted to reduce weight and improve lubrication. By mounting the cam a little way from the centre of the tappet face, and making the latter slightly domed, the tappet could be made to rotate slightly during operation and so wear more uniformly. The top end of the pushrod terminated in a cup which engaged a ball on a threaded member at one end of the rocker. A hardened pad at the other end of the rocker pressed against the top of the valve stem. Clearance was adjusted by raising or lowering the ball. The rockers themselves were fabricated or cast in one piece from cast iron and they ran on a hardened shaft and were generally of inverted T section. Their spacing on the rocker shaft was maintained by light springs or tubes and lubrication was by means of oil flowing along the hollow rocker shaft.

The valves themselves were made of heat-resistant steel but later, to reduce costs, the stems were made of a cheaper steel than the heads. On most engines a single roller chain was used to drive the camshaft and the chain was fitted with a chain damper or tensioner of one sort or another.

Overhead camshaft engines were used on a few high performance and sports cars like the Aston Martin and the Jaguar (figure 5.21), and only on a very few other cars, such as the Hillman Imp and the Vauxhall Victor. Pushrods were highly developed and proved to be reliable, and there was the difficulty of providing a cheap, quiet, and positive drive for overhead cams.

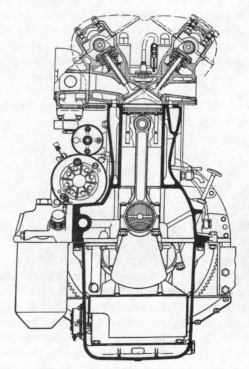

Figure 5.21 Jaguar twin-camshaft engine. By permission of Jaguar Cars Ltd.

As engine speeds increased the disadvantages of pushrods and the corresponding advantages of overhead cams became more and more apparent and in the seventies overhead cams began to displace pushrods. The reciprocating weights associated with overhead cams were less than for pushrods and the gear could be made more rigid so that OHC engines could run at higher speeds. More efficient heads could be used for a further increase in power. Overhead cams were also simpler and had fewer parts, and the design of the head was simplified, though it could be bulkier than when pushrods were used. Equally important, toothed belts had been demonstrated to be reliable so that the drive was no longer a problem. The internally toothed or cogged belt was made of rubber and reinforced with, for example, glass fibre, and faced with nylon. It was quiet and needed no lubrication, it was practically inextensible and it was long lasting.

Reducing stresses on cam and follower and providing lubrication to their surfaces remained rather of a problem, but detailed design improved lubrication and provided simple ways of adjusting clearances and so of reducing maintenance costs. It was easier to design an efficient head with twin cams than with a single cam but two overhead cams were more bulky, and of course more expensive, so that they were only used on expensive, high-performance cars.

Overhead cams could engage a flat tappet on the top of the valve stem as on the pre-war Morris (the tappet could be screwed in or out of the large-diameter valve stem to adjust for clearance) or operate through rockers. On most engines, however, little buckets or pistons were interposed between cam and valve stem. These buckets oscillated in bores within the cylinder head. The valves were therefore only subjected to axial loads. Clearances were adjusted by inserting shims between stem and bucket which was a complicated business, but Fiat in 1967 recessed the tops of the buckets and adjustment was made by slipping hardened steel shims into the recess which was much easier to do. Another approach was to use finger rockers which were pivoted at one end and engaged the valve stem at the other end. These were much lighter than bucket tappets and were easy to adjust. They had the further advantage that the leverage of the finger reduced the stress on the cam and opposing surface, which was important because the contact was subjected to a sliding force as well as to the normal load.

In years gone by some touring cars had been fitted with three or four valves per cylinder. Two valves have less inertia than a single valve of the same effective area, they give better breathing and swirl, and in the case of exhaust valves they run cooler. Also a centrally located sparking plug can be used in a compact combustion chamber giving a shorter flame path and smaller ignition advance.

It was not until 1960 when Honda used four instead of two valves per cylinder that it was realised that increasing the number of valves brought about improvements in the performance of contemporary engines that could outweigh the extra cost and complexity. Two inlet and one exhaust valves were used by some manufacturers; this improved fuel economy when the inlet valves were of different sizes, the smaller valve operating at low speeds and both operating at higher speeds.

Cylinder blocks

Aluminium alloy blocks are only about two thirds the weight of a cast iron block but the aluminium block usually has to have cast iron liners and the head has to have hard metal or cast iron valve inserts.

Aluminium blocks were used on high-performance and luxury cars and on rear-engined cars like the Hillman Imp on which it was important to reduce the weight on the rear wheels. Rolls-Royce gave up aluminium blocks mainly because aluminium did not damp out noise as effectively as cast iron. Special alloys and surface treatments were developed which enabled pistons to run directly in aluminium bores. The pistons were coated with iron or other metal. Porsche, in particular, used linerless blocks in Europe in the early seventies. Aluminium alloy cylinder heads were much more widely used, partly because they could be easier and cheaper to make than cast iron heads.

Pistons

The top face or crown of the piston was often no longer flat or domed but shaped according to the requirements of the combustion chamber. It could be angled to make a wedge-shaped combustion chamber, or made with cut-outs and recesses to give clearance to the valves, or in a Heron head carry a bowl-shaped recess which acted as the combustion chamber. Such pistons could give the designer weight and heat flow problems.

Split skirts had worked reasonably well when duties were light and as duties increased T-shaped slots had been effective but as compression ratios increased the T slot was barely adequate, for the slot weakened the skirt and the slit tended to close up permanently.

The piston continued to be made slightly oval and slightly tapered. The circumferential slot was retained in the bottom ring to direct heat away from the more heavily loaded parts of the piston. Steel inserts, belts and bands continued to be used in the USA to restrain the skirt from expanding in the wrong places. Light alloys of reduced thermal expansion were also widely used. Eventually only two compression rings were fitted with the rings machined to such a shape that they exerted uniform pressure all round their circumference. The scraper ring migrated from near the bottom of the skirt to a position above the gudgeon pin and so acted on more of the cylinder bore. The scraper ring was grooved so that it had two lands and the intervening area was slotted and the slots communicated with holes in the piston. A helical spring could be fitted between ring and piston to increase the pressure on the cylinder wall and more complicated arrangements were also used.

Crankshafts

Strokes and therefore crank throw radii became smaller and smaller, making for a stiffer crankshaft, eventually the crank pin began to overlap the crankshaft journal in section, and the crankshaft became even more rigid. Crank throws could be ribbed to decrease their weight without decreasing their strength and pins and bores drilled out to reduce weight. Crankshafts had to be accurately counterbalanced. The crankshaft was generally forged from carbon or alloy steel machined to approximate size and then hardened and finally the bearing surfaces ground to radius. Crankshafts were also cast. High speeds, which were sustained on motorways, showed that every throw on a four-cylinder engine should have its own main bearing, and similarly six-cylinder engines generally had seven bearings.

Thin-wall bearings soon became standard, but copper–lead, aluminium–tin and other alloy coatings had to be developed which would withstand high speeds and high loads, and which also had good fatigue properties.

Supercharging

One way to improve the volumetric efficiency of an engine is to pump the mixture into the cylinders, that is, to supercharge the engine and not rely on natural aspiration. Sir Dugald Clerk showed before 1900 that supercharging increased engine output and Chadwick won a number of competitive events in the USA in the period 1908–11 using supercharged engines. Supercharging was shown to be very effective on aircraft engines during the First World War and after that war it was used on sports and racing cars, notably by Fiat on GP cars in 1924. All GP cars were supercharged from 1925 to 1938 after which the regulations changed. In 1937 a supercharged Mercedes engine developed 646 hp.

Few superchargers, however, were used on other than racing cars. The supercharger had to be accurately made and was expensive, and the engine had to be robust enough to withstand the increased loads. If extra power was required it was generally cheaper and better to fit a bigger engine than to supercharge an existing one. A supercharged Mercedes was made in 1924 and the supercharger further developed in the thirties. It was apparently effective but it cut in with a noise like a siren and went out of production. These superchargers were of the positive-displacement type, that is, they used intermeshing lobes as in the Roots blower, and were driven by the engine.

In another type of supercharger, the turbocharger, (figure 5.22) a vane was driven by the exhaust gases and this vane drove a second vane that acted as the compressor. Turbochargers were simple and found to be very effective on diesel engines, so that they were adapted for installation on cars, for example, the Oldsmobile Jetfire appeared in the USA in 1962. Saab demonstrated their effectiveness in Europe and in 1976 showed that a 2 litre turbocharged engine had the performance of an unsupercharged 3 litre. Saab also introduced an electronic system which prevented knock and made the engine insensitive to variations in the quality of the fuel.

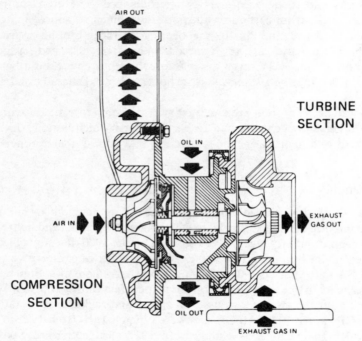

Figure 5.22 A typical turbocharger. By permission of Schwitzer Division of Household Manufacturing.

Turbocharging enabled the output of an engine to be increased by as much as 30–50 %, and that without greatly increasing the peak cylinder pressure, or much increasing the weight or size of the engine. It offered an attractive way of increasing the performance of the top car of a range with minimum alteration to the engine itself. It was also claimed that turbocharging reduced obnoxious emissions, and petrol consumption was less than that of a normal engine of similar power.

There were problems. The first were the mechanical problems of coping with the high temperatures (up to 1000 °C) and the very high speeds involved (up to 100 000 rpm). A second problem was that the turbocharger boost varied with speed, so that it could be too little at low speeds, and much too great at high speeds; the turbocharger therefore had to be carefully matched with the engine and some of the exhaust gases by-passed at higher speeds. Another problem was that there was a lag before the turbocharger responded to a change in throttle position but this could be reduced by minimising the inertia of the rotating parts.

The turbocharger itself is very simple; a typical unit consists of a centrifugal outflow compressor and a centripetal inflow turbine, one at each end of a shaft. Compressor and turbine are mounted in an aluminium and cast iron housing which is connected to carburettor and inlet manifold, and to the exhaust respectively. The shaft bearing is oilcooled or watercooled. A tailgate valve installed in the system bypasses more and more of the exhaust gases as the speed increases. An intercooler between compressor and engine can increase the volumetric efficiency of the turbocharger still further. Other improvements have been the introduction of a variable inlet valve and the use of ceramic rotors. It is likely that the conventional supercharger will be developed to compete with the turbocharger.

Compression ignition engines

The early oil engines had spark or hot-tube ignition, but Ackroyd Stuart mounted a small combustion chamber at the end of the cylinder and connected it to the latter through a narrow neck. Oil was sprayed into the combustion chamber during the suction stroke and during the compression stroke air was pumped into the combustion chamber, the mixture was then right and the temperature sufficient for the oil to ignite. The combustion chamber needed external heating by a burner when the engine was cold, but when it was running well no further heating was necessary. Similar engines were made by other manufacturers. These semi-diesels worked well and were used for many years.

Diesel in the nineties dispensed with the separate combustion chamber and used such high compression ratios that the air reached high enough temperatures for the fuel to ignite when it was blown at very high pressures into the cylinder. Pressures and temperatures were very high so that the engine was remarkably efficient. It used less fuel than any other contemporary engine, could be made to burn less refined fuels, even crude oils, which were of course much cheaper than refined petroleum spirit. The disadvantages of the diesel were that, because of the fuel injection equipment and the high pressures involved, it was expensive and heavy. Starting could also be a problem. Later, instead of blowing the fuel into the cylinder, a jerk pump could be used. The diesel engine was much more convenient than a steam engine of the same power and so the diesel displaced the small steam engine. It was adapted to rail locomotion in the thirties in the USA but it was not until the late fifties that the diesel engine and the electric motor began to displace the steam engine on the railways of the UK.

Many trucks and buses were converted to run on diesel engines in the UK in the thirties. These engines were specially developed for them but they were big and heavy. Mercedes-Benz fitted diesel engines to some of their cars in the twenties and the 260D model of 1935 had a 2.6 litre engine which produced 45 hp. Their 190D model of 1958 had an overhead cam engine (figure 5.23). Over the years a considerable number of diesel-engined cars were produced. However diesel engines cost more than petrol engines, for the pump and other parts had to be very accurately made; compression ratios were very high and so the engine tended to be noisy. It could be difficult to start, and the engine did not have the flexibility and refinement of the highly developed petrol engine. The diesel engine therefore continued to be mainly used on taxicabs and light delivery vans, and other high-mileage vehicles where the savings on fuel more than offset the greater first

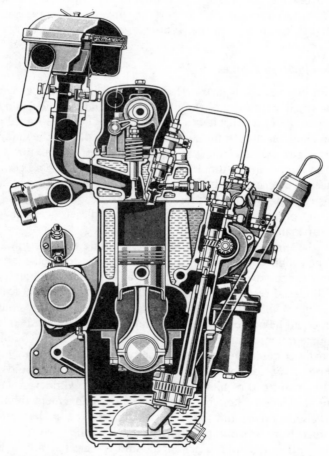

Figure 5.23 Mercedes-Benz 190D diesel engine of 1958. By permission of Daimler-Benz AG.

cost of the engine. Most heavy goods vehicles in the UK and Europe are powered by diesels.

The situation changed in the seventies and the oil crisis brought about a period of intense activity in the diesel field. There was some penetration of the USA market particularly by GM and four-, six-, and V8-cylinder engines were built, but despite considerable technical progress diesels had only a short-lived vogue. Low-price petrol became available again and in the mid-eighties GM ceased diesel production. However, because of their very low fuel consumption and good performances diesels continued to penetrate the market in Europe and Japan.

Small-car diesel engines are almost exclusively of the indirect injection type, that is, they use a separate small swirl or precombustion chamber, many based on the Ricardo Comet type that Ricardo developed for commercial vehicle engines in the thirties. These engines reached speeds of 5000 rpm and more which permitted the use of the same transmission as fitted to spark ignition cars. Similarly many engines made use of the same production facilities as their petrol-engined counterparts as many of the components were either the same or similar enough to use the same machining equipment. Emissions and noise were made generally acceptable and idling knock reduced by electronically controlling the fuel injection pump. The pump and distributor (generally combined) have to deliver to the injector nozzles, at very high pressure, the correct amount of fuel at the appropriate time, and as only very small amounts of fuel are involved precision engineering is necessary.

Turbocharging of diesels became a common practice and increased the output of the diesel until it was similar to that of a petrol engine of the same capacity. The turbocharged Mercedes-Benz 3 litre diesel of 1978 produced 113 hp. Direct injection will give even better fuel consumption but noise and emission problems are greater than with indirect injection, as are the costs.

Chapter 6

Carburation

The internal combustion engine converts into useful work the combustion reaction between the oxygen in air and the fuel, usually petrol. Petrol is a complex mixture of hydrocarbons and has to be mixed with air in the right proportion to ensure complete combustion in the engine; the stoichiometric air/fuel ratio is approximately 15 to 1 by weight. This ratio does not give maximum power or in general maximum economy from the engine; the mixture has to be rich for starting, less rich for slow running and weak for economical running but richer again for acceleration and high speed. The mixing of fuel and air, that is, carburation, has to meet these requirements and also at the present time ensure that after combustion noxious emissions in the exhaust gases are below levels specified by law.

Carburettors to 1914

In the early days carburettors were relatively simple but there was a great variety of them, with surface, wick and spray carburettors all being used.

Surface carburettors

The earliest way of forming a mixture of suitable proportions was by passing air bubbles through the spirit, or by passing air over the fluid. The corresponding carburettors were called bubbling or surface carburettors. Their features are so similar that they can be considered together.

A crude surface carburettor together with a mixer to control the strength of the mixture entering the engine was used by Benz in his first vehicle. An improved design (figure 6.1) consisted of a cylindrical vessel and to its top a pipe was connected from the petrol tank. A float regulated the supply of fuel in the feed tube by means of a needle valve. On the suction stroke of the engine, air was drawn through an adjustable valve to the enclosed chamber and a conical-shaped member inside the chamber prevented air from having direct access to the engine before mixing with fuel. The fuel at the base of the carburettor was heated by exhaust gas to improve vaporisation. Apart from minor changes such as easier throttle controls for the driver, the same basic design of carburettor was retained on all Benz models up to 1900. Daimler also used a surface carburettor on his motor bicycle engine and a modified form of the Daimler carburettor, in which air bubbled through petrol maintained at a fixed level by a float chamber attached to one side of it, was used by Delahaye until after the turn of the century.

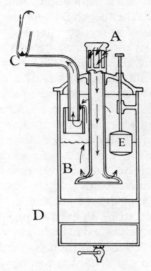

Figure 6.1 Benz surface carburettor: Air is drawn in through the inlet A and passes through petrol in B and on to the engine at C. Part of the exhaust gases heat the chamber at D. The petrol level is controlled by float E. By permission of Seeley Services and Co (from E C H Davis 1965 *Memories of Men and Motor Cars*).

A strong point in favour of surface carburettors was their simplicity. They were bulky, however, and they had the outstanding weakness that during operation the more volatile constituents of the fluid were carried off first, leaving the heavier residue which would evaporate more slowly, and in some cases solid particles would aggregate and even clog the carburettor. It was therefore difficult to maintain a correctly proportioned mixture over a long period and also over the speed range of the engine. The mixture could be erratic when the car travelled over a bumpy road. These drawbacks were responsible for their replacement by other types.

Wick carburettors

Another early form of carburettor was the wick type. In this carburettor as the level of the liquid fell the exposed surface of the wick increased so that the area of air in contact with fluid increased and the mixture strength remained fairly constant.

In the Petreano carburettor exhaust gases passed through a central pipe surrounded by a layer of asbestos fibre maintained moist by fuel. Four funnels inside the cylinder and lined with asbestos acted as baffles and promoted the mixing of air and vapour. The carburetted air was then transmitted to a lower chamber before entering the engine. The arrangement behaved very much like a distillation column in that all the heavier, unwanted, constituents of the fuel could be made to flow to the bottom of the cylinder and drained off at regular intervals.

Perhaps the best known carburettor of this type was that originated by Lanchester and used on his cars from 1897 to 1914. The original Lanchester carburettor (figure 6.2) consisted basically of a wick chamber, containing three layers of fibres, which was attached to the top of the petrol tank. Petrol was delivered into the lower part of the wick chamber by a hand pump operated periodically by the driver. Suitably sized overflow holes in the base of the wick chamber allowed any excess petrol to be returned to the tank. To speed up the mixing process hot air was drawn through the upper part of the wick and this mixed with the vaporised fuel to give an over-rich mixture. The latter was diluted with cold air from a separate supply and a barrel valve was used to regulate the mixture strength entering the engine. Later the pump was replaced by an exhaust pressure-fed system, and fuel regulator float and needle valve used to maintain a constant level in the wick chamber. A further change was to introduce a mechanically operated pump driven from the engine to overcome any trouble which could arise from a blocked valve in the pressure system. The Lanchester wick carburettor was extremely simple and very reliable in service compared with other forms in existence at that time, but in 1914 Lanchester fell into line with fashion and changed over to a spray carburettor. The spray carburettor offered greater possibilities for the future as it was a far less clumsy arrangement than the wick type, and could be made automatic in action.

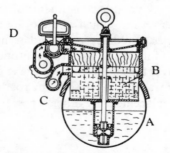

Figure 6.2 Lanchester wick carburettor: Air is drawn in through C and mixes with fuel pumped from the tank A into Chamber B containing the wicks. The inlet pipe at D is fitted with a heat-operated air valve. By permission of Seeley Services and Co (from E C H Davis 1965 *Memories of Men and Motor Cars*).

Spray carburettors: early types

Julius Hock used a primitive spray carburettor in 1873. Maybach invented the float chamber in 1886 and in 1893 he put the two together and fitted Daimler engines with spray carburettors. Butler had used a similar carburettor on his 1887 tricycle. Even in the early days of motoring such carburettors became the most popular type and after 1900 they were almost exclusively used.

All the spray carburettors operated on the same basic principle shown in figure 6.3. Here on the induction stroke air was drawn along a pipe and passed into a choke tube or venturi containing a spray jet. The air intake was controlled by a gauzed adjustable valve. As the air passed through the restricted area in the choke tube its velocity increased and the pressure fell below atmospheric pressure. Petrol then issued from the jet in a fine atomised spray, and mixed with air to give a combustible mixture. A float chamber, needle valve and weighted lever enabled a constant level of fuel to be maintained in the spray jet. It was usual to further improve vaporisation by allowing the jet to impinge on to a serrated inclined surface or to strike against a cone. Another improvement was to provide a secondary supply of air above the spray, which in the Daimler-Phoenix carburettor was controlled by rotating a cap containing holes to give a better controlled mixture.

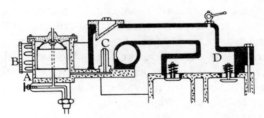

Figure 6.3 Daimler carburettor: A, the float chamber; B, the air intake; C, the single jet; D, induction valves of the two cylinders. By permission of Seeley Services and Co (from E C H Davis 1965 *Memories of Men and Motor Cars*).

Another well known carburettor was the Longuemare; the float feed closely resembled that of the Daimler-Maybach and fuel entered the mixing chamber via a plug having radial holes. A current of air was driven through the lower part of the chamber and its mixing with fuel was assisted by a layer of gauze before the mixture passed into the upper chamber and entered the engine. The amount of air entering the carburettor was adjusted by a cylindrical valve which could be turned by a lever at the top of the chamber.

As spray carburettors were sensitive to starting conditions, especially when cold, additional heating was provided by circulating hot exhaust gas round the carburettor.

When petrol evaporates heat is absorbed from the surrounding air, the temperature of which falls, hindering further evaporation, and if the ambient temperature is low enough water vapour can condense and even form ice. This cooling was to remain a problem for years. Once the engine was warm it was not so troublesome, the difficulty was when the engine was cold and starting up (see p 149).

Spray carburettors: later developments

These early carburettors worked reasonably well because of the limited range of engine speeds. As engine speeds increased it became necessary to design carburettors that would give good performance at both low and high speeds. The trouble was that the flow rates of air and petrol followed different physical laws and therefore the mixture could be right at one speed only on a simple carburettor, at lower speeds the mixture was not rich enough and at higher speeds it was too rich. The simple single-jet carburettor had other inherent failings as well. It could not provide the appropriate mixture for starting and low-speed running; combustion was inefficient when the engine started from cold and so a richer mixture was needed, and at very low speeds the pressure drop might not be sufficient for the jet to deliver any fuel. Another problem was that sudden opening of the throttle could cause a temporary weakening of the mixture when a richer mixture was really required. At maximum speed and full throttle again some enrichment was needed. Many devices to effect 'compensation' in a fixed-choke carburettor were therefore tried. Most of these originated in France and were to play a major part in modern carburation techniques. They mainly involved an extra air valve, or were hydraulic devices using compensating or air bleeder arrangements.

Extra air valve. At first, most carburettors incorporated an extra air valve operated by the driver, and so their effectiveness depended largely on the driver's skill. Krebs in 1902 went one stage further and designed one of the first successful semiautomatic carburettors. At high engine speed under light load and a nearly closed throttle the mixture strength was balanced at maximum depression. Under load with an opened throttle the extra air valve closed and gave a richer mixture. Designs incorporating an additional air valve based on Krebs' ideas became fairly common in France and were a big improvement on those that had to be controlled by hand.

Royce made a big advance in 1904 by designing a single-jet carburettor with an automatic air valve for mixture correction with varying air flows. This device, with later modifications, was incorporated in all Rolls-Royce carburettors until 1932, and worked so well that it would permit a car to pick up smoothly from low to high speed in top gear.

Air bleed. The Claudel Hobson carburettor (figure 6.4) represented a new development in design. It had a rotating cylindrical throttle shaped internally in such a way that when it was open it formed part of the choke tube and slotted on its lower surface in order to clear the jet nozzle when it was closed. The tube carrying the jet was surrounded by a second tube closed at its top and pierced with holes at its top and bottom. Fuel passed up the inner tube and was sprayed out through the holes in the outer tube but, as the air speed increased, more and more air was drawn in through the holes at the bottom of the outer tube, so weakening the mixture.

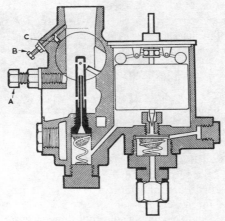

Figure 6.4 Early Claudel Hobson carburettor: A, air regulator; B, slow running regulator that restricts the size of tube C. By permission of Century Hutchinson Publishing Group Ltd (from E F Carter 1962 *Veteran Car Owners Manual*).

For slow running, a channel delivered fuel to an orifice near the edge of the nearly closed throttle, and for starting a flap valve strangled the air intake.

Many early carburettors had barrel throttle valves, but they could not be properly lubricated and so most had disappeared by 1914 or so.

In the Solex carburettor more and more air was bled into the jet as the air speed increased. The tube above the main jet was surrounded by two concentric cylinders (figure 6.5). The inner annulus was supplied with petrol through fuel holes just above the main jet, and as the air speed in the venturi increased, the level of the petrol fell until eventually it was below the level of the fuel holes, and the inner annulus emptied. Air then passed through holes in the outer cylinder, up the outer and down the inner annulus, and into the carburettor, and so the mixture could be automatically adjusted to the demands from the engine. Many carburettors worked on this principle, falling fuel levels exposing holes through which air was bled into the fuel from the jet.

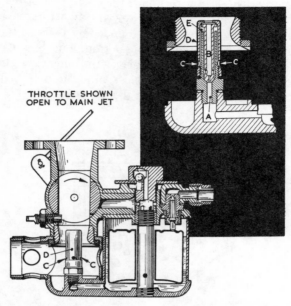

Figure 6.5 Solex carburettor and main jet: A, jet; B, well; C, air holes; D, annular space for upward flow; E, annular space for downward flow. By permission of Century Hutchinson Publishing Group Ltd (from E F Carter 1962 *Veteran Car Owners Manual*).

Compensating jet. One way of obtaining the correct mixture over a range of conditions was to use a compensating jet in addition to the main jet. In the Bavery system each jet supplied its own tube discharging in the centre of the venturi. The compensating jet was submerged and metered the flow of petrol into a compensating well which was open at its top end and which communicated with the compensating tube. At low speeds petrol passed through the compensating jet to fill the well and into the body of the carburettor, but as the speed increased the rate of flow was not sufficient for the well to remain filled and it was eventually emptied and air was drawn in, weakening the mixture. By suitably dimensioning the jets etc, the mixture could be kept constant and independent of speed. The Bavery system was the basis of the Zenith two-jet carburettor (figure 6.6). The Zenith carburettor was popular because of its easy starting, good slow running, simplicity of adjustment, and the absence of moving parts. The float chamber fed the main jet directly and the compensating jet was located in a secondary channel and limited the amount of petrol that could pass through the compensating nozzle; the latter was concentric with the nozzle of the main jet. For starting, a channel later led from the well to an orifice in the choke wall near the butterfly valve, so that when the latter was nearly closed petrol was drawn into the choke tube when the engine was turned over. Tickover speed was adjusted by rotating a sleeve in the well and so regulating the petrol flow. Below the main jet and compensating jet were two removable hollow nuts which collected any sediment formed in the carburettor.

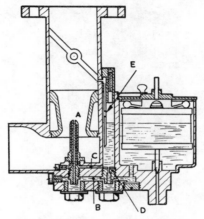

Figure 6.6 Early Zenith carburettor: A, jet; B, channel for main jet; C, channel for compensating jet; D, gauged hole; E, well. By permission of Century Hutchinson Publishing Group Ltd (from E F Carter 1962 *Veteran Car Owners Manual*).

Constant-vacuum carburettors

A further type, the variable-jet or constant-vacuum carburettor, maintained a constant air speed past the petrol jet, with the air passage and jet orifice areas being automatically adjusted to give the right mixture at all times; the size of the choke tube was controlled by the degree of vacuum in the inlet pipe, which increased as the engine speed increased.

A well known British carburettor of this type was the SU carburettor (figure 6.7). This had a jacketed vertical mixing chamber, into the bottom of which the petrol jet assembly entered at an angle of 45°. A piston slid up and down a cylindrical guide at 45° to the horizontal and it was connected to a disc that formed part of a pair of leather bellows, the chamber of which connected to the top part of the mixing chamber and so to the inlet manifold. A taper needle was mounted at the centre of the lower disc of the piston and penetrated into the petrol jet orifice. As the engine speed increased the bellows drew the piston and needle upwards, increasing both the air and the petrol flows.

Devices that made use of sliding choke members all suffered in the early days from excessive vibration caused by bad roads and poor suspensions, and this could have a greater influence on their operation than the demands made by the engine.

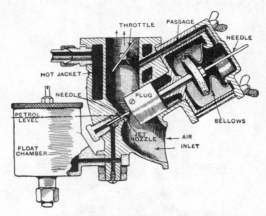

Figure 6.7 SU carburettor. By permission of Haymarket Publishing Co (from *Autocar* and *Autocar Handbook*).

Multi-jet carburettors

Another way of obtaining the required mixture over a range of conditions was to use a multi-jet carburettor. The Smith carburettor had, for example, four jets, one of which was a wall jet near the butterfly valve for starting. The other jets were fixed to a base and, with their separate choke tubes, formed a cylindrical body. The top ends of the chokes terminated in radial ports, the three ports being at different heights. A sliding cap fitted over the ports and was lifted by the suction of the engine to open the ports in turn, until all four were in operation when the engine was developing its maximum power. The jets were of different sizes. Thus the carburettor approximated the constant-vacuum variable-jet type. Later a five-jet model was made.

Various ways were tried to enrich the mixture for starting, for example, strangling the air intake, and possibly, at the same time increasing the area of the main jet manually by a needle valve arrangement, but the simplest way was to pass fuel through a channel to an orifice in the choke wall just on the engine side of the throttle when the latter was nearly closed. When the engine was turned over the resulting inlet manifold depression drew petrol into the induction pipe. The same channel and orifice could be used when the engine was idling and the throttle nearly closed. A pin, the 'tickler', was often mounted above the float chamber, and jiggling it made sure that all the channels and wells were full of petrol before the driver attempted to start the engine; he would also retard the ignition and advance it as the engine pulled.

American carburettors

American carburettors were generally more complicated than European ones with automatic operation of air supply being adopted in preference to fitting compensating jets. Stromberg carburettors were popular and there were three models, one had a single jet and concentric float construction and the other two models had single and double jets together with a glass tube float chamber. Later during the First World War a needle valve was used for metering fuel from a well that became empty when the flow through the main jet equalled that through a restriction valve.

Other arrangements like the Schlebler carburettor had a mechanical device so that the throttle opened a passage to the engine as well as releasing a needle from the jet orifice.

The Ford Model T carburettor was one of the automatic submerged jet, float feed type and had one adjustment and a needle valve. Two holes that communicated with a well provided the means of controlling the mixture flow at low or idling speeds. Above a vehicle speed of about 25 mph the vacuum effect on the two holes ceased and fuel mixture was taken from a main jet. After the engine was stopped fuel overflowed and filled the well and main air passage which permitted easy starting.

The above makes were typical of the progress made with early carburettor design in the USA before 1914. Carburation for cars received much attention in America during the First World War, owing to the poor quality fuel that had to be used. Most manufacturers bought carburettors from specialist firms and with the advent of Vee engines most used Zenith duplex carburettors, so that the carburation of each cylinder block was virtually independent of that of the other. Cadillac and Packard with their V8s and 12s, respectively, still used carburettors of their own design but this was against the trend.

Carburation generally was reasonably efficient on engines with a T head which had a limited range of engine speed. Fuel conditions were also simple from the carburation aspect as the fuel had a low density and the problems introduced later by the use of blended and cracked fuels were unknown. Engine speeds, however, increased and the use of the L instead of the T head permitted a more compact combustion space so that valve and port shapes had to be more carefully designed and manufactured to improve flow. With these changes existing carburettors tended to show their limitations, especially during idling, and the changeover to the main jet could only be made to occur smoothly by tricky control of the accelerator pedal.

The manifold area and choke area of the carburettor were both increased which could create blow-back at low engine speeds.

By the start of the First World War the power unit was flexible and reasonably economical although the specific fuel consumption was rarely better than 0.7 pints per bhp per hour. Even so, it was difficult for carburettors to be tuned to function satisfactorily on all makes of fuels as these varied considerably in quality.

Carburettors 1919–1939

The carburettors of the interwar period were in the main developments of designs which had been established before the war. Choke and jet sizes were more closely matched to engine requirements, and fuel consumption reduced by better atomisation, distribution and control of the mixture. Carburettors became much more efficient and reliable; they also became more complicated.

There were improvements in the detail design of the main jets. In the Zenith V-type model (figure 6.8) the flow from the main and compensating jets and from the capacity tube were combined to form an emulsion in a channel in the emulsion block, and issued from a common nozzle. Firms like Solex and Claudel-Hobson (who eventually gave up their barrel throttle) used more flexible and efficient emulsion tubes (figure 6.9); the name that was now given to the arrangement of concentric tubes through which air was bled into the fuel to form the emulsion that was drawn through the nozzle into the venturi.

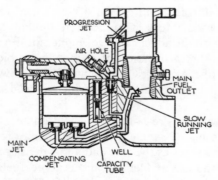

Figure 6.8 Sectional view of Zenith V-type carburettor. By permission of Haymarket Publishing Co (from *Autocar* and *Autocar Handbook*).

A number of manufacturers produced multi-venturi carburettors, Zenith for example, developed a triple diffuser in the twenties to cope with the less volatile and heavier fuels that were replacing the earlier, lighter fuels. The carburettor had the submerged jet of the earlier designs, but had three separate diffusers or choke tubes mounted one above the other and each bigger than the one below. After passing through the jet the petrol entered an annular space round the first, or primary diffuser, and passed into its air space through a series of radial holes. The diffusers resulted in increased turbulence and also more atomisation of the fuel, and so a more homogeneous mixture.

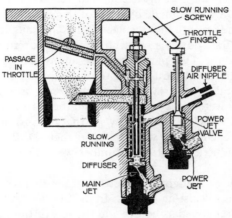

Figure 6.9 Claudel-Hobson carburettor showing diffuser well and passage in throttle. By permission of Haymarket Publishing Co (from *Autocar* and *Autocar Handbook*).

The leather bellows of the SU carburettor needed constant treatment with oil to maintain its flexibility, and so was replaced by a metal piston and cylinder, suction discs being fitted to the piston to maintain the vacuum in the cylinder. Movement of the piston was damped by a hydraulic damper so that the carburettor did not respond to sudden movements of the accelerator pedal. These carburettors retained the 45° angle of the piston air valve and needle; they were somewhat unusual in that the throttle butterflies were part of the shallow induction pipes as were the slow running jets.

In the early Smith multi-jet carburettors the jets were arranged at different angles to the main passage and some of the mixture was therefore reversed in direction when flowing to the induction manifold. Later designs had the choke tubes pointed towards the inlet manifold so that the mixture had an uninterrupted flow to the engine. A rotary air sleeve was also fitted to admit extra air to each choke tube and a sliding sleeve covered the wall jet and provided for easier starting.

Downdraught

Most early carburettors were mounted vertically with the air intake at the bottom, or horizontally, but late in the twenties the downdraught carburettor with the main air opening at the top appeared. The air was drawn downwards past the jet and the mixture downwards to the inlet valves. With this arrangement rather more mixture entered the cylinder which resulted in a higher mean effective pressure. It was usual to fit a larger choke size and this too gave better engine performance. Accessibility was improved and air cleaners and silencers more easily accommodated. This type of carburettor, developed in the USA, was made by most of the proprietary manufacturers in the thirties and eventually displaced the updraught and horizontal carburettors. At first, however, downdraught carburettors were regarded with suspicion in the UK partly because many cars had petrol tanks mounted aft of the dashboard above the carburettor and so there was a possibility of flooding the latter. This was not a real problem, for mechanical pumps were being introduced and the petrol tank transferred to the back of the vehicle. Humber were the first to use downdraught carburettors in the UK; they were fitted to their DX2 and DX3 models which were in production in 1933. Humber were also the first to use the economiser.

The change to downdraught carburettors generally involved making only small modifications to existing designs. In the SU carburettor, for example, the float chamber remained as before and the rest of the carburettor turned to be in a vertical position.

The use of diecasting in the early thirties enabled carburettor bodies and parts to be produced more cheaply and accurately.

Metering devices

The old-fashioned air valve reappeared in the form of an 'economiser'. This was a spring-loaded valve that opened when the depression was a maximum and admitted air into the carburettor. It was popular in the US in the twenties and in the UK somewhat later.

When the throttle of a carburettor was suddenly opened, the mixture was temporarily weakened and this was allowed for on many carburettors by providing accelerator pumps. Such a pump was particularly needed if the carburettor was fitted with an economiser. One type of pump was a simple piston cylinder device operated by the linkage to the accelerator pedal. The pump could be worked directly, or through a spring, when the piston acted as a dashpot and the fuel was delivered over a short period of time. Ingenious linkages were designed that reacted only to sudden movements of the pedal. In another type of pump, fuel was stored in a small chamber which was opened by a spring-loaded valve when the vacuum was suddenly increased.

The mixture also requires enriching when the engine is at maximum speed on a fully opened throttle. This was arranged by using the last few degrees of throttle movement to open a valve and bring the 'power' jet into action.

The performance of the carburettor was improved at low as well as at high speeds. The idling jet continued to discharge into an orifice in the choke wall, but to give a smoother transition from the idling to the main jet a second or progression hole could be drilled in line with and on the atmospheric side of the idling jet and fed from the same channel as the latter. This second jet came into action when the throttle was opened a little to ensure there were no 'flat spots'. The pressure differential was now not so great and some air was bled into the progression jet from the idling jet. Some carburettors were eventually to have a row of jets.

Idling speeds were adjusted by jets which metered fuel supply, by air bleeds, or by altering the minimum gap between throttle and choke tube.

Starting a cold engine could still be a problem and a number of ways were employed to obtain a rich mixture and to weaken it as the engine began to fire. Starting and idling systems could have channels and walls in common and use suitable air bleeds to weaken the mixture for idling, but generally the starting device took the form of a completely independent system.

Solex carburettors had, for example, a separate starting jet, well, and air bleed with orifice which was switched in and out by the driver. In later models petrol standing in the channel above the starting jet passed through a hole in a disc valve controlled by the driver, and then on to an orifice in the choke tube, so that when the engine turned over air was bled into the fuel. As the engine picked up further the driver rotated the disc and a smaller hole came into circuit and when the engine was pulling the disc was rotated further still, and the starting jet isolated. Later a sliding piston was substituted for the disc valve to obtain a progressive cutting of mixture from the starting jet. Zenith mounted a little venturi on the side of the choke tube on some of their carburettors and it opened into the latter through a valve at the usual site of the starter orifice. The venturi had its own jet, well, and air bleed, and was brought into action by opening the valve by a control on the instrument panel. The enriched mixture was obtained on the SU carburettor by lowering the jet assembly and so increasing the size of the jet orifice (figure 6.10).

A disadvantage of many starting devices was that the driver had to remember to switch them out when the engine was running properly, otherwise the performance of the engine could suffer badly and it would probably peter out.

Another approach was, instead of closing the throttle, to close a strangler on the air inlet. The strangler (or 'choke') was manually operated so that the driver had more control as he could operate both the strangler and the butterfly valve and could delay opening the strangler completely until the engine picked up. Again the driver could misuse the choke and so some were made semiautomatic, and a subsidiary blade valve on the flap valve was opened by the further pressure drop that occurred when the engine was firing. Later it was arranged that the strangler itself should open when the pressure fell. The driver was still supposed to open the strangler himself when the engine was firing properly, so to remove all responsibility from the driver automatic chokes were introduced which opened when the engine warmed up. However, cars are still made with manual chokes today.

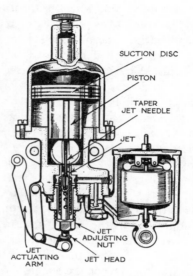

Figure 6.10 SU horizontal carburettor. By permission of Haymarket Publishing Co (from *Autocar* and *Autocar Handbook*).

Automatic chokes

An early automatic choke was that on the Stromberg B model introduced in America in 1932 and first fitted on the Oldsmobile six- and eight-cylinder cars. In the Stromberg C automatic choke, a thermostatic bimetallic coil spring was mounted close to the exhaust manifold and the torque from the spring closed the offset strangler when cold. As the unit warmed up the spring lost tension and the strangler valve opened progressively to provide the right mixture as the engine temperature varied. Later, the thermostat unit was partly housed on top of the carburettor and partly on the exhaust manifold to assist the automatic control.

Other models had the thermostat spring working directly on the end of the strangler spindle. Hot air was used to assist the automatic control.

Automatic control devices were very popular in the United States before 1939, but were little used in the UK. Humber was the first manufacturer to standardise on the automatic choke; they used a Stromberg device that had a thermostat mounted on a flat plate with a spring inside and a lever outside. The lever was connected via a ball joint arrangement to the strangler lever by a rod with a right- and left-hand thread at one or other end, so that the length could be altered for adjustment. There was another rod from the strangler lever down to the fast idler cam arrangement. Rootes made the device work well and later they used it on all models except the Minx. Most other UK companies had hand stranglers and did not change to automatic chokes until after the war.

What with starting, idling, progression, compensating and main jets, wells and channels, economisers, accelerating pumps, automatic chokes, etc, and the ingenuity of their designers, many carburettors became very complicated, too much so in the opinion of some critics.

American carburettors

American carburettors continued to be more refined and to have more moving parts than their European counterparts. Many American carburettors were fitted with air bleeder arrangements that had a small air leak on the side of the main jet; this assisted atomisation of the fuel. Most were also provided with some form of heating and had special pump devices to ensure adequate mixture when the throttle was quickly opened.

Many American carburettors were fitted with proper air strangler controls. Stromberg, for one, arranged a dashboard control that would close the strangler and at the same time slightly open the throttle from the idling position. A rich mixture was therefore obtained and a small automatic flap valve in the strangler could allow for increased air flow and automatically weaken the mixture at any higher speeds. Later, Stromberg and others introduced devices that controlled the air bleeder arrangements so that the jet was progressively partially relieved of the choke tube depression and provided automatic correction over a wide range of engine speeds.

Zenith handled Stromberg carburettors (figure 6.11) in Europe after 1936, and Zenith-Stromberg designs were found on English cars as well as large numbers of American cars. The DA design was basically the same as the American, except that the accelerating pump and economiser were separated. An offset semiautomatic strangler was retained together with a device known as the 'fast idle'. The latter consisted of a cam operated by a lever attached to the strangler lever, the cam was further linked to the throttle spindle by a rod. This arrangement allowed the throttle to be held open to a required amount on starting from cold. Later a DB model of the downdraught type was introduced, that was the DA model fitted with an automatic choke. A further refinement on the DBV model was an automatic choke and vacuum economiser instead of the mechanical arrangements used in the DA model.

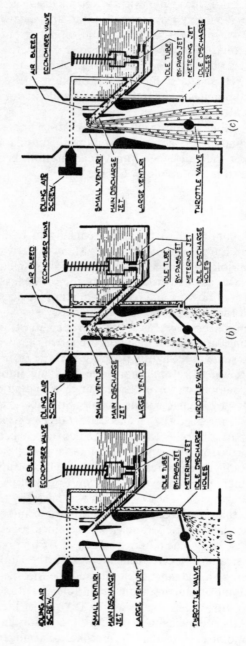

Figure 6.11 Zenith-Stromberg Models DA and DAV Series showing the operation of the metering elements: (*a*), on idle; (*b*), cruising on a lean mixture; (*c*), full throttle on maximum power mixture (bypass valve open). By permission of Chapman and Hall Ltd (from C H Fisher *Carburation*).

Other features of carburettors

Dual downdraught carburettors were also very popular. These had two barrels with a single throttle spindle on which were mounted two throttle valves. Each barrel had its own jets, etc, but a float chamber was common to both and a single economiser valve and jet served both barrels. The accelerating pump fed a separate jet in each barrel.

Dual carburettors were needed on big engines because if a single choke was made large enough not to interfere with the breathing of the engine, the pressure drop at low speeds could be too low for the carburettor to work properly. Consequently two smaller chokes were used instead of one larger one.

The fitting of air cleaners to protect the engine from excessive wear caused by dust and dirt drawn into the engine with the carburetted air became almost universal by 1939. Filters usually consisted of metal gauze, metal swarf or felt, either dry or soaked in oil. The cleaner was mostly fitted on top of the carburettor and provided with an air silencer which not only silenced the intake roar but also cleaned the incoming air to some extent. The cleaner was generally designed so as not to restrict air flow appreciably nor cause more than about 2 % loss of peak power.

Carburettors since 1945

Carburettor design had stabilised by 1939 and there was little further change, apart from improvements in detail design, until emission requirements in the sixties complicated matters. One interesting development was the introduction of the Zenith-Stromberg variable-choke carburettor in 1963; Zenith had hitherto made constant-choke instruments. It was a fairly straightforward design; the air valve piston was moved by a flexible diaphragm controlled by the engine depression, the piston was lifted slightly when starting but in later models fuel passed through a series of calibrated holes on a starter disc. An automatic starting device dependent on engine temperature was also produced. The carburettor was progressively modified in the light of emissions legislation. Constant-vacuum carburettors have only one jet so there are no flat spots, they can be better tuned to the engine and their performance kept closer to the optimum, thus reducing emissions. Incidentally Zenith-Stromberg and Solex joined forces in 1964.

There was also progress in the design of carburettors for bigger engines. Two carburettors or more could be fitted; they gave better distribution as well as better breathing but adjustment could be tricky. Two carburettors were used on smaller cars to give them a more sporting performance. Dual carburettors were used on bigger engines, the two choke tubes sharing a common float chamber and both throttles were operated together by the pedal linkage. A V8 could have two such carburettors. In the compound carburettor, one carburettor carried the starting and idling jets, the other the main jet only, and at low speeds only the throttle of the first carburettor was operated by the pedal but as the pedal effort increased a linkage opened the second throttle. Both choke tubes shared the one inlet pipe. Dual carburettors also improved the performance of cars in the medium capacity engine range (figure 6.12).

In steady running it is not too difficult to reduce CO and hydrocarbon emissions to very low values by, for example, running on a weaker mixture and adjusting the ignition timing accordingly, but it is very difficult to do so in transient conditions such as starting, accelerating, and decelerating, for the carburettor does not respond quickly enough. The mixture may be too rich for a few seconds, all the petrol will not be burnt and the emissions will increase.

Emissions control

To help meet emissions requirements carburettors were made to meter the fuel within closer tolerances. A weaker mixture was used which required the ignition to be retarded, and automatic chokes were employed so that the mixture weakened further as the engine warmed up. A weaker mixture was required when the car was decelerating than when running; this was provided by a gulp valve which opened when the throttle was closed and the engine depression intense, and bled air into the carburettor.

To help meet more stringent requirements carburettors were fitted with additional devices. One such device is the anti-dieseling solenoid used to stop the throttle valve at normal idle position when the engine is running and to shut off fuel flow when the engine stops. A dashpot may also be fitted to the carburettor to prevent sudden closing of the throttle and thereby avoid a rich air–fuel mixture. The vacuum in the intake manifold is also used to operate devices such as the exhaust gas recirculation valve (EGR), control of the heated air system etc, and better match the operating conditions. In four-barrel carburettors adjusted part throttle (APT) is sometimes provided by the fitting of an additional metering rod and a fixed metering-rod jet (figure 6.13) to enable a more accurate adjustment of the flow to be set initially.

Even though the required mixture might be leaving the carburettor this was no guarantee that all the cylinders served by it were receiving the same charge; for this was dependent on the shape and layout of the inlet manifold.

The Inlet Manifold

The mixture leaving the carburettor might have the required fuel/air ratio but much of it was in the form of droplets instead of the homogeneous

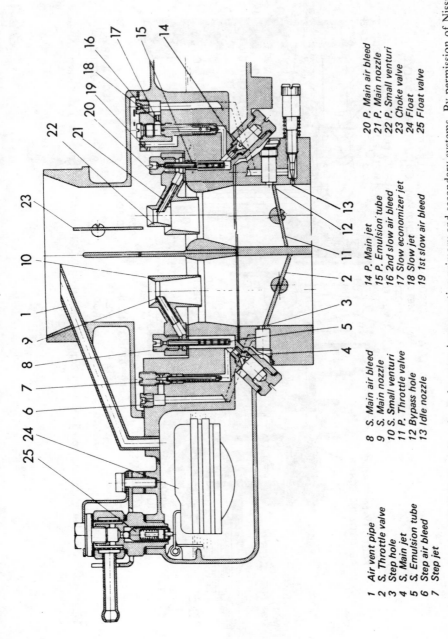

1 Air vent pipe	8 S. Main air bleed
2 S. Throttle valve	9 S. Main nozzle
3 Step hole	10 S. Small venturi
4 S. Main jet	11 P. Throttle valve
5 S. Emulsion tube	12 Bypass hole
6 Step air bleed	13 Idle nozzle
7 Step jet	

14 P. Main jet	20 P. Main air bleed
15 P. Emulsion tube	21 P. Main nozzle
16 2nd slow air bleed	22 P. Small venturi
17 Slow economizer jet	23 Choke valve
18 Slow jet	24 Float
19 1st slow air bleed	25 Float valve

Figure 6.12 Hitachi twin-barrel downdraught carburettor incorporating primary and secondary systems. By permission of Nissan.

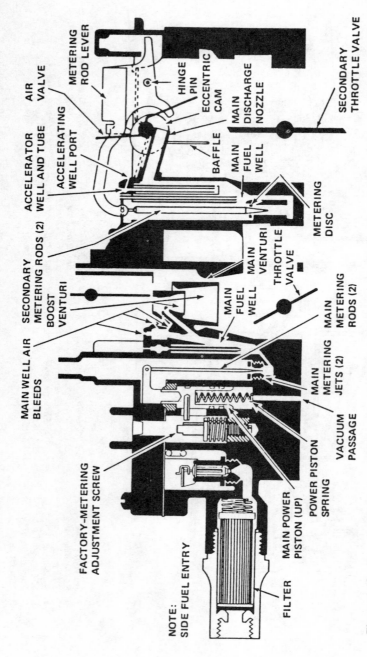

Figure 6.13 Four-barrel carburettor fitted with adjusted part throttle (Chevrolet). By permission of General Motors Corporation.

vapour really required. As vaporising the fuel absorbs heat from the surrounding air lowering its temperature and making further vaporisation more difficult, it was advisable to warm the inlet piping. Droplets could be so large that they could settle out, particularly when the airspeed was low, weakening the mixture. A further problem with multicylinder engines was to obtain equal distribution so that all cylinders got the same charge, and some did not run on too rich while others ran on too weak a mixture. Induction tract problems were not important in the early one- and two-cylinder engines; or, rather, the designer had more pressing problems to consider, but they became more serious when four-, six- and eight-cylinder engines were built.

So far as heating was concerned the carburettor and induction system generally had to depend on air heated by the exhaust manifolds and engine warming them, though on even some of the simpler engines a muff was mounted on the induction pipe and hot water from the cooling system or exhaust gas was passed through the muff.

Totally enclosed pipes were used on side valve engines in order to give the engine a neat appearance; only the horizontal carburettor bolted to the engine block could be seen. The pipes, however, could be badly cast and often the passages were too long, resulting in poor distribution, and, as the engine was slow in warming up, the car was slow in pulling away.

Internal piping could not easily be fitted to OHV engines so these had an outside manifold which resulted in short pipes and the manifold could be cast in high conductivity aluminium. Such manifolds enabled a quicker, smoother getaway and even side valve engines were eventually fitted with similar manifolds and thus overcame their reputation of being bad performers when cold.

Hot air heating, because of its variable nature, tended to wane in the twenties although Austin and Alfa-Romeo amongst a few others, for a time, placed the inlet of a vertical carburettor close to the engine to pick up warm air. Other problems with air heating were the loss of volumetric efficiency because of the reduced air density, and the delay in getting away quickly owing to slow engine warm-up. Excessive heating introduced the danger of raising the carburettor temperature above the boiling point of the more volatile fractions of the fuel and so cause vapour locks inside the carburettor.

Water-heated manifolds were almost exclusively fitted on large cars in the luxury class which often had one large carburettor to feed six cylinders. Usually a thermostatically controlled valve was fitted in the cooling system to provide a hot spot to heat up the mixture until the engine warm-up was completed. The acceleration behaviour from cold was therefore improved but in very cold weather the time to reach operating temperatures was longer than that of an exhaust-heated system.

The manifold could also be heated by the exhaust gases and this solution became particularly attractive when less volatile fuels came to be used in the twenties. Some fairly elaborate castings were needed, however, and there was the possibility that the manifold might be overheated. Some means of controlling the exhaust flow was therefore usually provided such as, for example, a thermostatically controlled flap valve. This layout was most favoured in the US and was well developed in the twenties.

In the UK it was more common to have a hot spot to supply heat at low speeds and then decrease its effect as the intake velocity increased. A good arrangement was that adopted by Standard on their six-cylinder models where the exhaust was carried past the back of the induction pipe on which impinged the spray from a horizontal carburettor and this ensured good volatilisation of the liquid. Sometimes as in the 14-45 Talbot a hot spot was simply obtained by casting a short section of the induction pipe integral with the exhaust manifold. Thermostatically controlled hot spots enabled a good performance to be attained over extremes of cold and the Americans became very adept at catering for these conditions.

Improvements in cylinder head design in the twenties were not matched by improvements in induction tract design, and it was not unusual to find flow restricted by right-angle bends and generally poor design. There was a trend to provide pipes that were well curved and of large diameter but, though they could give good breathing, it was found that with such piping fuel droplets could still settle out. On the other hand, suitably designed, short small-diameter angled pipes could encourage turbulence in critical areas and this turbulence could keep the droplets in suspension and encourage vaporisation so that a more homogeneous mixture would pass into the cylinders.

Manifold design improvements enabled satisfactory carburation of the small six-cylinder engine to be accomplished. Poor mixture distribution was no longer a problem, and effective hot-spot heating with suitable control to vary the amount of heating was available. This controlled air heating came from the use of better thermostatic elements. By this means stable carburation at low engine speeds under full throttle was maintained. Keeping a constant, correct air temperature resulted in smoother engine operation, good fuel economy and good pulling away from cold. However, many of the factors influencing efficiency were not well understood, and most induction systems were designed by trial and error. Often efficiency was sacrificed for a neat appearance but, even so, fairly efficient arrangements were well established by 1939.

Even in the sixties the design of the induction system was still very much a matter of trial and error, and the final design largely a matter of compromise. A layout that could suit one engine might give poor results on another.

On four-cylinder engines fitted with single carburettors, either two ports, each with a siamesed valve port cast in the cylinder block, or four ports feeding each individual valve, were used. If twin carburettors were used, they were fitted directly to siamesed ports in the two-port layout or to a short siamesed manifold passage in the four-port system. It was usual to connect a balance pipe between the manifold systems so that pulsations could be dampened out and idling conditions improved.

Manifolds for six-cylinder engines were mostly of the three- or four-port types. It was generally arranged that the three-port layout had each port feeding a siamesed valve port. The four-port manifold, the most popular, had the two main ports siamesed and the outer ports feeding single valves.

On straight eight-cylinder engines it was fairly general practice to use a dual carburettor, the earlier engines being treated as two separate four-cylinder engines. Later, one four-cylinder port in the centre was fed by one barrel of the carburettor and the other outer pairs were fed by the other barrel. Attempts were also made to try to keep the lengths of the gallery pipes the same. This was no problem with V8 engines and an arrangement used on Ford engines permitted alternate feeding of the two banks of cylinders. Each was fed by a single carburettor or a large double carburettor and this resulted in a fairly uniform mixture distribution.

At the turn of the century attempts were made to obtain more power from the engine by pumping mixture into the combustion chamber. This 'forced induction' or supercharging, however, did not reach the production stage until it was used on the 1924 Mercedes 15/70/100 car and also on later Mercedes vehicles. The main types of supercharger were the Rootes type of blower with its twin type rotor, and the Cozette with radially sliding vanes operating in an eccentric casing. Mercedes used a supercharger that blew through the carburettor but fitted a disc clutch at the end of the crankshaft which was engaged when the throttle pedal was suddenly depressed to bring the supercharger into action. Other high-performance vehicles compressed the mixture between the carburettor and the engine to permit the carburettor to function normally during supercharging. Although supercharging had a limited appeal in the thirties it was largely to disappear with the advent of fuel rationing during the Second World War.

Since that war extra vaporisation and better mixture distribution is usually achieved by means of a thermostatically controlled exhaust hot spot. Water induction systems can be more easily arranged in some engines such as horizontally opposed flat fours. For example, on the Alfa Sud water passed directly below the manifold passages and the engine gave good driveability with lean mixtures and low emissions. One disadvantage is that there is usually some power loss with the use of water-heated manifolds on an overhead camshaft engine. With the need to reduce emissions more care generally was required in the design of manifolds. One thing that became necessary was the need to use a thermostatically controlled air intake to maintain the temperature of the air between 35–45 °C to ensure the carburettor operated under more stable conditions.

Many manufacturers also used two or more carburettors to increase the power and efficiency of small engines but shaped the manifolds to ensure the air/fuel mixture in one cylinder did not interfere with that of another. Such manifolds were used, for example, on the Ford Cortina GT and in cars built by Toyota. Another interesting development was the three-branch manifold for the six-cylinder in-line Mercedes-Benz 2.8 litre engine in the seventies. This was fitted with a Solex carburettor having two primary and two secondary chokes. The layout was such that there was one primary and one secondary choke for each branch of the manifold to improve the mixture distribution.

Despite much ingenuity being displayed on manifold designs the charge distribution was still uneven, even with those that it was thought should give uniform flow to all cylinders. The mixture distribution also changed appreciably between part-load and full-load conditions. Similarly the carburettor could not respond rapidly and accurately enough, and catalytic convertors were needed to oxidise the CO and the unburnt hydrocarbons in the exhaust gases. Petrol injection on the other hand could, in principle, respond very rapidly to changing conditions, meter fuel very accurately, and give each cylinder the same charge. It had much more potential than the conventional carburettor—especially as more stringent emissions laws were to emerge.

Fuel Injection

Work was done in Germany and the USA before the War to replace carburettors by fuel injection systems on military and stunt aircraft. Unlike carburettors petrol injection worked whatever the attitude and acceleration of the aircraft. Fuel injection does not interfere with the breathing of the engine and enables each cylinder to obtain the same charge. For these reasons it was tried on racing cars in the fifties when it was found to be quite successful. In 1954 petrol injection was used on the Mercedes-Benz 300SL and on the 1957 Chevrolet Corvette but it was not used on UK cars until the 1967 Triumph TR5. The Mercedes-Benz system was based on diesel practice; a jerk pump passed fuel to injector nozzles near the inlet valve. Bendix developed the electro-injector system in the fifties. Fuel was pressurised and injected into the inlet ports by solenoid-operated valves, the amount being determined by the duration of the injection pulse. Injection was timed by contact points on the ignition distribution shaft. Sensors determined the position of the throttle, engine speed, intake inlet manifold pressure and temperature, and this information was fed to an electronic control unit which regulated the pulse time of the injectors accordingly. A cold start valve increased the fuel supply for starting. The electro-injector, however, was too expensive and never went into large-scale production.

Fuel injection

Fuel injection was complex and expensive, as well as being difficult and very expensive in service. It therefore did not arouse much interest until the limitations of the carburettor were shown when attempts were made to meet emissions requirements, at which point fuel injection appeared as a viable alternative to the conventional carburettor and inlet manifold.

Electronically controlled systems

Bosch took out patent rights from Bendix and in collaboration with Volkswagen brought out the D-Jetronic fuel injection system in 1967 (figure 6.14). This was a more sophisticated version of the electro-injector and took advantage of the developments in electronics.

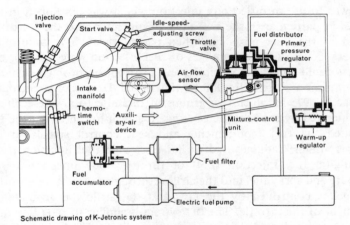

Figure 6.14 Bosch D-Jetronic petrol injection system. By permission of Robert Bosch GmbH.

The airflow into the cylinders is a much better indication of the fuel needs of the engine than the manifold depression, and in the L-Jetronic system of 1973 the airspeed was measured by the displacement of a circular disc or flap pivoted in an air funnel through which the incoming air passed; the flap carried a second flap which swung in a damping chamber. Again progress in electronics enabled a more powerful control unit to be used, and costs were reduced and reliability increased. Airflow speed was later also measured by a hot-wire anemometer, the greater the airspeed the lower the temperature of the wire.

In 1973 Bosch introduced the cheaper K-Jetronic system. The pressurised fuel was now metered mechanically and was sprayed in a continuous stream into the inlet valve port. The load on the engine was measured again by the displacement of a flap in the air funnel.

In later systems the electronic control unit (ECU) was converted to digital operation. Microprocessors are extremely powerful. Signals from air flow, engine and ambient temperature, engine speed, and pedal position are combined, and the microprocessor determines the fuel requirements and regulates the operation of the injectors accordingly. The microprocessor may also use the information from the sensors to control the ignition timing.

An important development in the mid-seventies was the use of the λ-probe to measure the air/fuel ratio indirectly by sensing the amount of oxygen in the exhaust gas. The signal from the probe is fed back to the control unit and action taken to adjust the mixture. Feedback control is enormously powerful.

A complete fuel injection system was expensive and a cheaper arrangement was to use throttle body injection which was developed in the early eighties in particular in the US. Fuel was injected into the 'throttle body' instead of the inlet ports and then passed through a manifold to the cylinders. Some of the advantages of fuel injection were lost, for example, droplets could condense on piping and distribution problems could arise.

An injection system similar to the Bosch D type was introduced by Cadillac in 1975 in which two groups of fuel injectors were each alternatively operated by the ECU. Sensors again sent information to the ECU that computed the fuel requirements and sent control signals to the fuel injectors. Later Cadillac used digital electronic fuel injection (DEFI) and then digital fuel injection (DFI) or throttle body injection to control, distribute and spray fuel into the airflow through a two-barrel throttle body. The fuel is controlled by the ECU and the airflow is regulated by the movement of the throttle. The DFI system is similar to DEFI but an oxygen sensor 'λ probe' is added to the exhaust manifold; information concerning the oxygen content can then be provided to the ECU monitoring the air/fuel ratio so that exhaust emissions can be kept within legislated limits.

Ford also replaced the carburettor by throttle body injection in the early eighties. Two pressure-actuated fuel metering valves are used which are incorporated in the throttle body on the intake manifold. A fuel pressure regulator is employed to maintain a constant fuel pressure to the injectors. Again sensors monitor engine operating conditions and signals sent to a system computer are used to ensure correct injector opening times. A signal from a bimetallic switch on the throttle body provides extra fuel during starting and warm-up.

Chrysler used a continuous-flow fuel injection system in their cars. This system uses a computer to control the electronic fuel injection and the electronic spark control system. The latter gives the correct voltage pulses to suit the engine operating conditions and minimises exhaust emissions while also ensuring better fuel economy and idling characteristics. The air/fuel ratio is controlled by varying the pressure on the fuel delivered from two sets of injection bars in the throttle body.

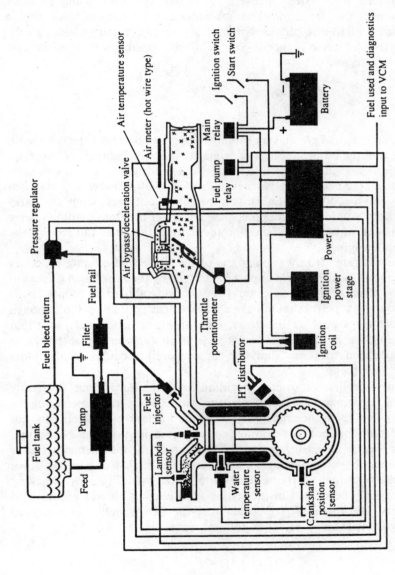

Figure 6.15 Lucas 9CV engine management system for the four-valve Jaguar XJ6. By permission of Jaguar Cars Ltd.

Not only are systems designed to meet fuel economy and exhaust emission legislation but electronic management systems are being developed generally to optimise engine running under all operating conditions. The Lucas Micro 9CV system (figure 6.15) is one developed using air mass measurement that takes into account water and air temperatures, engine speed and load to meet driver requirements, economy and emission levels. A high level of performance is also achieved, together with considerable inbuilt self-diagnosis.

Fuel

The oil industry had grown very rapidly in the last quarter of the nineteenth century, and the oil produced was primarily used for lighting and heating. Processing the crude oil was fairly straightforward; the oil was fractionally distilled and the middle cut, kerosene, distilled again, if necessary, and then cleaned and deodorised. Lubricants could be obtained from the tarry residue but its main use was as fuel oil. The lighter fractions, naphtha, were practically a waste byproduct, but these were the ones utilised by the automobile engine.

Crude oils came from various parts of the world and refining processes varied, and so petrol, a term apparently coined by the firm Capel Carless in the early years of the century, could vary considerably in its properties. It was thought that petrol should have a relative density of about 0.68, though this figure crept up over the years. One problem, incidentally, was that petrol could lose its more volatile constituents on standing so that the engine became difficult to start. Similarly the carburettor could use the more volatile constituents first.

The mass production of cars, beginning with the Ford Model T, greatly increased the demand for petrol, and this was met in part by increasing the yield of petrol from the crude oil. This could be done by widening the cut during distillation, but the resulting fuel was more prone to cause knocking. Another way of increasing the yield was to crack the oil thermally, that is distil it under high temperatures and suitable pressures, which breaks down longer chains of hydrocarbons into shorter, more volatile chains. Cracking had previously been used in processing kerosene, and it was extended to petrol just before the First World War; the petrol yield was increased typically from 15 to 30 % of the total product.

Knock

The output of an engine can be increased by increasing its compression ratio, but if the latter is increased beyond a limiting figure the engine 'knocks', that is, it gives a ringing sound, the output falls off, and the engine is likely to suffer damage. Knock became a serious problem during the First War because it limited the performance of aircraft engines. It was found that aviation fuel based on Far East crudes had better antiknock properties than fuels based on American crudes and this observation prompted Ricardo's systematic work on knock in the early twenties. He built a variable-compression single-cylinder engine to determine the compression ratios at which different fuels and hydrocarbons knocked. Of the various materials he tried heptane knocked at the lowest and toluene at the highest compression ratio, and so he determined the ratio of toluene to heptane needed to match a given sample of petrol and called it the toluene number of the sample. Later iso-octane was used instead of toluene to give the octane number or rating.

Ricardo showed that knock was caused by the expanding front of burning fuel increasing the pressure and temperature of the remaining fuel until it ignited explosively. Suitable design of the combustion chamber could delay knock, for example, the last gas to be ignited should be in the coolest part of the chamber, such as over the inlet valve, and the end gas should be confined in a thin layer between cool metal surfaces. It had earlier been found that the compression ratio could be increased by designing inlet ports, valves and combustion chambers to encourage turbulence in the mixture and so increase the rate of propagation of the flame front. Turbulent heads gave many side valve engines a new lease of life in the twenties.

Ricardo showed that aliphatics were bad, for example, and aromatics and alcohols good, for knock, and so fuels could be selected and blended, and refining procedures modified in order to increase the octane number of the fuel.

Thomas Midgley in the US was also trying to devise ways of making gasoline less prone to knock. He found that tetraethyl lead (TEL) was a remarkably effective antiknock agent and it was soon widely used in the US. It was not available in the UK until 1928; there were reservations for it could have been a health hazard, and it affected valve guides and some sparking plugs.

As a result of work of this kind the motorist soon had a bewildering choice of petrols. No 1 petrols could be sold unblended, blended with benzol, blended with alcohol (and possibly benzol) as well or could contain antiknock additives like TEL. There were also heavier petrols, including No 3 petrol, which were intended for commercial vehicles but could be satisfactory on the larger car engines. As most of the oil companies tried to cover the whole range of petrols there was a remarkable proliferation of brand names. The outbreak of war in 1939 put a stop to all this and only the one grade of petrol, pool petrol, was sold. Its octane number was only about 74.

To return to the thirties, cracking had increased the yield of petrol but still the demand for petrol increased faster than that for the other oil products and so attempts were made to increase the yield further still. Higher proportions of petrol were produced when the oil was cracked in the presence of suitable catalysts, and after some difficulty in making the process a commercial success, catalytic cracking was introduced. The resulting fuel had a better octane rating than the regular fuel plus TEL. Average octane ratings and compression ratios between the Wars are shown in the table 6.1. Other processes such as hydroforming and alkylation were developed, the latter giving high octane number aviation fuels. Catalytic reforming was used to obtain a high octane blending component.

Table 6.1.

Year	1925	1931	1935	1939
Octane number	45–50	63–75	72–78	74–83
Compression ratio	4	5.2	6.0	6.3

Aircraft engines during the Second World War were operating on 100 plus fuels. The cheapest and simplest way, however, to increase the octane rating was to add TEL, 0.4 to 0.6 g/litre, increasing the Research Octane Number (RON) by 6 or 7 units. Petrol rationing ended in the UK in May 1950, branded petrols were again on sale so that higher and higher octane fuels were produced, the octane number levelling off at about 97 RON for premium fuels in the early sixties.

The fuel had to meet other requirements as well as having the appropriate octane number. It should have good volatility over a wide range of temperatures so that it was satisfactory at very low temperatures but did not cause vapour locks at high temperatures. Freezing point depressants were added to prevent icing of the carburettor at very low temperatures, detergents to maintain engine cleanliness and other additives to prevent gum formation and corrosion.

Lead poisons catalysts and so lead-free petrol had to be used in the USA on cars fitted with catalytic convertors in order to meet emissions requirements. In Germany TEL was banned because it might harm children. The move to lead-free fuels has had an enormous effect on the refining industry. There were major investments in cracking and reforming plants, and advances in catalyst technology as the industry attempted to make good the loss in octane number.

Chapter 7

Ignition

Lenoir's gas engine of 1865 employed electric ignition, using batteries, a Ruhmkorff coil, and a sparking plug. The Ruhmkorff coil carried an automatic trembler make-and-break; the trembler was a flexible blade, fixed at one end on which an armature was mounted and with contacts at the free end. When current passed through the primary coil the armature was attracted to the iron core of the coil breaking the primary circuit. The armature then sprang back to remake the circuit and so the cycle repeated, and the resulting pulsating current induced a high-voltage current in the secondary coil.

Daimler used hot-tube ignition (p 18). A small platinum tube closed at one end was fitted in the combustion chamber of the cylinder, the closed end projected outside the cylinder and was heated to incandescence by a blow lamp of the Bunsen burner type. By moving the burner away from the cylinder end of the tube the firing could be retarded somewhat. The driver could therefore cope to some extent with changes in speed, particularly with slow running engines but the lamp could too easily be blown out, the system was inflexible, and it was dangerous.

Benz followed Lenoir, however, and used electric ignition. A rotary switch (figure 7.1) driven by the engine connected batteries to the primary of a trembler induction coil at the right instant, and the high-tension voltage induced in the secondary coil produced a series of sparks in the sparking plug. A condenser across the primary prevented arcing in the switch. The driver could alter the timing very simply by rotating the backplate carrying the stationary contact of the switch. A multicylinder engine could use a low-tension distributor, each contact feeding a separate trembler coil and sparking plug, or a low-tension switch and only one coil and a high-tension distributor to feed the sparking plugs. The Model T Ford was fitted with four trembler coils throughout its existence. de Dion did not use a trembler directly on the coil; instead the spring carrying the points of the contact breaker vibrated when the weighted end of the spring fell into the depression in the rotating cam (figure 7.2), the idea being that a series of sparks would be obtained at low speeds, making starting easier.

Instead of using a wiping contact and trembler coil, or the de Dion arrangement, a contact breaker and simple coil could be used (figure 7.3),

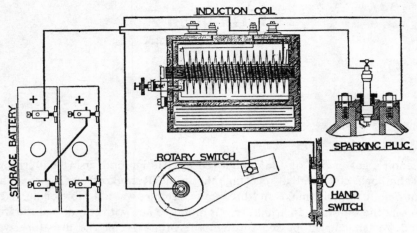

Figure 7.1 Diagram of Benz' electrical ignition apparatus.

though this was distrusted at first because it gave only a single spark, but it was eventually realised that this was generally sufficient.

Coil ignition had the disadvantage that, because the time during which current flowed in the primary decreased with speed, so also did the output. The early systems also had trouble with faulty insulation, and the contacts needed frequent cleaning and attention, but the biggest trouble was the batteries. The drain on dry cells was too great for them to be a practical proposition for most people, and accumulators were heavy, messy and susceptible to damage from vibration. They had to be lifted out of the car from time to time and recharged, one charge giving about 25 hours

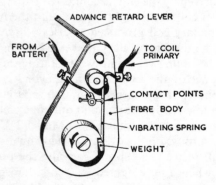

Figure 7.2 de Dion type of contact breaker; the spring was set vibrating by the rotating cam. By permission of Century Hutchinson Publishing Group Ltd (from E F Carter 1962 *Veteran Car Owners Manual*).

motoring in the early 1900s. Simms, however, in conjunction with Bosch adapted the magneto to the automobile engine and this did away with the need for a battery.

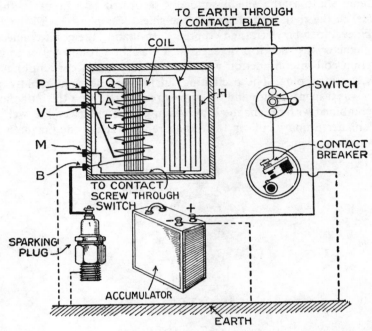

Figure 7.3 Ignition system with plain coil for a single-cylinder engine. A, primary wire coil; B, secondary wire terminals; E, secondary wire coil; H, condenser; M, battery terminal of secondary wire; P, battery terminal of primary wire; Q, soft iron wire core; V, contact breaker terminal of primary wire. By permission of Haymarket Publishing Co (from *Autocar* and *Autocar Handbook*).

Magneto Ignition

The magneto consisted essentially of an H-shaped armature wound with an armature coil and rotated between the pole pieces of a permanent magnet. At a critical orientation of the armature the magnetic flux linking pole pieces and armature suddenly changed configuration and in so doing induced an electromotive force (EMF) in the armature coil. Martini, of the Martini-Henry rifle, had used a spring to move the armature quickly through a quarter revolution, and used the EMF generated to ignite the charge of a gas engine. Other people, including Markus and Otto, also worked on the magneto. The early Simms-Bosch low-tension magneto consisted of a number of horseshoe-shaped magnets fitted with pole pieces; the H-shaped armature was, however, stationary and a soft iron shield was oscillated between armature and pole pieces; the resulting displacement of the lines of force induced an EMF in the coil wound on the armature. A cam driven by the same shaft as the shield suddenly separated two points within the cylinder at the right instant, and as the shield was phased with the cam, a spark developed between the two points. Insulating the point connected to the magneto was such a problem that it was soon found better to use an induction coil in conjunction with the magneto and a conventional jump spark sparking plug. The oscillating shield wore bearings rapidly so the shield was dispensed with and the armature rotated. Next the induction coil was combined with the magneto by winding a secondary coil as well as the now smaller primary coil on the armature, converting the magneto into a high-tension magneto.

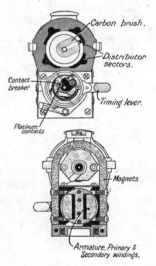

Figure 7.4 High-tension magneto. By permission of Haymarket Publishing Co (from *Autocar* and *Autocar Handbook*).

By mounting the distributor, contact breaker etc on the magneto, a compact self-contained, though rather bulky and certainly heavy, unit could be made (figure 7.4). The back plate carrying the points of the contact breaker could be rotated with respect to the body of the magneto and teeth on the circular back plate engaged in teeth on another circular back plate carrying the distributor contacts so that the timing could be retarded for starting and advanced as the speed increased, rotation being effected by a hand control on the steering wheel column.

By 1906 or so the high-tension magneto had largely replaced coil ignition particularly on four-cylinder engines. Many luxury makes, however, fitted coil as well as magneto ignition in order to take advantage of the easy starting of the coil. The usual arrangement was then for the circuit breaker, distributor and plugs to be shared, with the coil switched out automatically once the engine was running. Dual systems became usual in the USA. Sometimes completely dual systems were fitted, each with its own set of plugs.

Within a few years automatic timing was available on some magnetos; centrifugal weights engaging cams which rotated the contact breaker back plate. The maximum voltage, however, still occurred at the same angular displacement and it was not until much later that the centrifugal advance mechanism was inserted between the driving shaft from the engine and the shaft of the magneto itself, and so the maximum spark could be obtained irrespective of the ignition setting.

Despite its popularity the magneto had its drawbacks; it gave a poor spark at low speeds, which could make starting difficult, it was expensive, and its manufacture required skilled workmen. The latter was not a problem with Bosch who were precision engineers in the first place but was so in the USA, and to a lesser extent in the UK. Indeed the shortage of firms capable of making the magnetos which had previously been imported from Germany was a serious embarrassment during the First War, and a great deal of work had to be done to obtain an adequate supply of efficient machines.

Kettering

Because of these difficulties with the magneto, Kettering in the USA set out to improve coil ignition, which he did by attention to detail, and he improved things to such a degree that Henry Leland became interested and used Delco ignition on the 1910 Cadillacs. Delco stood for Dayton Engineering Laboratories Company, the little firm started by Kettering and his associates.

A little later Kettering developed an electric self-starting system for Leland and this too required a battery. For the self-starter to be practicable he had also to find a way of keeping the battery charged by means of a dynamo (generator in the USA) driven by the engine, and the Delco system was duly perfected (see p 177) and fitted to the 1911 Cadillacs.

Electric lighting became practicable once the battery could be recharged *in situ* (and suitable lamps had been invented). Once the car carried a battery it made sense to use coil ignition, but curiously the UK and Continent continued to use magneto ignition even after electric lighting had become standard. In the USA, however, magneto ignition rapidly lost ground after 1914 or so, and coil ignition was practically standard by 1920.

Magnetos in the twenties

The Old World persisted with the magneto for much longer, and indeed the production of magnetos quadrupled between 1921 and 1927. New types had been developed, the inductor magneto (mainly for aircraft engines) and the rotating magnet magneto, while the rotating armature type was improved. Laminated pole pieces gave a better spark, and spark jump instead of wipe contact was revived for the distributor. Spark jump contact reduced wear and contact problems and also made starting easier, for leakage could not occur in the spark plug area until the spark had occurred in the distributor. The simple rotating armature magneto gave only two impulses per revolution so had to run at high speeds on engines of more than four cylinders (1.5 times engine speed for six cylinders) but the newer types could run at lower speeds.

In 1920 the Japanese developed cobalt steel magnets which enabled the magneto to be made much smaller and lighter and, in particular, made the rotating magnet magneto a more practicable proposition. The cobalt steel alloys were followed over the years by even better magnetic alloys.

The magneto was expensive, and there could be mechanical complications in installing it, but it made a compact self-contained current source, whereas coil ignition depended on the battery which many people did not trust, particularly in the winter months. (Indeed many UK cars were fitted with emergency starting handles well into the fifties.) Consequently the magneto was still standard equipment in the UK in 1925 but then it lost ground rapidly, and was used only on about a quarter of the 1931 cars and in later years only on high-performance and racing cars, though some luxurious cars continued to fit dual systems. The magneto persisted on CVs much longer. It is still used on petrol-engined lawnmowers and marine engines so that the battery can be dispensed with.

Coil Ignition

Coil ignition simplified the drives from the engine; in the mid-twenties the contact breaker (figure 7.5) and distributor were generally mounted on the upper end of a vertical or near-vertical shaft driven by the camshaft through suitable gears. The lower end of the shaft extended downwards into the sump and drove the oil pump, an arrangement still used on some cars today. Timing could be varied with speed automatically in much the same way as on the magneto, and eventually automatic timing became universal. The control mechanism carried centrifugal weights which operated through a system of linkages and tension springs (figure 7.6), or which engaged contoured cams to rotate the back plate carrying the contact points. Similar systems are still in use.

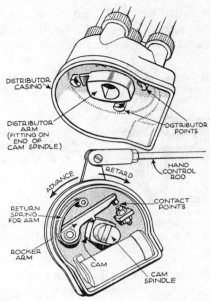

Figure 7.5 Contact breaker. By permission of Haymarket Publishing Co (from *Autocar* and *Autocar Handbook*).

The optimum timing should also vary with the load on the engine so that on some cars in the thirties the timing was made to vary automatically with the manifold pressure. The control unit (figure 7.7) consisted of a flexible diaphragm in a metal housing. One side of the diaphragm formed part of a vacuum chamber and was connected to the carburettor just behind the throttle butterfly valve, and acted against a spring. A central rod on the

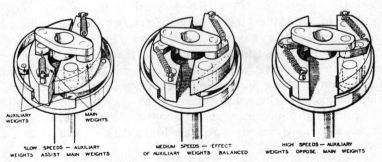

Figure 7.6 Automatic advance/retard by means of centrifugal weights. (*a*), Slow speeds—auxiliary weights assist main weights; (*b*), medium speeds—effect of auxiliary weights balanced; (*c*), high speeds—auxiliary weights oppose main weights. (By permission of Haymarket Publishing Co (from *Autocar* and *Autocar Handbook*).

diaphragm rotated the distributor body, but this arrangement gave wear and other problems so that in later designs the diaphragm rotated a light plate carrying the fixed contact of the contact breaker.

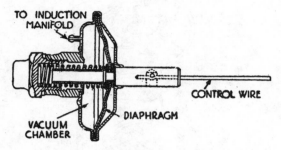

Figure 7.7 Automatic vacuum control mechanism. By permission of Haymarket Publishing Co (from *Autocar* and *Autocar Handbook*).

Automatic vacuum advance control became more widely used after the Second World War. Much later in the seventies some units incorporated a mechanism to retard the ignition under some circumstances and so reduce emissions. The vacuum was taken off downstream of the throttle and operated on an annular diaphragm that worked in opposition to the main diaphragm. The retard mechanism operated independently of the advance mechanism but was subordinate to it.

Coils were impregnated with wax, and sealed in position in their cans with pitch, so if the contact breaker was in the make position when the engine was stopped, but the ignition not switched off, the coil would overheat. To prevent this from happening a ballast resistor was used; this was a length of iron wire the resistance of which increased with temperature. Later the primary was wound outside the secondary coil so that it could cool more easily and in the fifties the coil was immersed in oil, reducing the temperature and doing away with the need for a ballast resistor.

A problem that arose in the thirties was that ignition systems could interfere with short-wave radio reception and, similarly, in the early fifties with television reception; interference suppressors, generally resistors, then became mandatory. The ignition system changed little in the following decades, although much was learned about the ignition and combustion processes, and improvements took place in the detail design of the various components.

Transistorised Systems

A major advance was the introduction of transistors and other solid state devices into ignition systems in the early sixties. In the earlier systems current from the contact breaker switched a transistor to complete the low-tension circuit. Because the points had to carry very little current they lasted longer and gave less trouble, and as the transistor could handle much larger currents than conventional points less inductance was needed in the primary winding of the coil, and the resulting higher build-up of flux gave higher voltages in the secondary winding. In later systems the points were done away with altogether, and the signal to actuate the transistor obtained by, for example, rotating ferrite rods mounted in a plastic disc past an electromagnetic pickup. In other systems a magnetic pulse unit was used in which specially shaped tips on a core mounted on the distributor shaft swept past poles on a coaxial magnetic ring inducing a voltage in a coil (figure 7.8), or blades on a rotor passed between poles on a magnetic pickup coil. The next stage was to process signals from the pickup, and from a sensor in the inlet manifold, using an electronic control unit; the latter adjusted the timing of the pulse and its steepness, size and duration, to suit the conditions obtaining. Transistor control gave stable and precise timing and there were no mechanical parts to wear, or get out of adjustment. The greater energy in the spark enabled unleaded fuel and higher compression ratios to be used, as well as leaner mixtures and higher engine speeds.

The motivation for transistorised systems came in the first place from

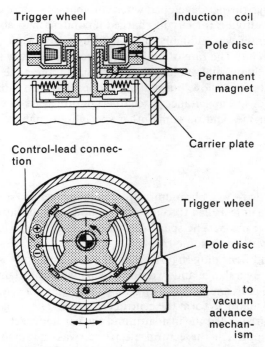

Figure 7.8 Magnetic pulse generator unit (Bosch). By permission of Robert Bosch GmbH.

racing. To obtain the maximum power from relatively small-capacity engines, racing car engines had to have a large number of small cylinders and to rev at high speeds. This meant there could be trouble with contact bounce, even when more than one set of points were used. Later the precise timing and flexibility of transistorised ignition made its use practically essential if American and, later, EEC emission regulations were to be met. Most European cars, however, still had mechanical contact breakers in 1980. More recently sophisticated electronic ignition systems have been developed in which the ignition system is integrated with other systems such as the fuel injection system or the antiknock system (see Chapter 6).

A stoichiometric mixture may be optimum for emissions control but it reduces the compression ratio that can be used unless some means is found to retard the spark when the engine is on the point of knocking. This can be done by mounting an accelerometer on the engine at a strategic position, filtering its output to extract the frequency corresponding to knock, and using the resulting signal to adjust the ignition timing.

Efforts are being directed to making smaller, lighter and more reliable components, and to reducing their cost. The more complicated systems are at present necessarily expensive and so are used only on more expensive cars, or when regulations can only be met by their use.

Another system, capacitance discharge ignition, which has had quite a long history, was further developed when suitable solid state devices became available. A multivibrator circuit charged a condenser to about 400 V and then the condenser was suddenly discharged by a solid state switch into the low-tension circuit. The rate of build-up in the secondary winding was very rapid indeed giving high voltages and reducing the time during which leakage currents could flow, but making the spark of too short duration for most applications. Capacitance discharge ignition was, however, used on some racing engines and on some other specialised engines.

Sparking Plugs

Lenoir embedded two platinum wires in a porcelain stem which was cemented into a hole drilled through a metal bolt, and the latter was screwed into the cylinder cover. The spark gap was about 0.05 in wide. Apparently these plugs gave Lenoir some trouble.

The very solid Benz plug (figure 7.1) had a cast iron body which was fixed to the cylinder by a flange and two bolts. A flanged porcelain insulator was clamped in the plug and sealed with lead packing, and a central rod passed through the insulator to form the central electrode. The plug carried a brass sleeve which surrounded the insulator and the earthed electrode, which was bent over to approach the central electrode, was mounted on the sleeve projecting about 1.25 in from the latter. The de Dion plug was rather

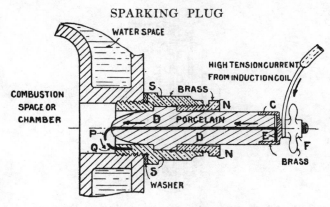

Figure 7.9 de Dion Bouton spark plug.

similar to the Benz plug but screwed into the cylinder head (figure 7.9). Porcelain was brittle, and liable to damage by thermal shock and so other materials were tried. Mica insulation was used in the early 1900s, mica being wrapped round the central electrode which was then pushed through a stack of mica discs with holes punched through their centres and the whole assembly clamped together (figure 7.10). The racing driver K Lee Guinness had so much trouble with the plugs of his Sunbeam about 1910 that he developed his own plugs and set up KLG Sparking Plugs Ltd to manufacture them.

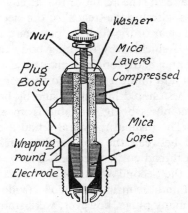

Figure 7.10 Spark plug with mica insulation. By permission of Haymarket Publishing Co (from *Autocar* and *Autocar Handbook*).

Mica and mica–porcelain composites were used for many years but mica was found to be attacked by the lead in leaded fuels when the latter came to be used. Other materials such as steatite were also used (figure 7.11).

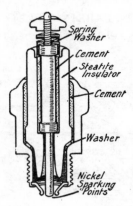

Figure 7.11 Spark plug with steatite insulation. By permission of Haymarket Publishing Co (from *Autocar* and *Autocar Handbook*).

One problem was to make the plug gas tight which was not easy because of the difference in thermal conductivity and thermal expansion of the porcelain and the metal body of the plug. A second problem was that oil and combustion products could be deposited on the insulator body, giving a path for leakage currents and so preventing an effective spark from being obtained. The solution to this was to try to run the insulator at temperatures high enough for oil and carbon to be burnt off and yet not high enough to damage the plug, or cause pre-ignition. This was done by shaping the nose of the insulator and altering the height of the latter with regard to the end of the body shell, that is, altering the 'reach' of the electrodes. If the insulator and points were well clear of the shell or 'petticoat' they would tend to run hot, and conversely if shielded by the plug body they would tend to run cool. Similarly if the plug had a long porcelain nose it could run hot whereas if the nose was short and blunt it tended to run cool. A designer therefore chose the plug that best suited his engine and he had a very wide choice, hundreds of different plugs being available from a number of manufacturers (figure 7.12). The insulator generally had a collar at its lower end which was clamped between the shell and a packing gland nut which engaged the internal thread cut in the shell. One or two gaskets sealed the join or gland. Until the Second World War the gland nut on many plugs could be screwed out and the insulator and electrode removed for cleaning. The components of many plugs, however, were simply cemented together, and a compression gasket between plug and cylinder head acted as a seal.

The central electrode was at first made of steel but soon oxidation-resistant nickel or nickel alloy were used. To reduce costs only the lower part of the electrode could be made of these more expensive materials. Sometimes the actual points were made of platinum or tungsten. Some plugs in the early days, particularly when dual ignition was used, had two sets of electrodes. A single central electrode together with a single right-angled earthed electrode mounted on the shell became the normal arrangement, however. But plugs could have more than one earthed electrode, which could take the form of an annulus concentric with the central electrode; the annulus carrying the points. Such plugs were made for many years; they gave longer electrode life but there was more risk of individual gaps being shorted by deposits.

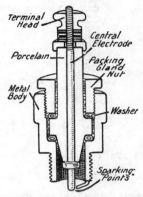

Figure 7.12 Typical porcelain spark plug. By permission of Haymarket Publishing Co (from *Autocar* and *Autocar Handbook*).

A major advance in the forties was the replacement of the porcelain insulator by sintered aluminium oxide insulators. The aluminium oxide had three times the strength of porcelain and ten times its resistivity. The gland nut was done away with and the top edge of the body shell rolled over to grip the insulator.

The diameter of the plug thread had hitherto been standardised on that of the de Dion plug of forty years earlier, but smaller 14 mm plugs that took up less space on the cylinder head began to displace the 18 mm plug.

Many plugs incorporated a carbon resistor in the central electrode; this helped to reduce radio interference and also limited the HT voltage, and so reduced erosion of the points. Sometimes the resistor was mounted in the plug lead. Instead of welding the central steel conductor and nickel alloy electrode together, they were later joined by a conducting glass element which also sealed wires and insulator body together.

Attempts were made in the sixties to adapt the surface discharge plug developed for gas turbines to the automobile engine, but with little success. Lower voltages were needed with these plugs and the sparks, which tracked across the surface instead of passing between points, tended to clean the surface. Emission control regulations made conditions more onerous for the sparking plug in the USA; it became necessary to increase the spark gap in order to burn the weaker mixtures.

Though its component materials and its construction may be different, the appearance of the contemporary plug (figure 7.13) is similar to that of the early plugs.

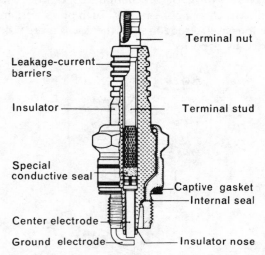

Figure 7.13 Contemporary spark plug. By permission of Robert Bosch GmbH.

Current Generation and Control

The dynamo and its associated circuitry had to be modified for it to be used on the automobile. One particular problem was that the output of the contemporary dynamo depended on its rotational speed, which in turn depended on the speed of the vehicle. At low speeds the voltage of the dynamo could be less than that of the battery and so current could flow back to the dynamo, possibly damaging it, and causing the battery to run flat. To prevent this from happening an automatic cutout was fitted. This was an electromagnetic relay (figure 7.14) wound with a high resistance coil of fine wire which was connected across the output of the dynamo. When the dynamo voltage was sufficient the relay closed and current passed through to the battery. This current also passed through a coil of a few turns of thick wire and so helped to hold the relay on, but if a reverse current flowed back to the battery the field of this coil opposed that of the shunt coil and the relay rapidly released.

It was not very easy to prevent the dynamo voltage and current from exceeding limiting values at high speeds. The rotational speed of some early dynamos was limited by mechanical means but, later, ways were used to limit the output directly. The constant-current system used a third brush in a manner that caused the magnetic field to weaken as the speed increased

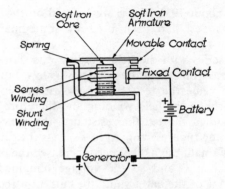

Figure 7.14 An automatic cutout circuit.

(figure 7.15). This system had been used in railway lighting. To prevent the battery from being overcharged the angular setting of the third brush could sometimes be varied manually. Another expedient was to have summer and winter settings, and to use the maximum charging rate when the lights were switched on; this was done by switching in and out a resistance in series with the field windings (figure 7.15). Though the third-brush system was simple and gave quite a constant voltage, the output of the dynamo decreased with decreasing battery voltage, and was therefore least when the battery most needed charging.

The third-brush system was replaced by constant-voltage vibrating

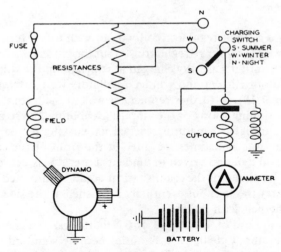

Figure 7.15 A third-brush control system. By permission of The Council of the Institution of Mechanical Engineers (from *Proceedings of the Institution of Automobile Engineers*).

armature systems about 1936. These were based on the Tyrrell system for generating stations and had been used by some manufacturers for a considerable time.

A resistance in series with the field winding was shorted out by a relay across the dynamo output when the voltage was low, but when the voltage had reached a set value the resistance was switched in and reduced the current flowing through the field winding and therefore the output of the dynamo, the relay then shorted the resistance, the cycle repeated and the armature set vibrating (figure 7.16). At low speeds the time the contact was open was relatively small, but as the speed increased the proportion of open to closed time increased and so the voltage remained fairly constant. However, the voltage of the battery and the current it required varied with its state of charge and to compensate for this some series turns were added to the shunt turns of the relay so that the voltage fell with load (figure 7.17).

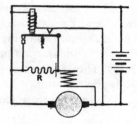

Figure 7.16 Circuit diagram of vibrator control of the twenties. By permission of Reed Business Publishing Ltd (from *Automobile Engineer*).

Sometimes a current regulator coil wound with a few turns was used in conjunction with the voltage regulator so that the battery was charged at the maximum rate until it was fully charged at which point the voltage regulator took over. Cutout current and voltage regulators were often mounted side by side (figure 7.18). A further refinement was to incorporate a bimetallic strip in the tensioning spring of the vibrating armature and so allow a higher rate of charge until the generator reached its working temperature.

The dynamo was sometimes mounted at the front of the crankshaft, or the magneto and dynamo driven in tandem. Later it was belt driven from a pulley on the front end of the crankshaft; it could be mounted on top of the engine and carry the fan, but eventually it finished at the side of the engine and shared the belt with the fan.

A move to the use of 12 instead of 6 V batteries, and therefore smaller currents began about 1930; 12 V batteries became standard by the fifties. The positive to earth arrangement was adopted in the mid-thirties to reduce corrosion, but in the sixties there was a reversion to the negative to earth convention to fall in line with Continental practice.

CURRENT GENERATION AND CONTROL 175

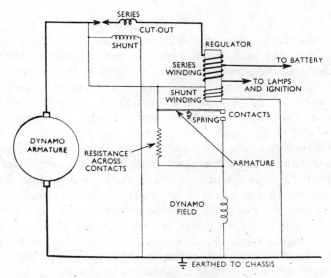

Figure 7.17 Circuit diagram of current–voltage regulator. By permission of Reed Business Publishing Ltd (from *Automobile Engineer*).

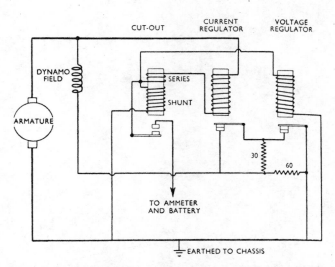

Figure 7.18 Circuit diagram of cutout, current and voltage regulators. By permission of Reed Business Publishing Ltd (from *Automobile Engineer*).

Alternators

Very simple alternators were used on motor cycles in the fifties and they appeared on cars with heavy electric loads in the early sixties. Despite its initially higher cost the alternator gradually displaced the dynamo. It had about twice the power-to-weight ratio and required less attention than the dynamo. It had no commutator so that it could be run at higher speeds than the dynamo and therefore it could be geared to develop current at low engine speeds. The alternator rotor carried an excitation coil wound coaxially with the axis of the rotor, and two claw-shaped poles, one at each end of the coil with the claws interpenetrating. They were magnetised north and south by the coil and, when rotated in the coaxial laminated stator, generated current in the stator coils which were connected with the excitation coils by slip rings. The three-phase current was rectified by six silicon diodes which also acted as cutouts, for they prevented current from the battery flowing back into the alternator. The output voltage was at first controlled in much the same way as that of the dynamo, but later a Zener diode and transistors were used; if the voltage exceeded the critical voltage of the diode the excitation current was interrupted. The output was self-regulating as far as current was concerned not rising above a set limit. By 1967 the components had been so miniaturised that the control circuitry fitted into a unit one inch square and 0.125 in thick.

Chapter 8
Self-Starters

Starting the car with a starting handle was quite a performance. It was always a nuisance and could be quite dangerous if the engine backfired, and so attempts were made to dispense with the starting handle. The engine was turned over by compressed air, or by charging the cylinders with acetylene and igniting it and various spring devices were tried but none of these was particularly successful.

Dynamotors

In 1910 a friend of Leland's was fatally injured cranking a Cadillac car for a lady. After this Leland began a crash programme, with the help of Kettering, to produce a self-starting device. Progress was so rapid that the self-starter was installed as standard equipment in the 1911 Cadillacs. A dynamotor was used, that is a combined generator (called a dynamo in the UK) and motor. When the dynamotor was used as the starting motor it was declutched from the engine and the driver slid a pinion along the motor spindle to engage a ring gear on the flywheel and switched on the motor; when the engine fired he withdrew the pinion, and the engine drove the generator via the clutch. A mercury voltage regulator limited the output of the generator. Series–parallel switching enabled the generator to charge the batteries at 6 V but for the latter to provide 24 V when the unit was run as a motor. As the motor ran for only small periods at a time it could take relatively large currents and this, together with the 20-to-1 gear ratio, enabled large torques to be developed to turn over the engine.

Once Cadillac had demonstrated the feasibility of electric starting other firms used the Delco system (figure 8.1) or brought out systems of their own. The early systems were bulky, heavy and expensive, but they were very successful, particularly in the USA. Many small cars, however, were sold in the UK without a starter well into the twenties; it was argued that the starter put an undue load on the battery, that it might jam, and that the cars were so easy to turn over by hand that a self-starter was not needed! Four-wheel brakes a little later met with even more opposition from some UK manufacturers.

The dynamotor was a specialised piece of equipment and needed skilled maintenance. In addition, the design requirements of a motor are very different to those of a dynamo, and the dynamotor was not a good compromise. A separate motor and dynamo were individually much more efficient, and so within a few years the great majority of US cars used two units.

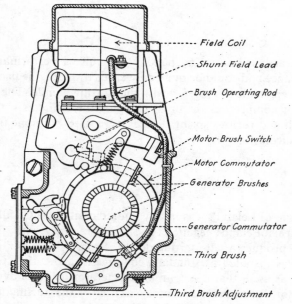

Figure 8.1 Delco dynamotor.

Dynamotors were still used by some UK manufacturers in the twenties and thirties, notably by Morris and Talbot. Instead of acting on the flywheel the dynamotor could be connected to the crankshaft by a chain and sprocket wheels when it ran at about one third engine speed, and on a small number of cars it was directly coupled to the crankshaft. As gear ratios were much smaller than when pinion and flywheel were used, bigger and more powerful motors were required. The dynamotor carried series and shunt windings, the former being excited when the unit was acting as a motor, the latter when it was acting as a dynamo, the series winding then supplementing the action of the third brush in regulating the output at high speeds.

Starter Motors

Most manufacturers, however, preferred to use a separate starter motor and dynamo. If the pinion of the starter motor was not withdrawn quickly enough the motor could be damaged by the resulting high revs so Bendix patented his automatic engagement and disengagement gear in 1912. His design is substantially that still used on many cars today. A coarse multiple thread (figure 8.2) was cut in a sleeve that slid along the spindle of the motor and was connected to the latter by a spring. The starter pinion was internally threaded so that when the sleeve rotated the pinion spun along the sleeve to engage the flywheel gear teeth, which were tapered to facilitate engagement, and drove the flywheel, the spring cushioning any shock. When the engine fired and picked up speed the flywheel tended to over-ride the pinion and consequently the latter was driven back along the spiral grooves and disengaged the flywheel teeth.

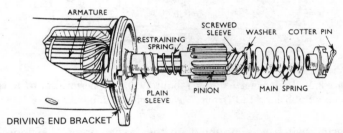

Figure 8.2 Inertia engaged starter drive (about 1950). By permission of Reed Business Publishing Ltd (from *The Automobile Engineer*).

To reduce the diameter of the pinion and so increase the gear ratio, a nut and barrel assembly ran on the spiral grooves in another popular gear, and the pinion was mounted on the end of the barrel. Different types of springs were used to cushion the shock of starting and some starters had a rubber coupling instead of springs. The starter could be of the inboard or outboard type depending on whether the pinion moved towards or away from the motor on engagement; the outboard starter generally carried a bracket with an extra bearing to provide support for the outer end of the shaft. The pre-engaged starter (figure 8.3) was introduced in the USA just before the Second World War and became popular. The pinion was engaged by the driver, the motor switched on, and when the engine was firing the pinion was withdrawn. An over-running clutch prevented damage to the motor if disengagement was delayed. Later a solenoid engaged the pinion and when the pinion was fully engaged a switch closed the motor circuit. The conventional starter was, however, so reliable and so cheap that the pre-engaged starter came into use slowly in the UK and was used at first only on larger, more expensive cars, though it was used on the 1968 Maxi.

In another arrangement the spindle carrying the armature and pinion could slide along the starter bearings; a spring kept the armature partly out of the field coils and the pinion out of mesh with the ring gear, but when the motor was switched on the field coils attracted the armature which slid along bringing the pinion into mesh. When the motor was switched off the spring pulled armature and pinion back.

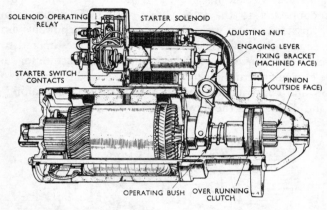

Figure 8.3 Pre-engaged starter (about 1950). By permission of Reed Business Publishing Ltd (from *The Automobile Engineer*).

The starter switch was mounted on the fascia, or on the floor by the driver's seat and the driver pushed down a spring-loaded plug with his heel. Switch and contacts had to be robust to handle the heavy current. This type of switch was replaced in the thirties by a solenoid-operated relay switch in circuit with the ignition key. If the engine stopped accidentally it could be arranged that the solenoid starter switch operated automatically, and when the engine got going again the current from the dynamo cut out the motor. One particular dynamotor acted as the starter motor if the engine stalled.

Chapter 9
Engine Lubrication

Developments to 1918

The need to separate rubbing metal surfaces by a lubricating film to prevent overheating, rapid wear and seizure was recognised from the earliest times. Not surprisingly, the lubrication practices adopted in the first internal combustion engines were based on many years of steam engine experience. In the earliest steam engines animal or vegetable oils were used to lubricate cylinder bores but the effect of steam temperatures was to cause partial decomposition of the oils into stearic and oleic acid and these changed into pitch-like substances that were unsuitable as lubricants. This difficulty was overcome to a large extent by the use of mineral oils that were not so easily decomposed by heat and these began to be imported in barrels in commercial quantities from America in 1867. Animal and vegetable oils were preferred for bearing lubrication.

Mineral oils

Mineral oils were generally either of the transparent type, amber in colour and obtained by distillation from petroleum, or of the opaque type, green in colour, and made by settling crude petroleum in tanks and suitably heat treating the residue to remove the lighter fractions. Vaseline-type oils were available which were obtained by filtering opaque oils at higher temperatures to remove bitumen. It was usual to subdivide the opaque oils into those suitable for bearings operating at ordinary temperatures and into those oils where the distillation had been continued to raise the flash point which made them better suited for bearings running at higher temperatures.

No attempt was made at standardisation and some properties of three early well known oils made by Price are shown in table 9.1. Viscosity was based on the time taken by 1000 grains of oil to flow through a small orifice in the testing apparatus at each temperature, sperm oil being taken as a standard of 100 at 70 °F. The setting point was the temperature the oil could withstand without solidifying, so that it would run to the bearings.

At first the most widely used oil was heavy gas engine oil. Engine lubrication was not a particularly critical factor as the earlier engines were not reliable, starting was uncertain and large oil flows were not necessary to keep pistons cool. Therefore it was only necessary to use oils that were suitable for steam engines and which were readily available, even if these oils did not treat the engines kindly. It was not unusual to decoke an engine after about 1000 miles running.

Table 9.1 Properties of various oils.

Type	Specific gravity	Viscosity at				Flash point (°F)	Setting point (°F)
		70°F	120°F	180°F	212°F		
Motorine A	0.89–0.895	2750	435	100	65	550	40
Motorine B	0.89–0.895	2000	340	85	55	500	40
Heavy gas engine oil	0.905–0.910	750	150	45	33	400	32

Compound oils

Compound oils were also available and these were later to have a big influence on lubricant development in motor car engines. These were developed purely by chance from an event which occurred in 1877 in Manchester at the Crossley Works. Here, on their horizontal gas engines because of dirty town gas it was essential to clear away deposits formed in the cylinders, otherwise these deposits would solidify overnight and the engines could not be restarted without first freeing the pistons in the cylinders by hand. One day an operator, by mistake, filled the cylinder lubricator with colza lamp oil and to his surprise the engine was free to turn over the next morning. The significance of this was soon realised and experiments by oil companies showed that a similar behaviour could be obtained if something like 10 % or so of fatty oil from animals and vegetables was blended into normal cylinder oil. This in effect resulted in a cleansing action, which lead to the widespread adoption of compound oils for gas engines. Blended oils had a lower coefficient of friction than mineral oils and their better lubricating properties under boundary conditions made them more suitable in bearings operating at higher rotational speeds.

This was particularly important as national competitions aimed at capturing the imagination of the motoring public were being keenly followed by the oil companies and as better engine design appeared so did compound oils begin to be used in racing car engines. At first they were exclusively used in vehicles prepared for sporting events but private motorists soon realised that blended oils had other virtues and commenced to use them. One major disadvantage with heavy oils was the hard combustion chamber deposits they formed, whereas engines using blended oils formed smaller amounts of softer deposits making easier cleaning, and decoking less frequent.

Compound oils were not popular with all users and some complained that they were responsible for the obnoxious odour of the exhaust gas. Most objections to their use naturally came from the American oil companies who supplied much of the uncompounded oil used in Great Britain and considered that the use of fatty oils was a serious threat to their business. This prompted a move towards standardising oils for car engines and in 1909 specifications were proposed by the Society of Automotive Engineers in America (SAE). These specifications recommended different values of viscosity at temperatures of 100 °F and 210 °F for summer and winter use, and for specific gravity, flash point and pour point. These specifications referred solely to mineral oil without additives. The American oil companies were discriminating against the use of non-mineral components in oils but there was some justification for doing this as certain users had resorted to adding bituminous substances and other agents to increase the apparent viscosity of oils. The SAE proposals were adopted generally by the oil industry in 1911 and the oil satisfying these proposals would be equivalent to today's SAE 20 grade oil.

Although straight mineral oils and compound oils were found suitable for car engines, problems were created when the same oils used by motor manufacturers were used in aircraft engines in the First World War. For one thing it became necessary to run the engine for a long warm-up period to reduce the oil viscosity to give sufficient power for take off. Partial seizure was always a possibility under these conditions.

Other oils required to be developed and it was found that castor oil, either pure or blended in mineral oil gave a lower coefficient of friction and reduced engine deposits compared with the thicker mineral oils. Such oils reduced the tendency for aircraft engines to seize and resulted in a better performance. Thus when the war was over other oils became available to the motor industry.

Lubricating systems

Splash. As regards lubricating systems these were very simple on early car engines as many had open crankcases, and cylinders and bearings had their own individual drip-feed lubricators. Other manufacturers such as Benz fed drops of oil to the cylinders, with the big end and crank bearings lubricated separately.

Later crankcases were enclosed to prevent oil from being wasted and thrown on to the passengers. Lubrication was then mainly by the 'splash' system where scoops on the crank dipped into a bath of oil in the crankcase and oil was thrown to all parts of the engine. The main bearings were lubricated by the general spray of oil assisted by catch troughs and small ducts gushing oil to them.

Some engines were provided with a small inspection cock for oil to just run when it was at the correct level in the sump. A few had a direct visual means to observe the oil level and others even used a wire suitably marked and placed through the filling hole to act as a crude form of dip stick.

In multicylinder engines there were generally partitions in the crankcase to prevent oil from flowing back to the rear cylinder and so missing the forward big end scoops when the vehicle travelled uphill. To minimise this risk some makers arranged for the oil supply to enter the front of the engine. Other makers like Renault had special equalising pipes with a tap fitted to feed each partitioned space. The sump was filled and kept topped up while the engine was running by a hand pump mounted on the dashboard and operated either by the driver or passenger.

The cylinder and main bearings were supplied with oil from sight feed lubricators by gravity feed, as were most of the other important components of the vehicle. Layouts consisted of a horizontal trough mounted on the dashboard from which a number of pipes were led off. Each pipe was provided with a small adjustable screw-down needle valve to control the flow of oil droplets which could be clearly seen through a glass pipe. It was often possible to place the oil tank high enough to supply oil to the sight feed lubricators by gravity feed and some interlinked the oil feed tap with the ignition switch so that oil flowed as soon as the engine started.

Later the oil supply was automatically fed to the lubricators and commenced when the engine started running so that the oil flow varied with engine speed. One popular means of raising the oil was by a small pump, usually of the plunger type, which was placed in the oil tank and driven by the engine. Sometimes engine exhaust pressure was used to force the oil from the tank, part of the exhaust gases passing through a non-return valve and then to the pressurised oil tank that was incorporated with a pressure relief valve. In some cases provision was made to stop the engine before a lubrication failure could result in engine damage by arranging for the exhaust pressure to feed to the carburettor as well. An oil supply failure resulted in pressure escaping to the atmosphere and the fuel supply was cut off. A pneumatic pump enabled oil to be fed to the engine if the exhaust by-pass valve failed. In another system hydraulic pressure from a water pump was used to feed the oil through to the drip feeds.

Other makers used a mechanical means to regulate the supply of oil, such as a chain of buckets to pick up oil from the tank and then to tip the oil into a trough as it passed over a sprocket. This system was called a dredger lubricator and was used on some early Mercedes cars to supply oil to the main and other bearings. Other parts of the engine were lubricated by a splash method.

The trough system was another popular form of splash system where the big end bearings were lubricated from oil contained in troughs, the trough under each big end being kept full to the brim from oil delivered by a pump, either internal or external to the engine. The overflow from the troughs and bearings drained back to the sump. In sleeve valve engines of Daimler and others, the trough could be raised or lowered with the throttle opening to vary the oil flow with engine speed.

In splash systems it was necessary to replace old oil every few hundred miles. Good practice was to flush out the lubricating system to remove foreign matter from the sump. One of the faults on some early cars was that the drain plug was sited at the base of the crankcase in such a position that anyone underneath the car who released the drain plug could not avoid being doused with oil. It was not surprising that failure of main bearings was rather more frequent than necessary.

Efficient lubrication was also very difficult to attain with drip lubricators if the setting was not right and it was all too easy to feed too much oil and then run out. Variations in oil viscosity had a profound influence on the performance of drip lubricators; in cold weather some even ceased to operate. Some improvement was attained by using a simple pump to keep the oil circulating.

In all systems great difficulty was experienced in attaining the correct oil level in the sump and it was considered safer to overlubricate. Excess oil then gave unpleasant smells from the exhaust and caused carbon deposits to form on piston crowns quickly.

These problems were not too serious when engine speeds were low but when engines became more reliable and attained higher speeds and had greater bearing loads it was essential to develop improved methods of lubrication to cope with the greater heat that was dissipated.

Forced feed. A natural development was to use a forced- or pressure-feed system so that the same oil in the sump was used repeatedly, being drawn off by a gear-type or eccentric-driven plunger pump usually positioned below the crankcase oil level.

Many engineers compromised by using a combined splash and forced-feed system, the pump directly supplying the main bearings, the oil leaking away to return by gravity to the oil chamber.

The first forced, albeit low-pressure, lubrication system was designed by Albert Pain in 1890 and was used with great success in a high-speed steam engine built by Bellis. The potential of this lubrication system was not realised by most internal combustion engineers, for apart from Lanchester most makers preferred to continue the use of the rather crude 'total loss' system. Lanchester realised the significance of the results of Beauchamp Towers' work on the friction in plain journal bearings and of the necessity to provide adequate quantities of lubricant to obtain a consistent frictional behaviour. He further appreciated the importance of shaft rotation to draw oil into the bearings to form a pressure wedge and give hydrodynamic lubrication as first outlined by Osborne Reynolds. Lanchester also understood the part played by the viscosity of the oil and the positioning of oil grooves. Efficient lubrication could therefore be made if oil was supplied at a rate sufficient to equal the losses due to side leakage from the bearing edges.

The first full forced system in cars, in the modern sense, was that of Lanchester in his 20 hp model of 1905 (although Lanchester had force-fed oil in the engines of his experimental cars) in which a pump supplied oil under a pressure of about $40 \, \text{lb in}^{-2}$ through the crankshaft and main bearings, and the gudgeon pin was also fed by the oil passing up inside a hollow cylindrical connecting rod. The gudgeon pins were also hollow so that oil could continue to flow through them to the cylinder walls. Despite the high oil pressure excess lubrication was avoided as the oil passages were small and the piston rather long and furthermore the small crank radius tended to keep down the amount thrown off from the webs.

Oil pressures

Oil pressures varied in different makes of engine, 5 to $10 \, \text{lb in}^{-2}$, was fairly common on cars made before the First World War, although experience had shown the higher the oil pressure the longer the life of the bearings. On some models a pressure relief valve was fitted to prevent pressure build-up when the oil was cold. Part of the oil delivered by the pump was then by-passed to the bearings and either drained directly back to the reservoir or went back via the timing gear case.

Typical of the lower pressure arrangement was that used by de Dion Bouton where oil from a gear pump was driven to a large-diameter pipe in the base of the crankcase and was afterwards transmitted to each main bearing. Oil spray inside the engine was used to lubricate the gudgeon pins and cylinders; overlubrication was prevented by sheet metal guards placed around every crank web. Another device used by the same company to avoid excessive lubrication was a thrower ring fitted to the top half of the main bearing to catch oil thrown from the crank web.

It was more usual to fit a scraper ring near the bottom of the piston skirt to remove excess oil from the cylinder bore. Some manufacturers drilled holes in the piston for oil to drain away to the piston interior. Others used both scraper rings and holes.

Filters

Many other systems were designed but apart from small details most were basically the same. One big difference between engines was the method of filtering used, as practical full-flow filters were not available. Some manufacturers like Daimler simply fixed a gauze across the whole of the bottom half of the crankcase for dripping oil to pass through and often this was the only filter in the oil system. Some like Lanchester had an extra screen over the pump intake. Others just used the gauze on the pump while some depended upon a screen submerged between the sump and the base of the crankcase. The effectiveness of the filter was limited (it would trap things about the size of nuts and bolts) as it was essential to ensure the grid dimensions would not restrict the pump supply at all times and the filter had to be accessible for cleaning.

Developments in the USA

Better lubrication systems were gradually becoming more popular in Europe at the outbreak of the First World War with the result that the plain splash system had virtually disappeared. Developments were generally slower in America and lubricating systems were comparatively simple. In 1916 splash systems were used on 50 % of engines in America; for only a small percentage of cars could travel at sufficient speeds on the poor roads to cause lubrication failures. The splash system either kept the oil level correct by adding oil at a rate to keep pace with the oil lost without using a pump, or a pump was used and the oil was kept recirculating.

A typical system was that used on the Model T Ford. This had a flywheel running in a well at the system's lowest point and oil was flung into an oil cup from where it was carried via a crankcase oil tube to the front of the engine where it lubricated the timing gears. The oil then flowed back to the sump filling small wells for the big end dippers to move in and so lubricate the cylinder walls. Oil was fed into the engine through the breather pipe and the oil level was aimed to be kept about halfway between the highest and lowest permissible levels.

About 23 % of US cars used a trough and pressure-feed system with the fully forced pressure-feed system being confined to good quality cars. 25 % of these were operating on all the V engines brought into production that year (1916), as an uneven oil distribution resulted if the splash system was used.

By 1918 the war had caused further changes and the lubricating systems in use, expressed as a percentage of the models introduced each year, in the war period, are shown in table 9.2. These figures indicate a big increase in the trough system owing to rising costs. Trough and pressure systems were also on the increase with an average oil pressure being about 10 lb in^{-2} in a warm engine. In pressure systems gear pumps were used almost exclusively because of reliability, instead of the plunger type with its check valves and more intermittent action which were mainly used in trough systems. At this time the best systems in America were as good as those developed in Europe before the war.

Table 9.2 The different types of lubrication systems used in the USA.

Year	Pressure (%)	Trough Pressure (%)	Trough (%)	Splash (%)
1915	37.5	16.0	0	46.5
1916	23.35	23.35	0	53.3
1917	30.0	35.0	0	35.0
1918	26.7	43.2	28.4	1.7

Developments Between the Wars

Oils

After the First World War the continued use of low-quality petrol restricted the specific performance of engines. Oils therefore were not required to withstand high temperatures and they were judged from their ease of starting and their ability to avoid carbon deposits.

Many brands of oil were available in the early twenties, most of which were imported into the UK from America, Rumania and Indonesia. The performance of these oils varied because of the chemical nature of the blending stocks. One very noted oil was Pennsylvania oil, consisting of paraffinic compounds with good oxidation resistance and viscosity/temperature characteristics which gave reasonable starting except at low temperature. The oil consumption was good but hard carbon deposits were formed in the combustion chamber. Other brands from America consisted of blends of mixed base stocks with less satisfactory viscosity/temperature behaviour which resulted in high oil consumption. These gave better starting in cold conditions because they contained no wax and further formed heavy but soft combustion chamber deposits.

Many extravagant claims were continually being made for the merits of different brands of oils, often by small companies. These mainly referred to the use of vegetable oils to give more desirable viscosity/temperature characteristics to the cheaper blended stocks.

Compound oils still enjoyed much popularity. The good lubricating properties of the added fatty oils was shown by Wells and Southcombe to be due to their containing free fatty acids. The compounding of mineral oils with small quantities of certain fatty acids was given the trade name of the 'Germ' process. Only small quantities were needed and 1 % of fatty acid reduced the oil friction by 30 % or so without any change in viscosity. Because of the many supporters of animal and vegetable oils Ricardo was financed jointly by Shell and the Air Ministry to carry out research on the suitability of lubricants for internal combustion engines.

The results of the investigations showed that under boundary lubrication conditions, bearings ran cooler when fatty oils were used because of their lower coefficient of friction, μ. They failed as lubricants at temperatures of about 350 °F, whereas mineral oils retained their lubricating properties at even higher cylinder wall temperatures. Straight mineral oils formed fewer carbon deposits and were not so prone to piston ring sticking. No difference in wear was observed between a straight mineral compound oil and mineral and vegetable oils. These results were not published immediately and later in the twenties similar results were confirmed at Farnborough but only after this did animal oils cease to be used as engine lubricants.

Before that time early lubricants continued to meet the requirements of engines of low performance. As engines gradually got more lively, lubricants gave more trouble as high temperatures degraded the oil to give excessive combustion chamber deposits, ring sticking and heavy sludging in the crankcase. The latter was a serious problem at low temperatures with poor petrol. Sludging was thought to be accelerated by the presence of air in the crankcase causing rapid oxidation of the oils, so many crankcases were virtually sealed and no ventilation was provided. An analysis of oil in crankcases revealed that it had been diluted up to 15 % by volume with fuel, and condensed water had contaminated the oil by 20 %. When oil was diluted the viscosity fell and oil consumption increased although it gave improved lubrication during starting. Water in the sump was a much more tricky problem to overcome for many designers thought it came from the cooling systems and spent much time redesigning these to prevent leakage. Some makers advised owners to drain water from the sump regularly. A better solution was again to compound oils to form stable emulsions and stop water from appearing in the sump. It was probably true to say at this time that manufacturers still knew much more about the design of engines than their lubrication. Many engineers still regarded oil as 'black magic'.

Chemists in the oil companies were beginning to realise that most of the serious problems such as ring sticking and carbonisation were caused by decomposition and oxidation of the oils. More stable oils were therefore required, but these were not developed until after systematic tests had been conducted. This was a slow process because laboratory tests did not simulate reliably the behaviour of oil under ordinary service conditions. Despite this difficulty oils of better stability were gradually developed by the laboratories.

Additives

For the first time, in 1926, soluble metallic additives appeared. These gave better oxidation stability and reduced ring sticking and carbonisation. Initially aluminium napthanate was used but, although it kept down oxidation, experience showed that it gave high piston ring and cylinder wear from increased abrasion. Soon afterwards calcium dichlorostearate was used which was effective in reducing carbonisation but which had corrosive tendencies at high operating temperatures, severely attacking little end bronze brushes.

Other additives were tried until they were stopped temporarily when a really outstanding step forward came from the introduction in 1929 of 'solvent refining' in the USA. By this means the aromatic components, ring compounds of poor thermal stability, are removed by the selective solubility of solvents, such as SO_2, or a proprietary solvent, Furfuraldehyde, giving a stock of a predominantly paraffinic nature of greatly enhanced oxidation and viscosity/temperature characteristics. Improved processes were further introduced for the removal of wax, which resulted in oils of lower pour points and lower low-temperature viscosity and made for easier starting in cold climates.

SAE classification

Along with these improvements proposals were made in America in 1926 by the Society of Automobile Engineers for rationalisation of oils by the SAE system for classification of crankcase lubricants: oils were graded in SAE numbers 10, 20, 30, 40, 50, 60 and 70 in terms of increasing viscosity. These were to be measured in Saybolt Universal Seconds, at 130 °F for the 10 and 20 grades and at 210 °F for the remaining numbers.

The SAE system was in general use by 1929 when further proposals were made by Dean and Davis for an arbitrary system for comparing the temperature/viscosity behaviour of oils from different sources and refining treatments. In this scheme the viscosity of a sample of oil measured at 100 °F was compared with that at the same temperature of two reference oils, one a high-viscosity reference oil of Pennsylvania origin given an arbitrary value of 100, the other a low-viscosity Texas Coastal Oil rated at approximately zero. Both oils had the same viscosity as the sample at 210 °F. If L is the viscosity at 100 °F of the oil of zero viscosity and H is the viscosity at 100 °F of the oil of 100 viscosity index, then the viscosity index VI of the sample oil is $100(L-U)/(L-H)$ where U is the viscosity of the oil sample in centiStokes at 100 °F. The higher the VI number the less is the variation of viscosity with temperature. With this system paraffinic base oil would have a high VI of about 100 and a mixed base oil a much lower VI value of between 30 and 50. Despite the arbitrary nature of the system it remains, apart from small changes, in existence today.

Improving engine life

One of the big problems in the twenties was caused by car manufacturers themselves who advised motorists to idle their engines for several minutes after a cold start to get the oil circulating properly. This practice resulted in high oil consumption and an engine life of 20–30 000 miles because of severe cylinder wear; this short engine life caused serious concern to manufacturers at a time when private motoring was becoming very popular. Much work was carried out by manufacturers to improve the life of their engines and some research was undertaken by the oil companies, but most of this was of a sporadic nature, uncoordinated and not capable of general application.

Ricardo suggested that chemical corrosion might be the most important factor responsible for high cylinder wear and suggested to W N Duff, the Director of RABMAM (Research Association of British Motor and Allied Manufacturers) that statistical data be obtained on the operating conditions under which cylinder wear occurred. Later in 1931 under its new Director, Dr C G Williams, RABMAM was asked by the motor industry to investigate on a more scientific basis the factors affecting cylinder wear and suggest ways of increasing engine life.

By carrying out cyclic engine tests wear could be accelerated alarmingly by a delayed warm-up, that is, by low cylinder wall temperatures. At low temperatures condensation of water on the cylinder walls would occur and if the cylinder bore temperature fell below the dew point chemical corrosion from various acids in the products of combustion were responsible for the high wear. Williams suspected, and this was later confirmed by others, that carbon dioxide in the form of carbonic acid was particularly important although other acid products of combustion, such as sulphur trioxide and formic acid were also responsible for corrosion.

This work spurred the motor industry into positive action; the first result of this was the introduction of thermostatic control of cooling water temperature which at that time had only been used on a few cars. This enabled the cylinder bore temperature to be brought rapidly above the dew point of the corrosive acids. Other firms made experimental piston rings and cylinder liners of austenitic cast iron because of its corrosion resistance. Good results were obtained with chromium-plated piston rings.

Other results of this work showed up differences in the behaviour of light and heavy oils. It was established that light oils did not increase wear, except in the presence of abrasives, but could give less wear by getting to the bores quicker. On the other hand, heavy oils could reduce wear in the presence of abrasives.

There still was not much cooperation between the oil companies and manufacturers in 1936 when the Caterpillar Company of America started producing high-speed diesel engines that were to have a pronounced influence on the future development of crankcase oils.

High-speed diesel engines were already well established in road transport in the UK, but deposits caused ring sticking which limited power output and therefore the engine bmep to about 80 lb in^{-2}. Engines were thus de-rated and gave a poor vehicle performance, but their good fuel consumption justified their popularity.

In America diesel engines had to be competitive with petrol engines in terms of specific performance and the Caterpillar Company insisted that oil companies should develop an oil that would withstand maximum performance from the engines. Caterpillar built special engines and developed tests to evaluate oils under controlled but more severe conditions than were likely to occur in service. Oils were assessed from the amount of 'lacquer' deposits on pistons, amount of wear and absence of ring sticking.

One result of the Caterpillar tests, and other work as well, confirmed that refining techniques in themselves would not bring greatly improved oils so that soluble chemical additives had been introduced into oils. Thus in 1939 refining became less important, and oils with metallic detergent oxidation inhibitors and alkaline anti-wear additives were classified as heavy duty oils. These coped successfully with diesel engine duties. Later their use was naturally extended to passenger car engines, to start a new phase in lubricant development.

Systems

After the First World War well established engines like those used on the Ford Model T and Daimler cars continued to use splash systems. Other makers of side valve engines also used splash systems but with all kinds of refinements to control the right oil level.

As motoring became more popular engines of greater specific output using overhead camshafts were gradually introduced and these required more sophisticated lubricating systems. The main problem was to provide the required amount of oil for the cams, rollers and rocker spindles without overlubricating the inlet valves. These OHC engines mostly had pressure lubrication systems.

It was quite popular to run the camshaft in its own chamber so that the cams received sufficient oil on each revolution. In addition, oil was force fed through a hollow camshaft and in cars like the Bentley external drainage was provided for oil to return to the sump. Some engines such as the Ansaldo had their camshaft placed below the tops of valve guides; it was then possible to have a good supply of oil on the top of the head for the cams to dip into, regardless of the slope on which the vehicle was moving, and without excess oil reaching the valve guides. On the Lanchester and AC engines the cams were lubricated by oil leaking from pressure-fed bearings of the rockers that were placed vertically above the camshaft.

It was more difficult to lubricate engines with a camshaft placed above the valve guides. On the Wolseley staggered-valve arrangement forced lubrication of the rocker spindles was used together with the provision of tiny copper tubes for oil to flow to each cam roller pin. In this system lubricant was directly supplied to the most important parts without oversplashing oil about the cam case. Excess oil from the head often escaped back to the sump via passages cast in the cylinder block including the one for the vertical driving shaft.

The Rhode OHC engine had a splash system, in principle not unlike that used on the Model T, in which oil was thrown from the flywheel through a detachable gauze filter into a pocket running the full length of the camshaft. After lubricating the camshaft the oil ran through four plugs with regulating holes to troughs in which the big ends dipped.

The pushrod and rocker-operated valve lubrication arrangements were anything but satisfactory, some even relied on the application of an oil can after the rocker cover had been removed. Others depended on oil mist from the crankcase reaching the pushrods. In these cases felt washers were provided to enable a film of oil on the working surfaces to be maintained. The valve gear cover usually had a filler opening and breather.

Later cup and ball joints were generally used on tappet rods and if the cups were inverted, holes drilled on the rocker arms allowed oil from their bearings to lubricate the ball joints. Some engines had copper tubes arranged for oil to drip into the tappet rod caps. By keeping the ball and cap well supplied with oil the valve gear was kept quiet.

By the middle twenties all new engines were moving over to forced lubrication to the main and big end bearings as splash lubrication was incapable of coping with the short bursts of speed often made on the better roads now beginning to be built. The higher friction from higher loads and rubbing speeds also required a rapid oil flow through the crankshaft and bearings to keep them cool.

Aluminium pistons were almost in exclusive use by this time since previous objections to their use such as piston slap with a cold or almost cold engine had largely been overcome. This was achieved by preventing excess oil working up past the pistons by fitting a special scraper ring or making a lower pressure ring behave like a scraper. New piston designs also played a part in eliminating piston slap.

Engines were now expected to work over a greater range of loads and speeds and in more varied climates than before. One problem was to provide some lubrication to cylinder walls and pistons while the engine was starting up.

Some companies like Armstrong Siddeley used plunger pumps but most makers preferred the gear type. The pump was mostly submerged in the sump to avoid the need of priming when pumps were mounted above the oil level.

Oil pressures were beginning to increase and on cars like the Beverley-Barnes oil was supplied under a pressure of 60 $lb\,in^{-2}$ to the crankshaft and big end bearings, and an oil pressure of 5 $lb\,in^{-2}$ was used to lubricate the valve rockers and overhead camshaft.

Filters

One of the biggest changes was the improvement in filters, as the old system of using a coarse filter could only separate out objects bigger than a split pin. This was in turn replaced by large, fairly fine mesh gauzes placed over the sump and finer mesh gauze filters inserted on the delivery side of the pump. Even this system would leave large amounts of abrasive matter in the oil that could damage bearings. Filters were gradually becoming more accessible and could be removed without having to drain all the oil from the sump.

The by-pass filter had been introduced in America in 1926, with coarse screens used on the suction side of the pumps, sufficient to stop large fragments damaging the pump, and a by-pass from the pressure side of the oil system led to a container full of packing made of asbestos, cotton, etc, on a wire gauze. These filters were very effective in separating out most of the fine particles in the oil and, unless they had large amounts of dirt to cope with, they kept the oil as clean as when much larger filters were inserted in the system.

These filters separated out most of the fine particles in the oil, but because of high resistance to flow only a small proportion of oil in the pressure system was allowed to pass through the filter and back to the sump. The oil was therefore filtered by a cumulative process. Oil filters of this type soon became standardised items on most American cars.

In the UK some cars used magnetic oil filters to collect particles of ferrous metals in addition to a fine-mesh gauze filter. Other cars used pressure oil filters. In one type, the Auto-klean filter, the filtering was achieved through the edges of a number of thin steel discs mounted on a rotatable spindle spaced by fixed interleaved scraper blades. By rotating the spindle which projected through the filter cover, accumulated dirt was scraped from the filter surfaces. These filters were not as effective as textile filters.

A number of cars were fitted with oil rectifiers which had the head of the filter heated by the exhaust to distil off any petrol that may have caused crankcase dilution.

In the late twenties many people still thought it advantageous to use thick oils to maintain an oil film and that thinning of oils should be prevented as far as possible. For a time it became quite common and it was even standardised by Renault to fit an oil radiator to keep the oil cool. Some cars like the Minerva arranged to cut out the oil flow to the radiator in cold weather.

Sleeve valve engines

Sleeve valve engines continued to provide difficulties in lubrication. One problem was to ensure adequate lubrication on starting as petrol had sometimes been initially deposited on the sleeves. By the late twenties Daimler engines had moved over to forced lubrication and had a gallery pipe passing along the bearings fed by a pump with a spring release valve. Initially oil was allowed to flow into troughs having narrow slots in the side, into which dippers on the big ends worked. On starting up the dippers splashed oil on to the sleeves and pistons, but when the oil was hot the oil supply decreased as a relief valve reduced the oil level in the troughs. This worked in a fashion but it was not very practical.

Oiling was always a problem on sleeve valve engines, either you got too little oil on the sleeves and they seized up, or you got too much and they produced smoke like an unswept chimney. Occasionally you got a good one but there was no telling why; neither could you tell why it was bad, it was a matter of luck more than anything else. Because of this erratic nature of sleeve valve engines it was not surprising that Daimlers bought the Lanchester Motor Co. to acquire ready developed poppet valve engines.

Other system features

During the thirties the design of engine lubrication systems settled down. It was practically universal to have a circulation system driven by a gear pump often placed above sump level and no pumps were now submerged. Generally the pump was driven by helical gears from the camshaft. Sometimes a vertical shaft was used, driving the oil pump at its lowest end immersed in the sump and driving the distributor, which was placed on top of the head, at its upper end. Other arrangements used a common drive with the shaft running diagonally across the engine; the ignition distributor projected out from one side of the crankcase and the oil pump was sited on the other side at a lower level in the crankcase. External pipes were mostly used to feed oil from the pump to the main bearings.

The main oil supply was usually governed by a valve, often of the ball type, placed in the delivery pipe to by-pass oil back to the inlet side of the pump if the oil pressure became too high.

Crankcase ventilation gradually became more common, the air circulation helped to cool the oil in the crankcase and reduce sludging. Increasing use was also made of baffles in the sump to stop oil surge.

On most popular cars external oil filters of fairly large size to cater for the whole oil supply, which gave improved cooling for the oil, tended to be increasingly used. The filtering element of the pressure filter was made of fabric and could be easily detached from the system.

With side valve engines sufficient oil was by-passed from the gallery pipe to drain on the chain drive of the camshaft and often supply the bearings of the auxiliary drive. On the overhead valve engines a supply was taken from the main pressure system to the rocker bearings and an adjustable needle valve was incorporated in the feed pipe to govern the oil flow. Suitable drain passages were provided to prevent overflooding of the valve gear.

More elaborate dual systems similar to those of the twenties were used on expensive cars. It was usual to have a high-pressure system feeding the crankshaft and a low-pressure system to lubricate the lightly stressed components. Each system had its own separate pressure filter and by-pass valve; the filters were very accessible and mounted between the two blocks in a V formation. Oil leaking from the camshaft bearings was used to lubricate the tappets. Such a system was designed by ex-Rolls-Royce engineers under W O Bentley for the 4.5 litre Lagonda and the neatness of engineering of the layout so impressed W O that he remarked that 'it was just as if it came out of the Derby works'.

Dry sump lubrication was used on only a few high-performance cars like the Aston Martin, Triumph 'Dolomite' and the expensive American White. Nearly all manufacturers preferred simplified lubrication systems and found that these would cope with the performances of most Continental and American cars of the thirties. When the Second World War started major progress was being achieved in oil technology itself, largely by the Americans, rather than improvements in the engineering side.

Developments Since 1946

Multigrade oils

As regards passenger cars the forties in the UK were the times of severe restriction from petrol rationing and pool petrol. Many vehicles were old and in somewhat indifferent mechanical condition so that as a result of short journeys and cold running, sludging appeared again as a serious problem.

Few steps were taken to overcome sludging when new engines with larger bores and smaller strokes appeared, although these engines were designed to take full advantage of oils of low viscosity even though motorists still adhered to the practice of changing oils to suit the season. This was because of the large number of older designs that were unsuitable for low-viscosity oils. The use of multigrade oils that had been available in the USA for a number of years was therefore held back until the appearance of engines of higher specific performance. The first 10W-30 appeared in 1956 quickly followed by 20W-40 and later by 20W-50. These oils were made up from a low-viscosity base oil, SAE 10W oil, and a viscosity index improver—a high molecular weight oil-soluble compound to give the required viscosity at high temperature (SAE 30 etc.). Other additives were included to give better oxidation stability, less deposit forming and sludging and lower pour points than the older single grades. They were in fact designed to meet the more severe operating conditions of the new engines from higher powers and speeds than existed in pre-war days and further, to be suitable for both summer and winter.

These new oils created havoc for the motorist who had an older car because of the much higher sludge deposits in older engines; the high detergency of these multigrade oils loosened up the sludge and ungummed the piston rings resulting in high oil consumption and sometimes complete engine failure.

Even more disastrous was the catastrophic wear of cams and tappets that resulted in some new engines. This wear came from the multifunctional additive zinc dialkyl dithiophosphate that caused severe pitting in chilled iron. This was well known in the USA where chilled iron for tappets had been replaced by hardened irons and the zinc dialkyl dithiophosphate now helped to control valve train wear.

The situation was different in the UK and Europe prior to the introduction of multigrade oils, for there was no serious problem with valve train wear which had been kept under control by careful design and selective metallurgy. Although there were differences between various engines, in the main chilled iron was most suitable to withstand the high stresses encountered in cams and tappets. Price was also in favour of chilled iron and much trouble resulted from the introduction of multigrade oil that might have been avoided with better cooperation between the petroleum and motor industries. As it was, the motor manufacturers introduced their own approval tests for crankcase oils, many of which had conflicting requirements and added to the oil suppliers' problems and delayed the general acceptance of new oils by some motor manufacturers.

After these preliminary troubles, multigrade oils have by general cooperation become established and have overcome the problems of high oil consumption and poor starting that occurred with some straight oils.

VI-improvers and other additives

Simple long-chain VI-improvers were used for many years, but they had the disadvantage that the molecules were eventually sheared and became ineffective. Synthetic oils could be made much more stable but were expensive. However, in the eighties lubricants were made which contained small heavily branched molecules (molecular weight about 300 compared with 500 for conventional lubricants and 100 000 or so for long-chain molecules) and these were combined with good base oils to make practicable semisynthetics. These could contain about 40 % synthetics and 60 % oil.

Besides the VI-additives other additives were developed and eventually lubricants could carry a dozen or more additives with the latter constituting 10 to 15 % by volume of the lubricant.

The VI-improvers had, as mentioned above, detergent properties, but detergents proper like calcium sulphonate could also be used. They originated before the war, but again were not widely used until the fifties. They combined with wear debris, etc, kept it from forming deposits and also neutralised acid contaminants into harmless salts. Dispersants such as poly-isobutenyl-succinimide kept contaminants in suspension so that they did not settle out as sludge or varnish.

The oil itself tended to oxidise at higher temperatures and anti-oxidants like zinc dithiophosphate reduced the formation of harmful acidic compounds and sludge. Water can condense on cylinder walls, then mix with the oil and the resulting mixture is churned in the crankcase to form a foam which is difficult to pump and which encourages oxidation of the oil, so antifoam agents like silicones were added to the oil to stop this from happening.

Waxes were frozen out of the oil during processing in order to lower its pour point, and so make starting easier in very cold weather. The pour point could also be lowered by the use of additives like polymethacrylate which control crystal formation at low temperatures.

The oil merely acts as a carrier for the additives. Antiwear agents form protective films which are repaired as quickly as they are damaged. Extreme-pressure lubricants form films at highly loaded areas. Sperm oil was used as a friction modifier, or boundary lubricant, until the seventies, when the whale was declared an endangered species and synthetics had to be developed as substitutes.

One additive could sometimes perform two functions, and conversely more than one additive might be needed for the one function. The various additives have to be compatible with one another, and so the formulation of an oil is a protracted and expensive procedure involving a great deal of testing. International specifications and standardisation of test procedures give some assistance to the engine designer. The additives eventually become depleted and the oil contaminated, so that it has to be periodically replaced, and there is always pressure to increase the mileage between servicing and oil changes; thus in 1960 the engine oil was changed about every 3000 miles, in 1970 about every 5000 miles and on some current cars changes are recommended every 10 000 miles. Similarly engines were decarbonised and the valves reground about every 5000 miles in the thirties and the engine might need a rebore after 20 to 30 000 miles, whereas in the late fifties decarbonisation might be needed after 15 000 miles and a rebore after 40 to 50 000 miles. The average driver nowadays does not expect ever to have to pay to have the engine of his car decarbonised or rebored. The metallurgist and the engineer can claim some credit for these improvements but the major credit must go to the big oil firms.

Another incentive for improvement is the increased duty the lubricant and lubricant system has to carry. Engine revs slowly increase, engines and sumps get more compact, and underbonnet temperatures increase. Turbocharged engines are a particular challenge.

Systems

With regard to the lubricating system itself, the rotor type of pump, in which a multilobed rotor rotates within a multilobed stator, became more popular than the gear-type pump.

Filters were of the by-pass type on small cars and full-flow type on larger models. The more efficient full-flow type needed a pressure relief valve, but eventually became widely used. The filter element of resin impregnated paper was replaced by a new filter during servicing, or a canister containing the filter replaced by a new canister. Lubricating cams and cam followers was, and still is, a problem, but again the use of additives has helped.

Other developments

Lubrication of the gearbox and back axle posed no great problems until the advent of the hypoid gear in the latter. Some sliding as well as rolling occurs with the hypoid and to cope with the sliding extreme-pressure lubricants were needed. The extreme-pressure lubricants were sulphur, chlorine or phosphorus compounds which reacted with the surfaces when they reached high temperatures at heavily loaded areas to form protective films just where they were needed.

The manual gearbox no longer needed its own little dipstick and drain plug, for the oil lasted the life of the engine and merely needed topping up.

The automatic gearboxes brought their own problems and special lubricants had to be developed for such boxes, indeed automatic gearbox lubricants have to meet quite exacting requirements. Before the Second World War there were numerous grease nipples on the chassis that had to be filled periodically using a grease gun. This was done automatically on some cars, lubricant being pumped out from a central point, but this was expensive to arrange and there was always the risk of a line not working. Automatic chassis lubrication, however, was used more successfully on commercial vehicles.

The number of points needing attention was gradually reduced by using sealed-for-life units packed with lubricant, and by substituting rubber or polymer bearings for metal ones. The greases themselves were improved by using, for example, lithium-based soaps which withstand higher temperatures than calcium-based soaps, so that these were used in wheel bearings, and by adding solid lubricants to the grease.

Chapter 10

The Cooling System

Benz mounted a small water tank (figure 5.1) over his engine and relied on thermal convection to circulate the water. The next step was to use a simple metal tube to dissipate the heat. The tube was of considerable diameter and, even if finned, was not very effective since it had a small surface-to-volume ratio and so had to be made very long to keep temperatures down. It was also very heavy when filled with water. Consequently, it was difficult to arrange a neat installation. One arrangement was to fold the tube like a jumping-jack firework. Smaller tubes would have required pumps to circulate the water and early water pumps gave a lot of trouble. The water tended to boil away but this was no problem when journeys were short, and the driver expected to fill up his water tank in the same way as he filled his petrol tank.

Radiators

When engines became bigger and more powerful, the single tubes could no longer cope, and by about 1900 many waterways in parallel had to be used. This arrangement was not particularly novel since it had already been used on condensers for steam engines on trams.

At first the tubes were horizontal and connected to vertical side tanks divided into chambers: a pump was necessary to circulate the water. Next the tubes were made vertical, connecting top and bottom tanks; on smaller cars natural convection could then be relied on to circulate the coolant. The tubes (figure 10.1(*a*)) were of copper or brass in order to resist corrosion. Several rows of tubes could be used and these were sometimes finned to increase heat dissipation. Instead of fins continuous helical strips or wire helices could be soldered to the tube. Often the fins were replaced by fin plates which ran continuously across the radiator. The multitube radiator with further modifications and developments has persisted to present times.

Daimler in 1898, instead of replacing the single tube by a series of smaller tubes, used a radiator based on the fire-tube boiler. Air passed through small horizontal tubes which were attached to vertical tube plates and the tubes were surrounded by water.

Instead of vertical tube plates Maybach used wire mesh ends, threading the ends of square tubes into the mesh and sealing each face by dipping the whole assembly in a solder bath. These radiators were very expensive but quite satisfactory, and though they soon became obsolete they were used by Mercedes until the late thirties. Even then new false radiator fronts were made to resemble the traditional Mercedes radiator.

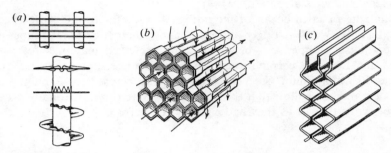

Figure 10.1 Types of radiator construction: (*a*), grilled tube; (*b*), honeycomb; (*c*), film tub, corrugated. By permission of Haymarket Publishing Co (from *Autocar* and *Autocar Handbook*).

The next stage in the evolution of the radiator was to expand the ends of round tubes into hexagons; they were then nested and the assembly joined and sealed again by dipping in solder (figure 10.1(*b*)). This type of radiator was called a honeycomb radiator in the UK whereas curiously the Mercedes arrangement of square tubes was called a honeycomb radiator in Germany. In the USA the term 'cellular' was used.

The introduction of the honeycomb and vertical multitube cores gave the designer scope to vary the shape of the radiator and its cowling, and to give it an attractive and distinctive appearance. The radiator began to serve as the hallmark of a particular make of car, so that many designs persisted for many years, even until after the Second World War. The Rolls-Royce radiator of the eighties is reminiscent of that of 1907, at least in outward appearance.

For many years cellular radiators were the most satisfactory available; they were light, had a large surface area and, though not very robust, were easily repaired. However, the area of metal in contact with air was similar to that in contact with water whereas the rate of heat transfer from metal to air is much less than that of water to metal. Cellular radiators were on many of the more expensive cars in the early twenties and were still on some cars in the thirties. Other cars continued to use finned or gilled-tube radiators and a typical tube radiator could have over a hundred 0.25 in tubes stacked in five rows and connected by numerous fins.

Film-type radiators began to displace the other types in the twenties. In these, two strips of metal slightly separated, and sealed at the edges, formed the waterway, and the strips were bent in a zig-zag fashion so that when adjacent pairs were joined together airways were formed between the pairs (figure 10.1(c)). The strips could be shaped to give a honeycomb appearance. Alternatively more sheet-metal strip could be sweated to the waterway strips to increase the amount of metal in contact with the air, with this airway strip pierced with louvres and slots to promote turbulence. Other arrangements were also used.

Later the practice began of mounting the radiator behind a decorative grill; its looks were no longer important and radiators became cheaper and more robust. Copper and brass were the preferred materials for water tubes, but they had become so expensive by the late sixties that they were replaced by aluminium. Steel strip continued to be used for finning. Some cores became so shallow that the water tubes were run horizontally instead of vertically, that is, as they had eighty years before.

Coolant

Water was obviously the most convenient liquid coolant. It has one major drawback; it expands when it freezes and so can crack the block. The radiator therefore had to be either drained overnight in very cold weather or the vehicle kept in a warm garage. Chemicals like calcium chloride which depress the freezing point of water were tried, but they could corrode and react with materials in the system. Alcohol reduced the boiling point and evaporated and a mixture of glycerine and water was very viscous. Ethylene glycol was eventually used but as it could form corrosive degradation products, anti-oxidants and other materials were added to it.

Another disadvantage of water is its low boiling point. However, the boiling point can be increased by pressurising the system; for example, an overpressure of 14 lb in^{-2} raises the boiling point by 12 °C. Loss of water and antifreeze can also be reduced by pressurising the system; altitude does not then affect matters, and the engine runs at a higher temperature making it more efficient. Some cars in the mid-thirties had spring-loaded valves in their filler caps which remained closed until the pressure had built up to 3 or 4 lb in^{-2}, so increasing the boiling point by about 5 °C. A subsidiary valve opened when the system cooled down; this admitted air back into the radiator to prevent a partial vacuum and buckling of the hosepipes.

Pressurised systems did not become standard, however, until well after the Second World War. As early as 1918 Cadillac condensed any steam formed in a small container and arranged for the water to be sucked back into the radiator when the latter cooled down. This was developed in the sixties into 'sealed for life' systems, the steam condensing in a separate sealed reservoir and passing back into the radiator. The header tank was kept full, so aeration and degradation of the antifreeze was minimised, and the system did not need periodical topping up.

Circulation

For many years natural convection was utilised to circulate the coolant—the so called thermosyphon system (figure 10.2.). This worked reasonably well and indeed was still used on some small cars after the Second World War. On larger cars, however, the circulation had to be boosted and from about 1915 impellors began to be used, more to supplement rather than supplant the thermosyphon effect. The impellors were paddle wheels usually driven by the fan spindle and these pushed water up into the header tank.

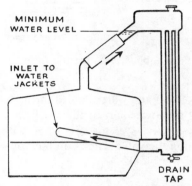

Figure 10.2 Thermosyphon system. By permission of Haymarket Publishing Co (from *Autocar* and *Autocar Handbook*).

Later centrifugal pumps were used, again at first on larger cars, which pumped water into the water jacket of the cylinder block. Because of the positive circulation the radiator could be made smaller and less water was needed. There was the further advantage that the radiator no longer had to be located so that the header tank was well above the cylinder block.

It was soon realised that to work efficiently the engine should be brought rapidly to operating temperature. In the thirties it was shown that most of the cylinder and ring wear occurred during the warming up period, making it even more important to reach operating temperatures quickly. As the pump slowed the warming-up process from about 1916 thermostats (figure 10.3) began to be fitted. These were generally metallic bellows filled with a volatile liquid such as ether; when the ether was cold the bellows blocked off the radiator, but when the ether vaporised it lifted a valve allowing hot water to pass to the radiator. Nowadays a wax filling is used in a plastic casing. With this simple arrangement pumps could develop very high pressures which made them more prone to leaks. To overcome this a by-pass valve and tube were later fitted; when the thermostat was closed the water was passed back into the engine (so, incidentally, keeping it at a more uniform temperature) relieving the pump and when the thermostat opened

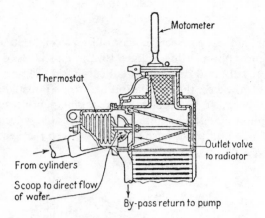

Figure 10.3 Early (about 1918) thermostat with by-pass return.

the by-pass was blocked off and water passed through to the radiator (figure 10.4). Overcooling of the lower part of the block can still be a problem. To overcome this some cooling systems are split, with one system for the cylinder head containing the valve seats, combustion chambers and exhaust ports and another for the block.

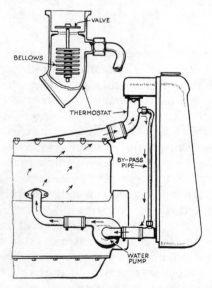

Figure 10.4 Later thermostat and associated water circulation. By permission of Haymarket Publishing Co (from *Autocar* and *Autocar Handbook*).

Radiator shutters were also used from about 1920 to bring the coolant to working temperature quickly. Some could be operated from the dashboard and others worked automatically, for example, by a system of levers operated by a bimetallic strip, or by a bellows filled with a volatile fluid immersed in the coolant. Shutters were used on the Rolls-Royce, Salmson and other cars. Radiator muffs were also widely used in very cold weather. Thermometers in the radiator caps which measured the temperature in the header tank were therefore helpful (See p 212).

Fans

A cooling system that could cope with ordinary running might overheat when the vehicle was stationary, or slogging uphill, and fans were being fitted at the turn of the century to prevent this. These became more necessary the larger the car. Indeed many small cars even in the thirties had no fan and relied on the relatively large heat capacity of their cooling systems.

For many years fans consisted merely of two- or four-paddle blades of sheet metal riveted to a central carrier or a spider mounted on a belt-driven pulley. Later assemblies were made of pressings attached to the spider or as castings in aluminium.

Bonnets had more or less vertical louvres cut into their sides to assist airflow through the radiator. At first the sales people disliked these but by the thirties they were often decorative rather than useful; they did not survive the styling changes of the forties.

The fan could be cowled to ensure that it drew air in through the radiator only, and did not merely churn up the air in the engine compartment. One problem was that ways of increasing fan efficiency, such as increasing the speed or the number of blades, also tended to increase fan noise. Uneven numbers of blades could be used, they could be unevenly spaced, or the blade tips could be radiused in attempts to reduce noise.

It was soon realised that fans could waste a lot of power and indeed their overall benefit was questioned from time to time. A number of arrangements have been tried to reduce their power consumption at higher speeds; for example, fans driven through viscous clutches that slipped at high speeds, flexible fans, variable-pitch fans, and fans driven by thermostatically-controlled electric motors, the latter being mounted directly on the radiator.

In one variable-pitch fan a central disc drove a second concentric annulus. At lower speeds a helical spring ensured that there was no slip between disc and annulus but at higher speeds relative motion occurred and caused the fan blades to pivot on the driven disc so reducing their angle of incidence.

Air Cooling

Air cooling is simple, the engine lighter, there are no freezing problems and operating temperatures are soon reached. On the other hand, aircooled engines are noisier, their compression ratio is lower, they generally need an oil cooler, the fins take up more space than water jackets so limiting the size of the engine and it is easier to get water than air to areas that tend to run hot, such as exhaust valve seats. The big American Lincolns were notable for using air cooling for many years, air from a large fan on the flywheel being ducted to the individual cylinders, but otherwise air cooling has generally been used on small cars, and particularly on cycle cars. The VW Beetle used air cooling very successfully and it was also used on a number of other small post-war cars. A powerful fan is generally required.

Chapter 11

Instruments

Instrument Panel

For some time the only gauges the driver had to watch were the sight feed lubricators. The petrol tank was pressurised if a simple gravity feed to the carburettor could not be arranged, so that a pressure gauge was next needed to indicate the pressure in the tank. (A simple sight glass could show the level of petrol in the tank when the latter was mounted in the scuttle.) When high-pressure lubrication was adopted a gauge appeared for the oil pressure. The speedometer was at first an expensive accessory bought by the enthusiastic driver; even in the twenties some cheap cars did not have a speedo.

The lubricators were mounted on a panel but otherwise the gauges were mounted wherever convenient. Similarly when knobs and switches began to appear so that the driver could adjust things when the car was in motion, these too were mounted where convenient, on or near the scuttle dash. Controls were mounted on the steering column. It was function, not appearance, that mattered. Eventually there were sufficient instruments and switches to warrant the use of an instrument board, which gave a much tidier arrangement. The dashboard, however, was rather far back from the driver and so a new dashboard was mounted just below the scuttle to act as the instrument panel or fascia (the latter is an architectural term for the long flat surface under eaves or cornices). Things could be mounted on the panel wherever convenient for the driver and designer but by the early twenties the stylists were at work to make an attractive display.

By the mid-twenties standard equipment included speedometer, odometer, ammeter, oil pressure gauge, and often a clock. The instruments were grouped compactly (figure 11.1), partly for appearance but also to enable them to be illuminated at night by one lamp which might be behind the panel, or, if in front of it, cowled to prevent light dazzling the driver. Radiator thermometers would be mounted on the radiator cap or on the instrument panel. Similarly petrol gauges could be mounted on or near the petrol tank or on the instrument panel. Tell-tale lights began to displace some of the instruments, a light coming on if pressure had fallen to a dangerous level, or if something was not functioning properly. The warning light was cheaper than the proper instrument but did not give as much information to an experienced driver as a gauge.

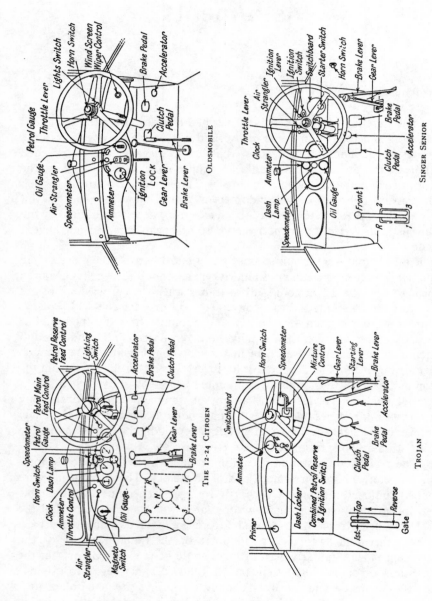

Figure 11.1 Instruments and controls on some cars of the mid-twenties.

Besides the speedometer (and associated odometer), radiator temperature, oil pressure and petrol gauges, clock and ammeter, cars in the thirties could also be fitted with an oil temperature gauge and, on some sports cars, a tachometer. Two or more instruments could be combined in one housing and share the same dial. There could be switches for the ignition, starter motor relay, lamps and dashboard lights. Knobs operated such things as the choke, shock absorber setting and the radiator shutters.

The number of instruments on the panel depended on the cost of the car. A speedometer was a legal necessity from 1927 and some cheap cars only had a speedometer, a few indicator lamps to show, for example, that the battery was charging, and very little else. Expensive cars could have a polished walnut panel carrying very large speedometer and tachometer dials, with smaller dials for the radiator temperature, oil temperature and pressure, battery charging rate and petrol tank contents, plus a multitude of knobs and switches in arrangements which were often hardly ergonomic. It was not unknown for the speedometer dial to be partly hidden by the steering wheel spokes and glare and reflections was often also a problem.

Behind the panel haphazard wires gave way to a wiring harness like the cabling in contemporary telephone switchboards, and after the Second World War to printed circuitry and later flexible printed circuits. Plastics spread from the bezels of the dials and switches until the whole fascia was plastic on all but the most expensive cars. The basic set of instruments has not altered greatly over the years but there has been a considerable increase in the number of warning lights—lights to show the trafficators are working, to show whether the brake pads have worn and need replacing and even to show that the safety belts have been fastened. Similarly more switches have appeared, even one to wind out the radio aerial. Current developments in electronics make it possible to display a great deal of information, such as average speed during a trip, instantaneous and average fuel consumption, and fuel range as well as the more usual information. Some information can be given orally and not visually by a speech synthesiser. These features cost money in the first place but are so many more things to go wrong so that more conservative drivers think of them as sales gimmicks.

Speedometers

Early speedometers were based on the centrifugal governor. A cable geared to one of the front wheels ran in a flexible casing and rotated a pair of bob weights; as the speed of the weights increased they moved outwards lifting a toothed collar which in turn rotated a gear wheel to which the pointer was attached. These early speedometers were large, impressive units which were quite expensive. Vibration was a problem and the drive often gave trouble. A later arrangement (figure 11.2) was to pivot an annular weight on the spindle of the meter; a spring held it at an angle to the spindle but when the latter was rotating the annulus ran at closer and closer to a right angle to the spindle as the speed was increased, and this angular movement was transmitted to the pointer.

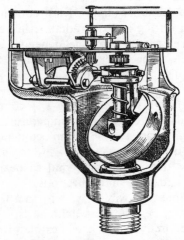

Figure 11.2 Centrifugal speedometer operated by inclined weight. By permission of Haymarket Publishing Co (from *Autocar* and *Autocar Handbook*).

By 1914 there were a number of types of speedometer in use including centrifugal, electromagnetic, and chronometric meters.

As centrifugal force depends on velocity squared the scales of centrifugal speedometers were non-linear and the figures tended to crowd together at the higher speeds. Centrifugal meters also could not be geared down to make things easier for the driver, and they had a number of pivots and moving parts that could wear. Electromagnetic speedometers (figure 11.3) were simpler and cheaper and eventually displaced them. In the early magnetic speedos, a rotating annular permanent magnet dragged a graduated light metal drum round against a spring and graduations on the drum were read through a window. Later the aluminium rotor rotated between a steel field plate or cup, and a carrier carrying one or two bar magnets. The rotor was restrained by a hairspring and its movement transmitted to a pointer. The rotor was sometimes mechanically damped. Speedos were designed generally to be most accurate near 30 mph, the legal speed limit in built-up areas; if they overestimated the top speed most car manufacturers were not unduly perturbed.

In the chronometric speedometer (figure 11.4) the pointer was geared to one of the road wheels, so that the distance the pointer moved round the dial in a short fixed period of time was proportional to the road speed. The

INSTRUMENT PANEL

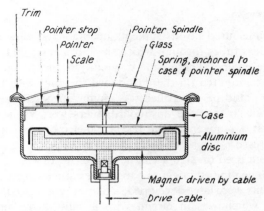

Figure 11.3 Early electromagnetic speedometer. By permission of Haymarket Publishing Co (from *Autocar* and *Autocar Handbook*).

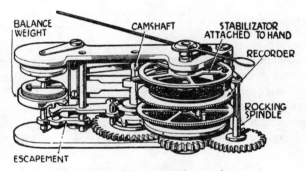

Figure 11.4 Chronometric speedometer.

meter incorporated a clock mechanism to keep the measuring period constant and the main spring of the clock was kept wound by the speedometer drive. The Bonniksen meter had two pointers, one travelled round and then stayed in position to indicate the speed while the second was moving round, and when the second stopped, the first pointer returned to zero to repeat the cycle. The later Jaeger instrument had only one pointer and to overcome the jerky movement the timing interval was made so small that the needle moved smoothly. Chronometric speedometers were expensive, precision instruments, and chronometric tachometers are still used on racing cars.

The speedometer drive was originally taken off one of the front wheels but the gears could become clogged with mud, so that in the twenties it became usual to drive the speedo from the gearbox. Electric drives make installation easier, but the mechanical drive has been made so simple and reliable that it is still in general use.

Oil pressure gauges

Benz and Daimler used no instruments on their first cars except for sight feed oilers—glass tubes through which the lubricant could be seen passing through to the engine. These were originally mounted on the engine but soon they were mounted on the dashboard together with taps so that the lubricant flow could be controlled. As engine speeds increased oil pumps were introduced so that the dashboard oilers were replaced by pressure gauges. At first a plunger on the fascia panel moved against a spring to indicate the pressure but soon Bourdon-type gauges were in use. Bourdon gauges had been used on steam engines since the 1860s.

The tube connecting engine and gauge was rather vulnerable to damage, so electrical units were later developed; a diaphragm moved the contact arm of a rheostat which was in circuit with a calibrated voltmeter and the battery. The voltmeter was modified so that the reading of the pointer was not affected by the state of the battery (see section on petrol gauge below).

Radiator temperature

The temperature of the coolant was measured by thermometers of the liquid-in-glass type or by units that incorporated a bimetallic strip, deflection of the strip being transmitted by quadrant and pinion to a pointer. These meters were mounted on the radiator cap. The radiator cap thermometer was replaced by a metal bulb filled with a volatile liquid immersed in the coolant and a Bourdon gauge on the instrument panel (figure 11.5), the two being connected by a capillary tube (which was difficult to instal and

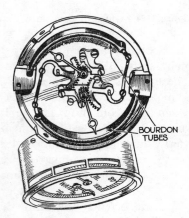

Figure 11.5 Radiator temperature and oil pressure gauges operated by Bourdon tubes and contained in the one housing—early thirties. By permission of Haymarket Publishing Co (from *Autocar* and *Autocar Handbook*).

easy to damage). It was some time before the thermometer was a standard fitting on the instrument panel.

After the Second World War electrical remote reading instruments were developed and eventually the temperature of the coolant was sensed by a semiconductor unit the resistance of which was a function of its temperature. The semiconductor resistor was in circuit with a moving-coil ammeter and a constant-voltage source, which was often the instrument voltage source that also supplied the petrol gauge.

Petrol gauge

The driver at first measured the amount of petrol in the tank with a dip stick in just the same way as he now measures the amount of oil in the sump. Alternatively a simple glassed tube attached to the tank showed the petrol level directly. A development of the latter was to mount a U-tube manometer on the dashboard and plug it into the bottom of the tank; the U tube was filled with a coloured fluid to make it easier to read. Another arrangement (figure 11.6) was to fit a float and lever in the tank and run a chain from the lever through a tube to the rack and pinion mechanism of a gauge on the dashboard. These meters were intended for tanks mounted under the scuttle, and when the tank was shifted to the rear of the car in the late twenties remotely-operated gauges became necessary. The float and lever now moved the contact arm (figure 11.7) of a rheostat in circuit with the battery and an indicating instrument. The latter was a voltmeter with two windings, one, the control winding, took the place of the hairspring of an ordinary meter with the two windings being connected in such a way that

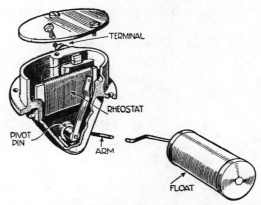

Figure 11.6 Petrol gauge, with float connected mechanically to indicator on fascia. By permission of Haymarket Publishing Co (from *Autocar* and *Autocar Handbook*).

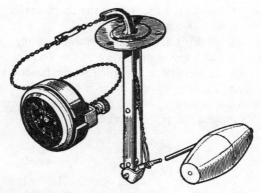

Figure 11.7 Petrol gauge with the float operating rheostat arm. By permission of Haymarket Publishing Co (from *Autocar* and *Autocar Handbook*).

the voltage of the battery did not affect the deflection of the pointer; a decrease in battery voltage, for example, not only decreased the deflecting force produced by the deflection winding but also decreased the controlling force, and the ratio of the two forces determined the position of the armature and therefore of the pointer. The reading tended to fluctuate when the petrol surged in the tank on corners and on hills. This was eventually overcome by replacing the meter by a unit containing a heating coil and bimetallic strip. The deflection of the strip, and therefore the reading of the pointer, depended on the current in the heating coil which in turn depended on the resistance of the sensor and the supply voltage, so that the latter had to be kept constant by a voltage regulator (see p 172). This idea was taken a stage further by utilising the thermal expansion of a length of wire to move the pointer and most contemporary meters are of this type.

Chapter 12

Clutches

The engine has to be started with the vehicle stationary and does not develop reasonable torque until it is rotating at some speed. Provision has, therefore, to be made to isolate engine from wheels, not only when starting but also when changing gear. Also, something has to be done to bring driving and driven shaft to the same speed relatively gradually to avoid shock and stalling. These factors mean that some slip has to be built into the transmission. The simplest way of solving these problems is to use a friction drive between engine and gearbox. Some motor bicycle manufacturers just ignored the problem—the bikes had no clutch and only one gear, so the driver had to make a running start and jump aboard when he got the engine going!

Benz, however, used belt drive on all his cars and so clutches were not necessary, for a jockey pulley could be used to tighten the belt on its pulleys. Daimler used friction clutches on his first two experimental vehicles though later vehicles, including the first made at Cannstadt, were fitted with belt drives.

Cone Clutches

The earliest Panhard-Levassor car was fitted with a plate clutch, but the poor performance and short life of the clutch soon resulted in its replacement by a cone clutch similar to the type that had been used for many years in workshops. The female member of the clutch was fitted to the motor shaft, and the leather-faced male cone was fixed to the forward end of a longitudinal shaft which slid in its bearings and was forced against the female part by a spring, and drawn out of contact by a yoke operated by a pedal.

The simple cone clutch worked well and was used for years, indeed well into the twenties, but it had its problems; the leather had to be kept clean and kept flexible with castor or collan oil and treated with fullers earth, etc, if it tended to slip. If not looked after it was prone to be fierce, or to slip, or to burn out; indeed the early clutch was a bit of a trial. Metal-to-metal surfaces were also used; these had to be accurately finished but even so were liable to be fierce; they ran in oil and therefore μ was very low and a large clamping force was needed.

Another major problem was to make the engagement as smooth as possible, and various modifications were introduced to take up the torque smoothly as the mating surfaces came into contact. One way was to introduce slight crimping into the driven member; in this case the extra flexibility gave a certain amount of cushioning as the load was applied. Another way was to use leather with cork inserts which took up the initial load on the clutch and made the engagement smoother. Sometimes a layer of rubber was inserted between the hub and outer rim and the extra flexibility was also useful in damping out torsional vibrations from the engine. Springs were often mounted between the leather and cone, or spring-loaded plungers used to help give a smooth take-up. Some makers preferred to use two clutches, a small metal cone for initial take-up followed by a larger leather-lined cone to complete the engagement and absorb most of the heat dissipated.

By 1910 or so, fabric facings based on contemporary brake linings came into use and these made smoother engagements possible; they were flexible, were not affected by the temperatures likely to be encountered, had reasonably high and consistent μ values and were a great improvement over leather facings.

Another problem was that heavy cones, because of their inertia, made changing gear more difficult. They were therefore cut away or assembled from mild steel segments; aluminium cones were also used. Another arrangement which reduced the weight of the cone was to grip a light cone between two heavier members. Clutch brakes or stops were also fitted: these were usually hardwood pads or later spring-loaded leather pads which contacted and slowed down the cone when it was withdrawn. Other types were also used and isolated examples persisted until about 1930. The real problem with clutch brakes was that, although a quick up-change could be made, the brakes generally interfered with the clutch shaft motion while changing down when the clutch pedal was pressed down hard, and this often resulted in a terribly noisy clashing of gears. The brake stop position was therefore often adjusted to allow the clutch to disengage before the clutch pedal was fully up. This could make it difficult for drivers to drive cars other than their own. Gear changing was never an easy matter up to the outbreak of the First World War.

One fault with the early clutches was that the clutch spring reacted on the transmission. This could be overcome by fitting a reverse cone external to the flywheel with the drive cone forward of this cone; the spring then reacted on the flywheel through a thrust bearing, provision being made to adjust the spring tension by a screwed cup. Linkages were complicated and the leather facing not very accessible in the early designs. However, easy clutch removal for servicing was possible with some designs (figure 12.1).

Another way to overcome the reaction problem was to use a hollow shaft for the cone shafts, the shaft enclosing a shaft attached to the flywheel, and this shaft carried the clutch spring.

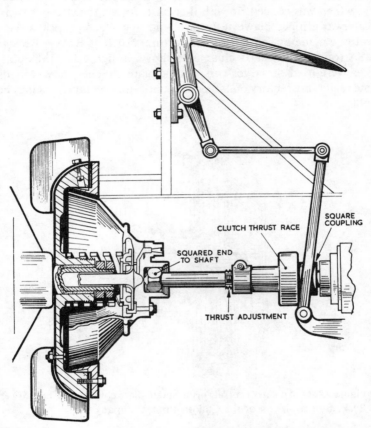

Figure 12.1 Cone clutch on a 7 hp Renault of 1908. By permission of Century Hutchinson Publishing Group Ltd (from E F Carter 1962 *Veteran Car Owners Manual*).

Plate Clutches

A second type of clutch, the single-plate clutch, was similar to the Panhard type of clutch but with the mating surfaces normal to the axis of rotation. Clamping force had to be greater because there was no wedging effect, as with the cone clutch. In the early models an annular recess was machined into the flywheel and a grooved ring screwed into the opening of the recess. Three or more radial levers pivoted in the groove and the other ends of these levers fitted into a grooved sleeve on the shaft. Near their outer ends the levers bore against an annular ring. When the pedal was depressed the sleeve was moved back against the clutch spring, pressure taken off the ring and the plate was free to rotate. Consequently when the pedal was released the spring, which was concentric with the shaft, forced the sleeve forward and the plate was gripped between flywheel and ring. Such proprietary clutches began to be made by specialist manufacturers such as Borg & Beck Ltd, in the USA (one of their early clutches is shown in figure 12.2). Double-plate clutches were used on larger cars. Single-plate clutches, however, did not become really satisfactory on larger cars until suitable fabric facings became available.

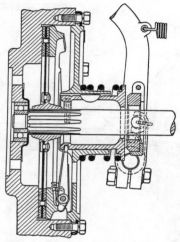

Figure 12.2 An early central spring single-plate clutch made by Borg & Beck. By permission of the Chilton Book Company.

Multiplate Clutches

Because of the limitations of the single-plate clutch, it was usual to fit multiplate clutches. These were basically modifications of the Weston clutch, which was built up from alternate discs of wood and wrought iron; it had been widely used on industrial equipment in Victorian times. Each additional plate increased the friction force, so for the same torque a smaller diameter clutch or a reduced clamping force, was needed. The rate of working per unit area was also much reduced. Multiplate clutches on cars both in Europe and in the USA generally used steel and bronze as the mating surfaces. The surfaces usually ran in oil and as a result a drag torque could be generated even when the clamping force was released. Small springs were sometimes inserted between the plates, or the plates could be perforated, to overcome this. Despite modifications like these drag was still a problem so Hele-Shaw used corrugated instead of flat plates. The plates disengaged readily so that the clutch slipped less than flat-plate clutches, so overheating causing buckled plates, burnt oil, etc was not a problem. In the Hele-Shaw clutch (figure 12.3) half the discs rotated with the flywheel, the other half, which interleaved the first half, rotated with the drive shaft, but both sets were splined so that they could be slid parallel to the shaft. The discs were thin, and the corrugations, besides assisting disengagement, increased their areas. To engage the clutch the stack of discs were forced together by the clutch spring to rotate as one mass. The clutch ran in oil but though the μ was low, the unit was small and compact; it was also very costly.

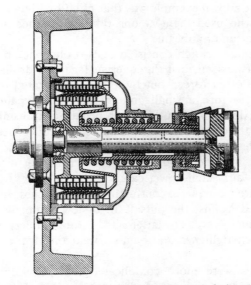

Figure 12.3 Hele-Shaw multiplate clutch. By permission of Century Hutchinson Publishing Group Ltd (from E F Carter 1962 *Veteran Car Owners Manual*).

Multidisc clutches were popular on high-powered and more expensive cars and many of these clutches had between twenty and sixty plates up to diameters of 8 in depending on engine horsepower. Many firms made their own clutches and developed their own special designs to ensure quick release of the plates.

Multiplate clutches were fairly extensively used in America but, unlike in

Europe, cork was often used as the friction material. The cork, in the form of inserts, was pressed into circular holes in the steel discs and the clutches were operated dry or in oil. In oil the cork had a coefficient of friction between 0.15 and 0.20 when rubbing against steel; it gave a smooth action with a good life and such a wet clutch was still used by Hudson in 1953.

Other Clutch Designs

Other types of clutches were based on internal and external contracting brakes. In the former type blocks were pushed out to engage the drum; centrifugal force, however, could make disengagement difficult, and a complicated linkage was needed. Such clutches were not widely used; perhaps the best known example was that fitted by Metallurgique although other makers who used them at one time or another included Bianchi, Mercedes, Rover and Sunbeam.

The Metallurgique clutch consisted of two phosphor bronze shoes that were forced outwards against the inner surface of a drum in the flywheel. The shoes were often forced out by levers that rotated rollers slightly so causing screws to turn a small amount and this expanded the friction segments. An axial load from the clutch pedal was transmitted to a sliding sleeve to act on the collars. Sometimes adjustable rollers actuated by a cone linked to the clutch pedal were used to force apart the lever to move the shoes outwards. These expanding clutches were powerful enough to work successfully on fairly high-power cars.

Even more rarely used were contracting band clutches. In one type a steel band was forced against the outside of a steel drum by wedge plungers carried by a sliding sleeve. The latter acted through rollers on a link bearing on toggles which tightened the bands round the drum in a progressive manner.

Band clutches were more complicated and required more accurate adjustment than cone clutches, and it was particularly difficult to ensure that rotational effects were suitably balanced out. Contracting band clutches were used on some Mors models until 1914.

Another type of clutch that had some followers for a time, including Mercedes (who moved over to a cone clutch in 1914), was the coil or scroll clutch. In this clutch the torque was obtained from the tightening action of a powerful steel spring acting against a drum. The spring was anchored to either the driving or driven part of the clutch and under rotation tended to wind itself up and grip the drum and so drive the car. Hydraulic and electromagnetic clutches were also tried but had little success.

The situation before 1914 was that the cone clutch was by far the most popular type but single-plate and multiplate clutches were also in use, the latter mainly on heavy expensive cars.

Developments From 1919 to 1939

The single-plate clutch was simple, withstood rough treatment and had a relatively consistent performance when used with asbestos-based clutch facings which had higher and steadier coefficients of friction over a wider range of temperature than earlier friction materials. The single-plate clutch thus gained in popularity, largely as a result of makers such as Borg & Beck Ltd and Long etc in the USA who were responsible for much of the development work. They were used on something like 80 percent of American cars by 1930.

In the twenties the single central spring Borg & Beck clutch was introduced into the UK, the agents being Automotive Products Ltd, who later designed and manufactured Borg & Beck clutches. One of the features of the Borg & Beck clutch was the T slots punched in the periphery of the steel rim of the driven discs to create a series of segments to which the friction facings were riveted with rivet heads on alternate sides of the facing which gave a cushioning effect. The driven plate was connected to its shaft by a star-shaped moulded rubber centre which acted as a sort of shock absorber.

Later it became common to fit a number of springs arranged circumferentially around the pressure plate to apply the clamping load. Smaller springs were needed and it was generally arranged that the pressure of the springs could be adjusted to make the pressure on the facing uniform. The release mechanism consisted of levers acting through struts on the pressure plate lugs, the struts being provided with knife edges to minimise friction. Rearward movement of the pressure plate against the helical springs disengaged the clutch. The driven plate was modified by including spiral springs in the plate instead of rubber to give greater control over flexibility and the cushioning characteristics with better durability. Fichtel and Sachs, another major manufacturer, commenced making a range of clutches in Germany in the thirties. Initially their spring-actuated clutches were less sophisticated than Borg & Beck's, being fitted with spring-loaded release levers having pivoted anchor pins in the cover plate. Carbon thrust bearings were introduced and actuation was by means of a cross shaft and pivoted fork linkages on the clutch pedal, pedal efforts being of the order 20–25 lb. In the earlier thirties many vehicle manufacturers began to make increasing use of these proprietary single-plate clutches.

Once established the basic principles of the clutch remained unchanged for many years though there was considerable improvement in the detail design (figure 12.4).

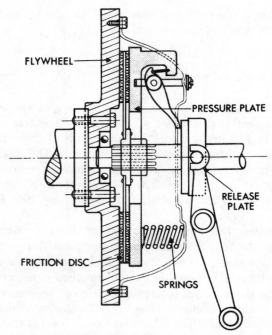

Figure 12.4 Typical single-plate clutch; the load on the clutch is exerted by springs. By permission of Mintex-Don Ltd.

Automatic clutches

Many attempts were made to make the operation of the clutch automatic. For example in a Newton clutch of the thirties weights mounted on the flywheel moved outward by centrifugal force when the engine speed increased, exerting a load on the pressure plate. The clutch could also be controlled in the normal way. Another arrangement used by Bendix was to use the excess of atmospheric over inlet manifold pressure to operate a piston to keep the clutch engaged. More in the mainstream of development was the Gillett system; a pump driven by the engine delivered oil to a cylinder, the piston of which operated the clutch under the action of valves controlled by the accelerator and a governor driven by the engine. The clutch was disengaged when the engine speed was low, and engaged when the throttle was opened.

Some Talbot cars had, in addition to a centrifugal clutch, an arrangement which enabled the car to descend hills without the engine coasting. Two heavy shoes held together by springs were mounted on an extension of the engine crankshaft and rotated in a drum attached to the gear shaft, and at a specified speed torque was transmitted without slip. Within the brake shoe assembly the free wheel unit was mounted. This was a spring-controlled roller on an inclined plane device which offered no resistance when the clutch was driving but came into operation when the gear shaft was tending to rotate faster than the crankshaft.

Fluid couplings

Letting the clutch out suddenly stressed transmission and passengers; on the other hand too much slip could overheat and damage the clutch facing. Dr Fottinger, a German engineer, invented a fluid coupling unit for ships engines in 1905, and hydraulic coupling for cars was adopted by Daimler in 1930 (figure 12.5) which gave a very smooth engagement. It operated quite independently of the driver, so an auxiliary friction clutch had to be fitted for use when changing gear. The 'Fluid Flywheel' was particularly effective when combined with the Wilson preselective gearbox and eventually became vital as an adjunct to the automatic gearbox.

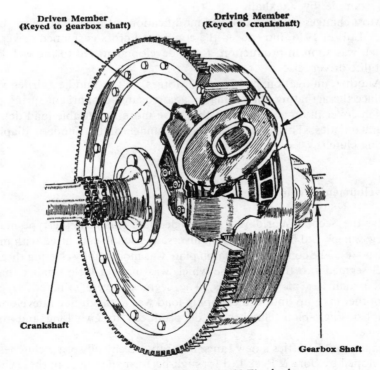

The Daimler Fluid Flywheel.

Figure 12.5 A Daimler fluid flywheel. By permission of Chapman and Hall Ltd (from A W Judge *Motor Manuals*).

The flywheel consisted of two rotors, one of which was embodied in a casing bolted to the flywheel and the other mounted on a flanged shaft connected to the gearbox. The rim of each rotor was of semi-circular section and divided into cells by radial webs. The casing was partly filled with oil, and as the speed of the driving member was increased the fluid was given angular momentum and pumped from the driving to the driven member which was set rotating. Slip occurred between the rotating members which was related to the torque transmitted; the difference in torque between input and output members was small at high speeds. Such couplings were developed by the Daimler group for use with planetary transmissions and were the forerunner of many fluid transmissions developed in the USA.

The mainstream of development, however, was in single-plate coil spring clutches. One weakness of these clutches was the loss of clamping load from the effects of centrifugal force acting on the springs at high engine speeds and to compensate for this effect Long in the USA included centrifugal weights on the release levers. Such clutches were favoured by some vehicle manufacturers and made by Borg and Beck Ltd in the UK and were used, for example, by Vauxhall.

Most changes involved detailed modifications of the driven plate. Borg & Beck Ltd, in 1936, introduced the compound cross-cushioned driven plate which was soon in production. Later developments led to the well known Borglite driven disc.

Another innovation by General Motors in the middle thirties was to replace the actuating coil spring by a corrugated diaphragm spring on the clutch cover in an arrangement that gave more favourable load deflection characteristics. This design was a forerunner of the modern diaphragm spring clutch.

Developments From 1946

After the Second World War the single-plate clutch of the type made by Borg & Beck Ltd became almost universally used in cars fitted with manual gearboxes. The compound cushion plate was modified by riveting the spring steel segments to the main disc which was made a little smaller than the inside diameter of the facings. The segments were crimped to give a smoother take-up until full clamping load was applied. Such developments enabled single-plate clutches to be produced to give long and reliable service.

In the middle fifties a new range of strap-driven coil spring clutches were developed by Borg & Beck Ltd for high-performance cars. In these clutches the drive to the pressure plate is taken by a number of tangentially arranged spring steel straps that deflect on release of the clutch and enable concentricity of the pressure plate to be retained. Accurate balance was ensured throughout full travel of the clutch as there is no relative movement between pressure plate and cover.

Mechanical actuation continued to be used for some time after the war but this gradually gave way to hydraulic actuation which was completely unaffected by power unit movement. Fluid under pressure was transmitted to a slave cylinder provided with a pushrod that operated the clutch release lever which was pivoted in the bell housing. On American and Continental cars lubricated sleeve-mounted ball thrust bearings were favoured for the clutch withdrawal mechanism instead of the carbon rings fitted on UK cars. The latter were to adopt thrust ball bearings widely in the sixties, the use of which could also eliminate the need for adjustment in the linkages.

Friction facings were also improved, the earlier millboard facings of limited mechanical strength were replaced by moulded and wound yarn facings which had a higher coefficient of friction and lower compressibility.

Diaphragm spring clutches

In the sixties the diaphragm spring clutch, well proved in the USA, and on the Continent, began to be used in the UK. This clutch overcame the

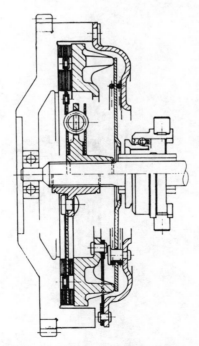

Figure 12.6 Fichtel and Sachs type M diaphragm spring clutch. By permission of Fichtel and Sachs AG.

weaknesses of the coil spring clutch by using a conical diaphragm spring to engage directly on the thrust member. This resulted in a reduction in the number of components in the cover assembly and a more balanced design. The clamping load was also maintained throughout the life of the facing unlike that of a coil spring.

In the early Borg & Beck design the spring plate was caused to flex about fulcrum rings. In the engaged position the resulting forces flattened the spring and exerted a clamping load on lugs on the pressure plate. In the disengaged position the whole spring flexed about the foremost fulcrum ring to lift the pressure plate. Laycock also introduced into the UK a design based on Hausserman patents in Germany, and first fitted on the Hillman Imp in 1963. Other manufacturers also made diaphragm spring clutches (figure 12.6), the principles involved being much the same although detailed differences existed in the designs. Strap-driven diaphragm spring clutches were also introduced. With the advent of the diaphragm spring clutch, pedal efforts were reduced and the torque capacity of these clutches was about 70 per cent greater than that of a coil spring clutch of the same size as used in the mid-thirties.

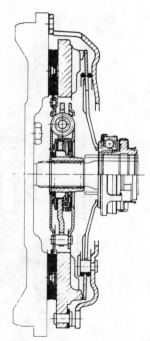

Figure 12.7 Automotive Products DST clutch. By permission of Automotive Products plc.

Once established, diaphragm spring clutches were subjected to continuous modifications to reduce the number and weight of components and hence the inertia of the rotating parts. One such innovation developed by Automotive Products Ltd in their Borg & Beck clutches in the late seventies was to introduce a new cover pressing with tabs in the central region that passed through gaps in the diaphragm spring and then bent backwards to grip the diaphragm spring between two fulcrum rings. The straps that transmitted the drive to the pressure plate were also riveted to the cover. This design, the DST clutch (figure 12.7), was also more durable, and it was thinner than previous clutches making it particularly suitable for transverse engine installations.

To reduce lost motion in the actuation system, pull-type designs have been introduced which result in less deflection of the cover and better utilisation of pedal travel compared with conventional push designs. Special release mechanisms were required with some designs, whereas with conventional designs the well established hydraulic operation is common although some arrangements are fitted with cable-operated clutches because of their flexibility in overcoming installation problems. The latter are provided with an automatic adjustment mechanism to cater for wear of the facings.

Friction facing materials have also continued to be developed to meet the increasing demands made upon them. Better fade and wear characteristics and burst strengths have been obtained with materials containing random asbestos fibre and metal inclusions, making them more suitable for high-speed engines than wound yarn facings. Similar desirable characteristics have also been obtained with non-asbestos facings.

Chapter 13

Gear Change Mechanisms

As the internal combustion engine develops low torque at low speed and maximum torque at high speeds a means had to be devised to connect the engine to the driving wheels to give torques to cater for the resistance loads that are encountered while driving the vehicle. Thus a torque multiplying mechanism is required to provide the tractive effort to accelerate the car and keep the engine speed within an economical range. A reversing mechanism is also required as the engine rotates in only one direction.

Early known ways of transmitting power were by belt transmission, by means of friction drives and by the use of toothed gears.

Belt Transmissions

Benz obtained intermediate gear ratios by means of a belt transmission based on contemporary workshop practice. In his first car a pulley was mounted on the end of the shaft used to provide the valve motion, and a belt transmitted engine torque to a countershaft, divided into two parts, by a fixed pulley. Next to this was a loose or jockey pulley which could spin freely when the engine was ticking over normally. Within the pulleys was a differential gear. A lever on a separate shaft was used to operate a fork and move the belt across the two pulleys by means of a bevel which could turn a bevel pinion mounted on a short spindle and a crank bar; when the control lever was pushed forward the crank pin was turned via the gearing and the belt was moved across to the fixed pulley and one gear ratio was obtained. When the control lever was pulled back the belt was returned to the loose pulley and simultaneously a brake block was applied to the fixed pulley on the countershaft.

Because the vehicle lacked climbing ability, Benz soon introduced an arrangement which would provide a second gear ratio by means of a two-speed gear operated by chains. Much difficulty was experienced with this system and it was replaced by the well known but less efficient twin-belt change mechanism fitted to most production models. Here, two fixed and two jockey pulleys were mounted on the countershaft which also carried the differential gear. By operating two handles, belt striking gear could be made to shift either crossed belt horizontally from its fast and loose pulleys to give two differential gear ratios.

Later modifications included the fitting of a sun and planet gear termed a 'Crypto' gear for hill climbing. The 'Crypto' was engaged by moving the belt to a loose pulley and applying a band brake to the outer surface of the gear case. The loose pulley then drove the fast pulley at a very slow speed and once in gear one remained in it. The car could then have climbed anywhere provided the driver lived long enough!

The Benz-type transmission was also used on other early production cars. A few refinements were added by some makers, for example, Mors in the middle 1890s included a gearing system that enabled the driver to adjust the belt tension from his driving position. Similarly, Stephens devised a system which could maintain an even tension on the belts when the car was in motion. An added attraction was that the belt arrangement was designed to allow the vehicle to coast down hills if the engine was switched off. At the bottom of the hill the belt transmission could be reapplied to provide a clutch action to restart the engine.

Surprisingly the Cannstatt-Daimler vehicle of 1896 was also fitted with a belt transmission despite Daimler having previously experimented with a sliding spur gear. In this vehicle the pulleys were arranged in pairs on either side of the engine and four belts used to drive four pulleys mounted on a countershaft. Each belt passed over a jockey pulley which could be raised or lowered to provide a tight or slack belt (figure 13.1). This was effected by a handle connected to a wheel, the lower half of which was toothed, which was keyed to a shaft so that it was free to slide to either side against spring pressure. When this wheel was moved to one side the teeth engaged with a pinion keyed on to another shaft to which was attached two pairs of crank arms. By turning the wheel either one or other of the jockeys could be lifted to provide a tight belt and result in a different gear ratio. Similarly, if the wheel was moved to the other side two further gear ratios could be obtained. When the wheel lay between the pinions a neutral gear was obtained. The operating belt mechanism was very similar in principle to a gate change as used in gearboxes. No reversing gear existed on this vehicle.

Another, and perhaps the best belt-driven system, was designed by Bollée and employed on the Darracq light car. It consisted of a cone of pulleys on the crankshaft with motion transmitted to a second drive shaft behind the rear axle on which was fitted a similar five-cone pulley, a clutch and a pinion gearing with a differential gear. The belt could be shifted from pulley to pulley by a fork on an inclined axis at the front and rear ends. The fork was moved by a hand wheel below which was a disc with five small projections. By lifting this disc and turning it to either of the positions the movement was transmitted by chains to a sprocket wheel on a rack to give a horizontal motion to the belt striker fork, the bearing of which was inclined to enable the fork to follow the inclination of either set of cone pulleys. In this way four forward gear ratios could be engaged. An epicyclic gear near the flywheel was used as a reverse gear.

The good features of belt drives were that they were simple, reasonably inexpensive, quiet, and able to withstand sudden changes of speed without there being any risk of damage to components under shock loads; indeed, if a belt was fitted the reverse gear could be used as a brake.

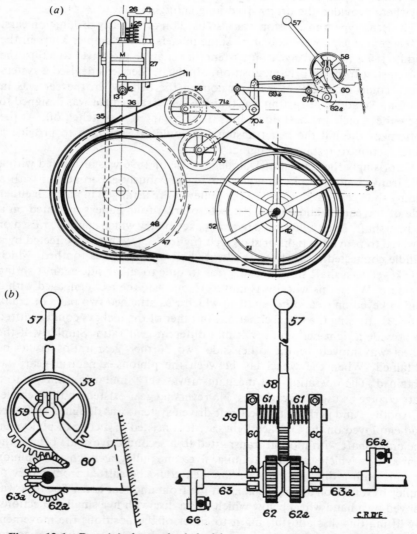

Figure 13.1 Part (*a*) shows the belt drive on early Daimler cars; part (*b*) shows the change speed control that allowed the gear lever to move both sideways, backwards and forwards as in a gate change to obtain four gears.

The big disadvantage of belts was that they could slip which resulted in a severe loss of power. Other undesirable effects resulted from the belts stretching which meant they frequently needed shortening to keep them taut. All belt systems also occupied considerable space, for the greater their length the better the performance. Because of these overwhelming disadvantages belt drives gradually lost favour with the public. They had to disappear if progress was to be made and vehicles run at higher speeds and have a livelier performance. It was really remarkable that they lasted so long.

Friction Drives

One method of transmitting power that was a fairly popular alternative to belt transmissions was the friction drive. This device generally consisted of a cast iron or steel disc on the driving shaft from the engine which was loaded against another movable wheel placed at right angles to the disc. In low gear the driven wheel was placed near the centre of the driving disc, whilst a higher gear was obtained by moving the driven wheel further out towards the periphery of the driving disc. It was thus possible to obtain a continuously variable output speed or torque for a constant engine speed. Reverse could be obtained by moving the driven wheel to the opposite side of the driving disc to that giving a forward drive.

Provided no slip between disc and wheel rim occurred, this relatively simple device worked reasonably well and could give a smooth and noiseless drive. As it was impossible to obtain line contact between the contacting members the grinding action between the rotating members frequently caused flats to be worn on the driven wheel at some stage, resulting in excessive slip and a horrible grating, whining noise from the drive. This would occur despite all combinations of materials, including friction materials, being tried on the driven member. There was also a limit to the power that could be transmitted.

To be effective this type of transmission required a delicate touch from the driver. Even in 1901 Holroyd Smith in a paper to the *I.Mech.E.* discussed the general scheme of friction disc transmission gear in the hope of preventing others wasting their time and money in a like manner. It still, however, reappeared at odd intervals and found most favour on twin-cylinder cycle cars such as the Crescent, Pyramid and Violet-Bogey designs which all incorporated a final chain drive to the rear axle. Perhaps the best example was that used in the GWK light car in 1911 which persisted until after the First World War when the vehicle was fitted with a larger four-cylinder engine. Even on this vehicle it was not too reliable and unpracticable compared with the more positive drive arrangements based on the gear wheel.

Later in the thirties designs such as the Hayes and General Motors infinitely variable transmissions used curved friction surfaces to reduce the size of friction drives. The latter design was extensively tested by General Motors prior to the introduction of the Hydramatic transmission but the high stresses still created metallurgical problems that gave insufficient life to the driving surfaces if the unit was to be compact and efficient.

Gearboxes

Other mechanical methods using change speed gears were soon introduced to overcome the weaknesses of belt transmissions and friction drives. In the first types gear wheels of different sizes went into and out of mesh to give the required ratios. In the period up to 1914 other arrangements were introduced; the most popular had pairs of gear wheels of various sizes in constant mesh which were locked to the shaft, that is, made 'live' when needed by clutches either of the 'dog' or friction type. Another type used epicyclic gears which were engaged by band brakes gripping and releasing various rotating parts. Most gear boxes with minor detail modifications fell into these categories.

Sliding gears

The outstanding innovation here was the introduction of the sliding speed gearbox as fitted to one of Panhard and Levassor's models produced towards the end of 1891. The Panhard layout comprised of a vertical engine mounted in front of the car, driving through a clutch, a gearbox and a differential gear shaft and this system set the trend for the future.

The gearbox consisted of a shaft connected in line with the crankshaft through a friction clutch that engaged the surface on the rear of the flywheel. The shaft carried three sliding gear wheels which could mesh with three gears mounted on an upper second shaft and so provide three gear ratios. Each sliding gear was moved by a lever whose lower end connected with a horizontal link bearing against a collar. Two pedals were fitted, one to control the clutch and the other one to actuate a band brake on the second shaft. At the end of the second shaft a bevel wheel transmitted drive to a transverse shaft and thence by a chain to the centre of the rear axle. The gear wheels ran free and unprotected in the open air and their only lubrication was by occasionally daubing them with grease. An early change was to replace this set up by one on which the bevel pinion on the second shaft engaged with one or other of two bevel pinions on a differential to give either forward or reverse motion.

It was not easy to effect a gear change without producing crunching sounds and damaged gears but as Levassor said of the mechanism 'it is brutal but it works' and despite F Lanchester's appropriate comment 'it is successful precisely in proportion to the extent to which it is not used' it was the best general layout for further development.

By 1900 the sliding gear had become firmly established, although initially it was regarded as a temporary mechanism until something more sophisticated appeared. The gears were now enclosed in a casing, usually made of aluminium and called a gearbox unit. Fig 13.2 shows a copy of the Panhard gearbox that was fitted to the 1897 4.5 hp Daimler. This large gearbox was fitted with plain bearings and had a countershaft or layshaft about the same length as the driving shaft and rather narrow gears were bolted to flanges formed solid with the shaft.

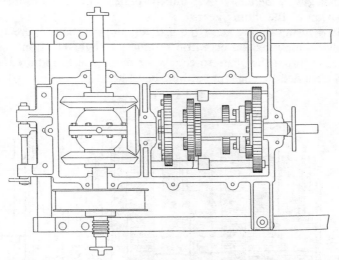

Figure 13.2 Panhard-type gearbox used on the 1897 4.5 hp Daimler car. Bevel gears were used to drive the chain sprocket shaft.

The Panhard sliding gearbox had several severe drawbacks: the long square shafts tended to whip and could run twisted slightly out of true which made it difficult to change gear. The gears were noisy and wore badly. As a result of these faults various other arrangements were designed.

One type had two separate shorter sliding sleeves carrying the forward gears and another sleeve for the reverse gear. The shaft diameter was also increased and deflections thereby reduced. The gears themselves were operated by a lever working in a quadrant that had three parallel slots with short transverse slots joining them and this layout became the gate change box. Each slot corresponded to one gear position.

A shift fork was used to operate each sliding pair of wheels individually.

To ensure the three forks could be moved by the one gear lever, a gate control was introduced. This enabled a shortened lever to engage with one or other of the three rods that carried the sliding forks. The gate made it impossible to engage more than one gear at a time. The familiar 'H' form was introduced for four-speed gearboxes.

In another popular type of sliding sleeve gearbox the drive went from the front of the engine clutch shaft to the layshaft and then transmitted power back to the driven shaft in line with the engine. In top gear direct power was transmitted through dogs to the tailshaft or propeller shaft.

The sliding gearshafts at first were square but later they were either castellated or splined, and milled out of the solid shaft. Soon improvements to multisplined shafts were made by using case-hardened steels with ground splines that gave a better fit over milled splines, and much more accurate centralisation of the sliding gears.

Instead of using sliding gears the first Renault gearbox (figure 13.3) had two layshafts mounted on eccentric spindles; a slight rotation in a radial direction permitted the gears to engage or disengage as required. Reverse gear was obtained by engaging a third bevel gear.

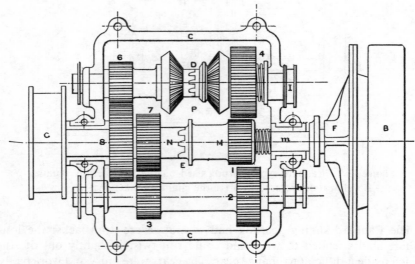

Figure 13.3 Renault gearbox of 1898, the first to provide a direct drive. The layshafts were left out of mesh and when needed one or other moved sideways into mesh with the main shaft. A third bevel gear meshed with the bevel gear shown to give reverse gear.

Constant-mesh gearboxes

Constant-mesh gearboxes in which pairs of gears ran loose on their shafts and were fixed as required by means of sliding keys were soon introduced by a few manufacturers but this layout was too complicated to become popular.

It was more usual at first to use expanding clutch type constant-mesh gearboxes, the best known early example being that of de Dion Bouton. The clutches and gear wheels were mounted on the driven shaft and by expanding one or other clutch the drive gradually became locked to transmit torque to the rear axle. The clutches were mostly operated by a lever mounted on the steering column although on some models a low gear could be engaged by operating a foot pedal. A gear change was possible provided the clutch did not slip and such designs really suffered if roughly handled and this often happened. Reverse gear was not in constant mesh and was engaged by bringing an intermediate gear into mesh with a driving pinion and expanding one of the clutches.

By 1905 de Dion Bouton moved over to the sliding gear type of gearbox. As the gearbox was by now considered to be practical, attempts were made to effect a gear change more easily. An early attempt at this was in the Linley self-change gearbox introduced by Commer Cars in 1905 (figure 13.4). This was a constant-mesh gearbox using a dog engagement. Spring-loaded forks on the gearbox allowed the shifts to be made from one gear to the next more easily than in most gearboxes at that time. A large rubber coupling was fitted to take some shock out of the gear engagement. Progress by about 1907 was such that the conventional gearbox using a pair of constant-mesh head gears was becoming fairly well accepted.

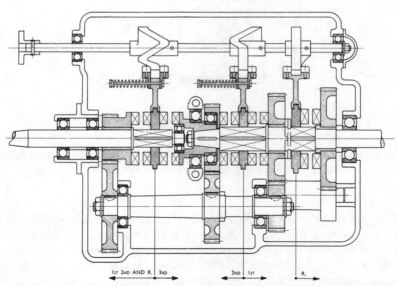

Figure 13.4 Commer gearbox 1905. By permission of The Council of the Institution of Mechanical Engineers (from *Fifty Years of Transmissions*).

Epicyclic gears

Epicyclic gearing giving one low speed, a direct action coupling engine to drive shaft for higher speed, and a reverse, was used by several early car makers. The epicyclic gears were brought into action by a contracting band brake on a drum containing the gearing. This arrangement was favoured by the Americans and it was usual to have a first gear ratio of about 3 to 1. Duryea used epicyclic gearing on his early cars but the satellite pinions rotated at too high a speed for the poor quality bearings fitted which soon failed. Cadillac used a similar system except that two cone clutches were fitted which each enabled a gear change to be made. The erratic behaviour of cone clutches made gear changing a frightening business. Many American manufacturers tried to eliminate the need for gear changing by using two-speed epicyclic gears, particularly Ford in his Model T (figure 13.5).

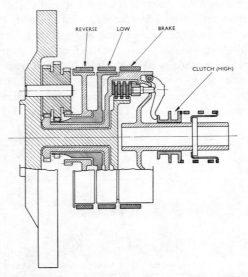

Figure 13.5 Model T Ford gearbox. By permission of The Council of the Institution of Mechanical Engineers (from *Fifty Years of Transmissions*).

The gear ratios of the simple epicyclic type were totally unsuitable for hill climbing so that Lanchester who had used this system in his first cars preferred to introduce an intermediate gear ratio to give the vehicle better climbing ability. Lanchester's solution was the compound epicyclic gear train that forms the basis of the automatic transmissions used today. This consisted of using the outer annulus of the first speed as the planet carrier of the second speed to provide an additional speed ratio. As fitted on the production car the transmission was driven from the lower of two crankshafts and a cone clutch was fitted to give direct drive on top gear. Later the cone clutch was replaced by a multidisc clutch and the compounded epicyclic transmission with various modifications remained in use on all Lanchester models.

Other important mechanical details in the compound epicyclic gearing were the use of cageless roller bearings for the planetary gear bearings, and splines for attaching the rotating elements to their shaft to transmit the drive. Roller bearings and splines were special innovations introduced by Lanchester which were used on his experimental vehicles. Furthermore, he invented new methods of manufacturing rollers and of cutting splines, and generally improving the product. All this occurred round the turn of the century.

Epicyclic gearing simplified the procedure for gear changing. Another manufacturer using epicyclics was Adams of Bedford who had a slogan 'pedals to push—that's all'; the driver simply pushed one pedal and went into first gear, the next pedal for middle gear, then into top and there was a pedal for reverse gear.

Design features of gearboxes

Despite the obvious attractions of epicyclic gearing most Continental and American manufacturers preferred the cheaper constant-mesh gearbox with coaxial input and output shafts which became firmly established practice before the First World War. It was usual to have four forward speeds on most cars except in America where large cars had four gears but medium-sized cars had three speeds and small cars had only two speeds. High gear ratios were not used owing to the poor state of the roads. Most three-speed boxes had short shafts, often square sectioned, but the trend was towards using splined shafts and roller bearings instead of ball bearings.

Lubrication problems were now less serious as oil could be maintained in the gearbox at a level high enough to contact the teeth of the smallest gear wheel. In some of the early rear-mounted boxes there were no seals and oil frequently worked its way along the transmission shaft bearing onto transmission brake drums as well as spraying the rear tyres. This leakage was soon largely eliminated by the fitting of oil-retaining spirals, sling rings or felt washers and lubrication was greatly improved.

The location of the gearbox changed considerably. In the early days the units were often independently mounted separate from both the clutch and propeller shaft, and in another popular arrangement the differential and tail shaft were included in the gearbox. The outer ends of the tail shaft were supported in bearings and usually attached to the longitudinal subframes or to cross members of the main frame. Other layouts had the gearbox directly attached to the rear axle but as the gearbox become neater and lighter it was transferred to unit mounting with the engine, as first adopted by Motor-bloc. This simple arrangement avoided the necessity of having an intermediate shaft and led to an improved change mechanism, although first a number of detailed modifications had to be made before it became the standard layout.

Most of the problems associated with early gearboxes were due to poor quality gear steels of low tensile strength that resulted in teeth failures. Great attention was therefore paid to produce direct hardening steels, and air and oil hardening nickel chrome steels became popular. The idea was to try to obtain the correct tooth profile by grinding the teeth after heat treatment to overcome distortion. The real metallurgical development was the use of case-hardened steels with their better surface hardness to give better wear resistance. Even then difficulties were experienced from improper heat treatment of gears that resulted in excessive teeth wear. Teeth also often bent over and gave a dreadfully noisy gear change. Gradually these troubles were overcome as better materials and production techniques became available.

Gear tooth design progressed slowly and many so called standard profiles were designed with too large tolerances; they were badly undercut and weak. Stub teeth were thought to be stronger and were preferred by some manufacturers. These stub teeth turned out to have unsuitable wearing surfaces, difficulties in adjustment and noisy running which led to a return to deeper teeth of better shape to combat the noise problem.

The involuted form of tooth was largely retained and the angle of inclination of the tangent of the path of the point of contact changed little from 20°. Most improvements were largely in detail and cutters were made to obtain greater accuracy during manufacture. Hobbing produced a much closer approximation to the true tooth form and caused fewer errors in the final product.

Gear sizes varied enormously between manufacturers; they were not related to engine horsepower, extremes were 0.625 in wide on cars of 80 hp and 1 in wide on cars of 10 hp with shaft centre spacing varying in much the same way. European practice was to have gears about 0.75 in wide and shaft centres from 3.5 to 4 in on engines of 3 in cylinder bore with suitable increases being made for larger bore engines. Those with 4.5 in cylinder bores had gears about 1.25 in wide and 4.5 to 5 in shaft centres.

Similarly a great variation existed in the pitch of gear teeth. Europeans seemed to settle for teeth from 6 to 8 diametral pitch but the Americans had much coarser teeth from 2 to 3 diametral pitch and had the shafts placed closer together.

Excessive play in early shafts gave great impetus to ball and roller bearing development to reduce bearing wear and give added rigidity to gearboxes. An additional central bearing between the sliding members was introduced on the Rolls-Royce in 1910. Progress was such that the compactness of the designs gave rise to less size and weight. Lipped roller bearings were introduced for pilot bearings, and journal bearings were developed to cater for increased axial loads. Needle roller bearings were used on floating constant-mesh gears.

Requirements

The requirements that a gearbox should meet had been more or less fixed by about 1910. One was that the gear ratio of the transmission should be such that when the engine was running near the speed at which maximum power was developed, the car should be travelling at maximum speed. This roughly corresponded to taking a fully laden car up a hill of about 1 in 15 in top gear. A second requirement was that the engine should pull from rest a fully laden car up a gradient of at least 1 in 4. These conditions were met by having a first speed ratio of about 4 to 1 with a top gear of 1 to 1 and the intermediate ratios were then arranged to be in geometrical progression. Once established these gear ratios were not in fact to change much for many years.

On Grand Prix cars gearboxes were designed to have a smaller range of ratios than on production cars. After Ernest Henry had successfully introduced his four-valve OHC engines in 1911 it was usual to have a first gear ratio of around 2 to 1 in a four-ratio gearbox. This arrangement made it possible to maintain peak powers in all gears while at high speed and thereby ensure good performance.

Developments in the USA

Steady development continued in America during the First World War; the three-speed gearbox became practically universal though Fiat, Pierce-Arrow, Locomobile and White used the four-speed box and Ford stuck to his planetary system.

A steady change in the location of the gearbox took place as can be seen from table 13.1 showing the percentages of models with their gearboxes in different positions.

By 1916 the majority of American cars were built on the unit system of combining engine, clutch and gearbox in one unit which was very convenient for factory production. The SAE even began to standardise on dimensions for crankcase ends and gearbox fronts to facilitate the matching of these components.

Gearboxes were mainly made of cast iron with the layshaft placed below the main shaft. The case was not divided and the shafts could be inserted through the top opening. If roller bearings were fitted in the layshaft a plate was generally fixed to the bottom of the gearbox to make it possible to remove the layshaft and reverse the idler pinion without disturbing the rest of the transmission. Detail refinements of this type were continually being made in America.

Table 13.1 Change of gearbox locations during the First World War in the USA.

Year	Unit power plant	Unit with axle	Amidships
1915	32.5	18.2	49.3
1916	63.5	15.3	21.2
1917	77.0	9.0	14.0
1918	74.2	7.4	17.4

The majority of cars had a gear change lever centrally mounted directly on top of the transmission. Striking rods were inserted at the top of the box and the lid carried the lever in a ball and socket joint. The lower end of the lever was also ball-shaped and worked directly in the striker slots without the need for interlocking mechanisms. This arrangement simplified the linkage and was easily adjusted and lubricated.

The centrally mounted gear stick followed from the USA moving over to the universal adoption of left-hand steering so that the central position was the natural place for the driver when driving on the right-hand side of the road.

Gearbox Developments From 1919–1939

After the First World War, three-speed boxes were used in the UK on light cycle cars up to 22 cwt laden weight, with four-speed boxes on heavier vehicles. Gradually four-speed boxes increased in popularity and three fifths of the vehicles in the 1927 British Motor Show had four-speed boxes and this trend continued.

In the early twenties most gearboxes had detachable plates at the end of the box. It was common practice to spigot the plate on the main shaft centre and dowels or bolts were relied on for the alignment of the layshaft bearings. The tendency was to move over to mounting the layshaft on ball bearings along American lines instead of using the plain bearings formerly in fashion.

The immediate post-war years also saw a trend in Europe towards a more compact gearbox unit that was fixed to the engine crankcase rear face, again following American practice. A further improvement was the reduction in inertia of the rotating parts in the gearbox as increased engine speeds with pre-war designs required the fitting of costly clutch stops to reduce the angular momentum of the clutch rapidly if gear changes were to be made reasonably quickly.

Central mounting of the gear change lever also became more usual as a less complicated linkage was needed than when the lever was placed at the driver's side. Some manufacturers like Austin used a stiff cross-sliding lever working in a gate, although the stiff construction was not as suitable for the average motorist as a light, springy ball-jointed change lever. The latter was virtually universal by the late twenties.

Separate gearbox units did not entirely disappear in the twenties and Renault for one continued their practice of mounting the gearbox at the front end of a torque tube supported on a ball joint surrounding the nose of the box and carried by a cast steel cross member. The gear lever was mounted on the cross member and picked up two selector forks in the plane of the ball joint. Bugatti for another had the gearbox connected to the rear axle in the straight eight as did several other heavier cars with long wheel bases. Rolls-Royce continued with a separate gearbox but moved over to a unit engine and gearbox in the 1929 Phantom II as did several other makers including Renault by 1930.

Noise

With certain mounting arrangements of the gearbox, gear noise tended to be amplified in saloon cars and this became a serious problem which was tackled in several ways. Attention was paid to the accuracy of gear tooth finishing and in some gearboxes where no machining was done after hardening, it was usual to use air hardening steels for gears subject to severe duties, especially for light cars. Most manufacturers preferred to grind nickel steel gear teeth as a finishing operation after case hardening which became common practice by the late twenties.

The tooth form came in for consideration and helical gears were found to be more silent than spur gears. Helical gears were superior to straight spur gears basically due to the conditions that exist during a tooth engagement. Both single (figure 13.6) and double helical gears became very popular and the double type eliminated end forces. As considerable end thrusts could be accommodated by better bearings single helical gears became more widespread towards the late twenties. The use of profile-ground helical gears together with increased use of intermediate bearings and stiffening of shafts went a long way towards solving the noise problem. Some manufacturers also moved over to cast iron housings owing to their greater damping capacity compared with light alloys.

An outstanding development that prevented noise from being transmitted to the passengers was to mount the integral engine and gearbox unit on rubber mountings. The full benefits of these flexible mountings were not, however, to be realised until the thirties.

Meanwhile in the late twenties, much attention was paid to obtaining silence of the indirect gears particularly in respect to the third speed of a four-speed gearbox. The idea was to have a quiet third speed to encourage the driver to make use of it. The back axle ratio was made higher than normal so that the difference between top and third speeds was much less than formerly used and high speeds could be obtained in third gear. The use of a dog clutch engagement made gear changing from top to third much easier than with clash gears.

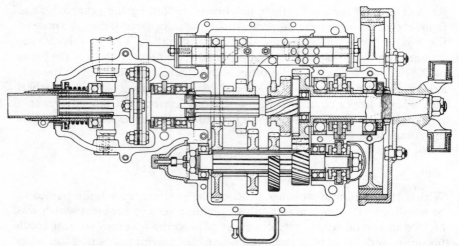

Figure 13.6 Gearbox of 18/40 hp Renault car. By permission of Reed Business Publishing Ltd (from *The Automobile Engineer*).

The overgeared top concept had been tried in the early days but excessive noise had prevented it from being successful. It was reintroduced by the Americans with a method of using integral gears for the third speed. On the Chrysler an offset sleeve, which formed the intermediate member for the third speed drive, was permanently geared to the clutch shaft engagement of third speed by the drive annulus.

A basic design change had thus occurred by 1930 with the introduction of the 'silent third' or 'twin top'. This resulted in the third gear being transferred from the centre of the box to the rear and kept in constant mesh with the layshaft. Single helical gears provided the constant-mesh gears and a conventional dog clutch for direct drive. The first and second speed gears were of the sliding spur type placed in the centre of the box.

The silent third with a high top gear did not, however, solve the transmission problem. The limiting factors were that cars did not possess sufficient acceleration in top gear, and when in third gear the engine speeds were rather high if good acceleration was required at speeds around

40 mph. This was a more critical factor with small-engined European cars and resulted in a reduction in the top gear ratio and simplifying the gear meshing operation to facilitate easy and silent gear changing.

One method that avoided the use of the clutch was the Maybach multirange gearbox fitted to the Mercedes-Benz. A separate gearbox was mounted behind the main gearbox giving two very closely set alternative ratios, attained by spring-loaded chamfered dog clutches, the gears sliding into mesh when the accelerator pedal was released. This latter operation caused a servo motor to throw over dead centre a toggle under spring pressure and enabled a rapid clutch engagement to be made.

Free-wheel devices

The free-wheel clutch reappeared as an aid to gear changing; it also allowed silent coasting and reduced petrol consumption. The free-wheel clutch was usually mounted behind the gearbox and disconnected the gearbox output shaft from the road wheel drive mechanism. The engine then slowed down and gear changes could be made without the use of the clutch pedal. When changes were attempted without the engine speed dropping sufficiently the free wheel did not always behave properly.

The free wheel depended on the ball and wedge principle; when the outer member was driven through the rollers they became wedged between the two members and the whole unit became locked up as one. During over-run the outer member revolved in the opposite direction relative to the inner member, to free the rollers and no drive was transmitted to the main member. By dispensing with the roller springs and using a roller retainer ring to locate and space the rollers equally the size of the unit could be made about 20 % smaller.

One device used in the late twenties was the Millam free wheel in which the driven shaft of the gearbox carried a cylindrical member with radial slots into which fitted rollers. Another, fitted as an extra to Lea-Francis cars, was the Humfrey-Sandberg that had inclined rollers. The Rover free-wheel unit was quite popular for a number of years and had an additional control so that the driver could slide a dog to mesh with the gear ring of the outer member and so lock the free wheel. The technique of gear changing then became no better than that of a clash box.

Most free-wheel devices used rollers but Willys Overland used an AJS gearbox with a free-wheel clutch operative in top gear only and which could be locked up if desired. The clutch consisted of a helical spring placed within the driving sleeves. The unwinding action of the spring took the drive and thus reduced the shock load in the transmission.

Free-wheel devices were fairly common in the USA but they were far from foolproof. At high speeds the gear inertia was often too high to give a comfortable quick change. The public did not like free wheels because of the reduced engine braking. Cars seemed to run away when they were going down slight slopes that previously could be catered for by engine braking alone.

Preselector gearboxes

At the same time as free-wheel devices were popular other transmissions were being developed to try to make gear changing considerably easier. One of the best known was the Wilson preselector gearbox fitted as optional extras on the 20 and 30 hp Armstrong Siddeley models in 1928. The gears were preselected by means of a lever mounted on the steering column, the actual gear changes being made by a pedal which replaced the clutch pedal. This gearbox was a development of the early Wilson-Pilcher unit that was later considerably developed at Vauxhall before Wilson became connected with Armstrong Siddeley.

The gearbox consisted of four epicyclic trains compounded together and a direct drive clutch. These were mounted side by side on the input and output shafts that were carried in four ball bearings fixed to end covers of the gearbox. The low-speed gears were engaged by band brakes that locked the corresponding gear drums. The bands were anchored at one end and contracted by a toggle mechanism, in which a pull rod, acting on the free end of the band was activated by a strut engaging a pivoted bar acted on by a powerful spring. Automatic adjustment for lining wear was made by nuts that rotated slightly at each stroke of the gear operating mechanism. Top speed was engaged by a cone clutch.

The box was lubricated by a thin oil circulated by a plunger pump. This oil was changed after about every 2000 miles as the lining material worn off the brakes formed a sludge that tended to interfere with the automatic adjustment device. Gear ratios were 3.67 to 1 in first, 2.27 to 1 in second, 1.46 to 1 in third and 1 to 1 in top, and the gearbox was remarkably quiet in operation.

The early gearboxes were not entirely foolproof as it was possible to set the gear lever down from top to second gear and introduce a fierce engagement in the transmission. The brake bands were also of small size and it was important for them not to slip excessively. As a substitute for the clutch and ordinary gearbox Percy Martin of Daimler suggested combining a Sinclair hydraulic coupling with the Wilson epicyclic gearbox to get the best out of each and this resulted in the Daimler Fluid Flywheel Transmission.

The fluid flywheel was a development of the Fottinger hydraulic coupling widely used in marine applications which was adopted for motor car transmissions by Sinclair in 1926. The device was filled with glycerine and the torque transmitted was related to the difference in speed (apparent as slip) between the driving and driven members. Below about 600–700 rpm the power transmitted was small but above that speed the power increased rapidly with a decrease in slip giving a smooth take-up. The slip was very small at high speeds.

As the torque was low at low speeds there was a need to use the disengaging pedal of the Wilson gear as the car was brought to rest without the engine stalling. The fluid flywheel, furthermore, relieved the friction bands and clutch of the epicyclic gear from slipping appreciably when starting off from rest. This allowed a gradual take off up to speed with a considerable reduction in the shock load on the gear teeth. Furthermore, it provided a close approximation to an infinitely variable gear.

In the early thirties the Wilson gearbox combined with the Daimler fluid flywheel was used on all cars manufactured under the control of Daimler and included the Lanchester and BSA models. Armstrong Siddeley used the Wilson gearbox almost exclusively with a manual clutch. Wilson gearboxes were also fitted to Talbot, Sunbeam, Lagonda and Crossley cars amongst others.

The ENV Company started manufacturing Wilson gearboxes which were fitted to the MG 'Magnette' cars, for example, and used a dry plate clutch between engine and gearbox. The first movement of the clutch pedal withdrew the engine clutch and further pedal travel compressed the gearbox operating spring. The latter operation ensured that in the top gear position there was no danger of the bands being released before the plate clutch. Another development was the use of a centrifugal clutch in combination with a Wilson gearbox to give a gradual take-up of the load without shock.

The Salerno transmission system to simplify gear changing was introduced in 1932 and was fitted to some Riley models. The Salerno transmission had a hydraulic coupling and used a clutch stop. This arrangement had the big advantage of avoiding overstressing of the gear teeth, since a gear was not engaged until the appropriate torque had been attained.

Another planetary gear mechanism that was used to a limited extent in the thirties was the Cotal box, found on Delage and Delahaye cars in conjunction with a single dry plate clutch. In this box electromagnet clutches were used to select the particular gear members to give the required gear ratios. The control was very simple, a small lever mounted in a quadrant gear selector or a series of push buttons mounted on the steering column. The transmission provided for four forward gear ratios and for coasting. Reverse was obtained by a separate gear with a separate lever for its control.

The Wilson gearbox was the most popular planetary gearbox; it continued to be developed with improvements being made to the controls to reduce pedal effort and travel, and generally make the box less cumbersome. Another problem that was overcome was that of the top gear cone clutch being prone to stick. This became dangerous when the fluid flywheel was in use because if the clutch stuck, and one preselected reverse gear and then accelerated the engine, the car would drive straight forward. This failing was removed when Pomeroy designed a multidisc clutch to replace the cone clutch.

Automatic transmissions

The Americans worked on slightly different lines and introduced two automatic change speed mechanisms in the early thirties. One, the Reo, was a two-speed gear with a wide difference between ratios. The lower gear was attained by an internal gear with an offset intermediate wheel that meshed with a driven annulus. A free-wheel clutch withstood the torque reaction of the intermediate member in low gear, and high gear was provided by a multiplate clutch engaged by fly-out weights. A change down from high to low gear did not occur unless the throttle was opened. This arrangement was only suitable in conjunction with a large engine capable of dealing with most conditions.

The other was the Chrysler arrangement with an automatic overdrive epicyclic gear used with a standard gearbox. The epicyclic was helical and the end thrust on the sun wheel and annulus was taken by a ball race on the tail shaft that carried it while the sun wheel was fixed to the casing. The overdrive was engaged by centrifugal force from fly-out weights each fixed with a locking arrangement to prevent hunting. Each engagement was positive and made when synchronisation occurred.

It was only when General Motors introduced a semiautomatic transmission on Oldsmobile cars in 1937 and Buicks did so in 1938 that mass production of three-speed epicyclics really got under way. This transmission consisted of two planetary gears, two multidisc clutches, two band clutches and two oil pumps to permit hydraulic control of the planetary gears. This arrangement still used a conventional clutch to provide the initial shift, neutral and reverse gear but in 1939 the fluid coupling was incorporated and the Hydramatic fully automatic transmission introduced.

The Hydramatic, the first successful passenger car automatic transmission, consisted of a fluid coupling, two epicyclic gear trains, two multidisc clutches and two hydraulic pumps to operate the epicyclic gear annulus brake bands (figure 13.7). This system allowed four forward speeds and reverse to be obtained. In forward drive the shifting of gears was done without interrupting the drive by the simultaneous action of the friction bands and clutches to produce an acceptable transition from one gear ratio to the next. The automatic gear changes were influenced by the degree of throttle opening obtained on the accelerator pedal. In third and fourth gears part of the power was shunted around the coupling thus increasing efficiency. For improved acceleration the selector lever was moved to 'Lo' and the transmission remained in first or second gear only. Reverse gear was obtained by manual engagement of a pawl with a reverse ring gear. In the parked position and the engine stationary the drive shaft was locked by moving the transmission control lever in reverse and thereby provided an effective parking brake.

Prior to this all the methods introduced to facilitate gear changing became insignificant compared with the synchromesh mechanism invented by the Cadillac division of General Motors and fitted in the 1928 Cadillac. This mechanism at last permitted easy and silent gear changing with a conventional gearbox.

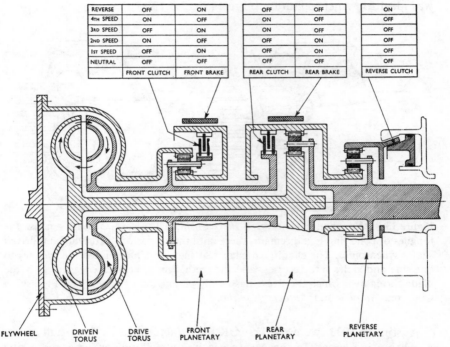

Figure 13.7 Hydramatic drive. By permission of The Council of the Institution of Mechanical Engineers (from *Fifty Years of Transmissions*).

Synchromesh

The synchroniser device consisted in principle of a mechanical unit that enabled the two gears to be meshed to be brought to the same rotational speed before the actual meshing took place. It was usual to use two clutch cones, one external and the other internal, and which was the sliding member. The first movement of the gear lever brought the driven internal cone member into contact with the outer cone member of the free gear so that the friction clutch brought the two members to run at the same speed. Further movement of the gear lever moved the gear wheel into mesh with an internal gear machined inside the driven gear in a smooth and silent manner. After engagement a spring device released the clutch and moved the internal cone member to its original position.

The Cadillac original synchromesh unit was of the constant-load type and the gears could be clashed if the gear lever was moved too hard in an attempt to make too rapid a change. This unit was introduced into Vauxhall Motors in 1931 and fitted to the Cadet three-speed gearbox in 1932. The Vauxhall arrangement was of the inertia lock type which overcame the problem associated with the Cadillac type and provided a gear change that was ideal for synchronisation of gears (figure 13.8).

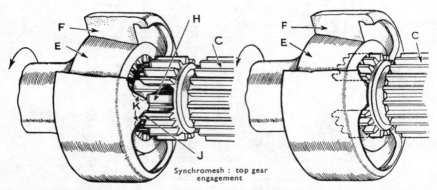

Figure 13.8 Vauxhall synchromesh mechanism. The sleeve C bears against the tongues of cone ring F and remains there until the speeds of the driving and driven shafts synchronise. The chamfered edge J of the sleeve has pushed the tongues to one side and slides into recesses, and the teeth sleeve becomes engaged. H is the blade spring over which the sleeve passes. By permission of Haymarket Publishing Co (from *Autocar* and *Autocar Handbook*).

Austin, in 1933, were the first UK manufacturers after Vauxhall to use synchronising cones. In the Austin gearbox a striking fork engaged a ring attached to projecting lugs on the sliding member splined to the gear shaft that carried internal teeth on each end. Inside the ring was a cylinder having internal coned faces which matched friction cones on the gear wheels to be meshed. Spring-loaded balls also engaged a shallow groove on the cylinder. The first movement of the gear lever brought the synchronising cones together and caused the gears to run at the same speed. Further movement of the gear lever overcame the resistance on the balls and completed the meshing of the gears.

At this time the inertia lock type was not used on many cars; most manufacturers preferred the constant-load type made by the Warner Gear Co in the USA. This consisted of bronze friction cones coming into contact under an inertial load from the movement of the gear lever from a neutral position. The member was moved by an inner toothed gear clutch actuated by spring-loaded plungers inserted in an internal groove in the clutch. Further movement of the gear lever caused the plungers to ride out of the groove and permitted the collar to slide on and allow the gears to engage positively. This arrangement was popular on many makes of car from Renault to Rolls-Royce.

The firms that used preselector gearboxes continued to use them but for most some form of synchronisation seemed to provide the most satisfactory solution to the gear change problem. Throughout the thirties there was a continuous trend towards synchromesh gears even if great diversity in practice existed. One problem with their use was that they required a bearing on each side of the gears to ensure good life and quietness during running. This necessitated some radical changes to the gearbox to accommodate extra bearings to support the shafts and ensure rigidity of the gearbox.

For these reasons synchromesh gears were only used on top and second of the three-speed boxes which were widely used in America and this was quite satisfactory on cars with a high power-to-weight ratio. On the Continent and in the UK some cars like Rolls-Royce and Bentley only used synchromesh on third and fourth for quite a time before extending it to other gears. Other firms like Alvis, and Hillman on the Minx, used synchromesh on all four gears. Some proprietary gearbox manufacturers like Moss and ZF used synchromesh on second, third and top although some ZF boxes had it on all four speeds but this arrangement was not common.

Other features

Another trend of importance in the USA in the late thirties was the moving of the gear lever from the floor to the dashboard or steering column. As Maurice Olley once remarked 'the gear stick couldn't find a place to live, so it dashed up the steering column and fixed itself there'. Methods of making a gear change with less force were also introduced. Studebaker, for example, had a gear lever that controlled a mechanism that selected the right striking rod in the gearbox, the gear being engaged with assistance from a vacuum cylinder. Hudson even used a combined electric and vacuum servo control to effect a shockless gear operation with little effort. Most manufacturers, however, still retained the centrally mounted gear lever until the outbreak of the Second World War.

By then the design of manual gearboxes themselves had settled down and helical gears were used for the constant-mesh arrangement with synchromesh, and although many double helical gears were used the trend was towards single helical gears. These were produced more easily and with greater accuracy as they allowed greater scope for gear cutting and final finishing. Helical gears were more silent than spur gears which was made possible by better control of the heat treatment process which resulted in so little distortion of the wheels that there was no need for them to be subsequently ground.

Developments From 1946–1960

Gearboxes

After the war three-speed gear boxes with synchromesh on second and third gears were used on American cars and also on UK Fords and Vauxhalls. The UK generally preferred four-speed boxes with a close ratio third to achieve maximum acceleration. In the USA four-speed boxes made a reappearance after many years and one of the most popular was the Borg Warner T-10 transmission used in the 1957 Corvette. Single helical gears were practically standard, with different helix angles for various gears to obtain a balance of end thrust on the layshaft. The layshaft ran on roller bearings with plain thrust washers. Greater detail was given to the lubrication system to ensure adequate lubrication of the plane journal and thrust faces of all gears. Continuous improvements in design and manufacturing techniques enabled greater loads and stresses to be carried, resulting in more efficient and compact units.

On most rear-wheel drive cars the gearbox main shaft was extended to the rear with lubricated splines on which the coupling flange could slide, the extended diameter of the hub being carried on a needle roller bearing. Low floor lines were obtained by tilting engine and gearbox downwards. Some vehicles, such as Lancia Aurelia and later Aston Martin cars, combined the gearbox and final drive with the rear axle. Many others combined the engine, gearbox and final drive at the rear. With front-wheel drive vehicles real progress for the future was achieved with the Mini, which had its gearbox moved below into the sump and arranged parallel to the engine.

On other than sports cars there was a move towards bench-type front seats so that steering column change became very general. Three-speed boxes usually had two striking forks with two striking rods going to the gearbox, each coupled to one fork. On four-speed boxes selection was obtained by means of a bell crank at the lower end of the column, engaging a sliding change speed rod. Other arrangements were also used and the shift ability of these devices varied enormously, especially after wear of components. They reached their lowest ebb when used with five forward speed boxes and it was a pleasure to see the gear lever revert to the floor.

Synchromesh was in general use but not widely applied on bottom gear although one exception in the UK was on the Standard Vanguard. Improvements to synchromesh systems were made to reduce operating loads on the gear lever and give quicker and more efficient gear changes.

With the cone type, cone areas were increased and in some cases cones were lightly loaded, but the contact pressure was supplemented by a baulking system that kept the dogs apart until they revolved at the same speed to give more rapid synchronisation and eliminate noisy gear changes. In the UK the use of spring-loaded balls in the inner cone member was widely used. The alternative baulk ring method of Porsche was introduced in 1947 and had the advantage of compactness as well as effectiveness. This had a synchronising clutch with internal gear teeth to enable it to engage with mating teeth on the gear to be engaged. During sliding the clutch mated with a baulk or synchronising ring to change its speed to that of the driven shaft and further movement brought the two gears into full engagement. A modified form using an anchored internal band brake together with a synchronising ring and thrust block was fitted on some BMC cars.

True five-speed gearboxes appeared in the early fifties but were mainly confined to sporting-type vehicles like Ferrari and Pegaso who pioneered their use, with Alfa Romeo and Aston Martin being manufacturers who adopted them as standard on saloon cars. In 1952, the Fiat 1900 used a friction clutch in series with a fluid flywheel and five-speed gearbox. Many companies, however, preferred the cheaper alternative of obtaining an extra gear or gears from a proprietary overdrive attached to the rear of a standard gearbox.

Overdrives

The widespread use of overdrives to reduce crankshaft speed and improve durability and fuel consumption was one of the features of the UK and Continental manufacturers in the fifties. Overdrives first appeared in the USA where units continued to be used again at different times until the seventies. Most units used epicyclic gearing. One of the most popular was made by Borg Warner which consisted of a planet carrier ring connected to the gearbox output shaft while the ring gear was attached to the outer race of a free-wheel unit and hence to the output shaft. When the sun wheel was arrested by a pawl actuated by a solenoid the planet carrier driven by the gearbox shaft rotated about the sun wheel taking the drive through the planet pinions to turn the annular gear and the output shaft to the propeller shaft. As the annulus rotated more slowly than the planet carrier, a lower gear ratio was obtained than in direct drive. A dog tooth ring on the planet gear could be made to engage teeth inside the planet carrier during axial movement of the sun wheel and enabled the epicyclic gear to be locked. The overdrive was then locked out and allowed over-run braking on long descents. Below vehicle speeds of 28 mph a free wheel operated in direct drive enabled gear changes to be made without the use of the clutch. For rapid re-engagement of direct drive for top gear acceleration, a kickdown switch was operated by depression of the accelerator pedal.

Another widely used design in Europe was the Laycock de Normanville (figure 13.9), which was available in various versions for vehicles of different engine capacity during the fifties. In these units a two-way cone clutch was attached to an extension of the sun wheel and was free to rotate on the input shaft. The clutch could engage either the casing or on to the annulus. In the latter position, the gear train was locked in direct drive. In

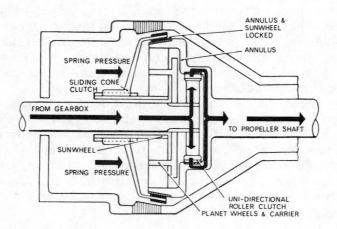

IN DIRECT DRIVE

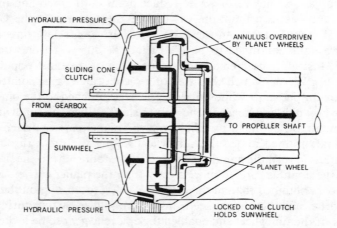

IN OVERDRIVE

Figure 13.9 GKN Laycock overdrive. Direct drive—cone clutch held to rear by spring pressure locking gear train. Overdrive—sun drive held stationary with planet wheels revolving round it, rotating the annulus. By permission of Laycock Engineering Ltd.

overdrive, the clutch cone was moved from the annulus against the casing to hold the sun wheel stationary. As the planet carrier was driven by the input shaft the planet wheels consequently revolve around the stationary sun wheels, and the annulus rotated at greater speed, that is, overdriven. Solenoid-operated selector valves were used to control the hydraulically operated pistons actuating the cone clutch. These overdrives were usually restricted to top gear operation on powerful cars, but on smaller vehicles units could be made to operate on any of the three gears, which was very confusing to the driver.

Semiautomatic transmissions

Automatic gear changing in this period was widely established in the USA but its acceptance on smaller cars was much slower in Europe where semiautomatic gearboxes continued to be developed. The pre-war Cotal electrical gearbox continued to be used until the demise of companies like Delage and Delahaye. Daimler in the UK persisted with the fluid flywheel with the Wilson preselector as did Armstrong Siddeley who for a time actuated the gearbox bands by electrical means. Armstrong Siddeley also used a centrifugal dry clutch as a means of coupling instead of the fluid flywheel. In the fifties attempts were made to obtain two-pedal cars by using an automatic clutch with a conventional synchromesh gearbox. Manumatic and Newton drives were examples and they used a centrifugal clutch and a vacuum-operated withdrawal mechanism. Systems like these were affected by fast idling speeds with a cold engine and also they had a tendency for clutch disengagement when running downhill on a closed throttle. More successful were the Daimler-Benz Hydrak and Fichtel & Sachs Saxomet transmissions. The Hydrak fitted to some later Mercedes-Benz types 219 and 220S consisted of a fluid coupling for take-up, and a single-plate dry clutch operated automatically by a vacuum servo unit to break the drive during gear changes. A free-wheel unit mounted on the engine side of the fluid coupling allowed the engine to over-run the driven members but locking action of the free wheel prevented the driven part from over-running the engine. Thus engine braking could be provided below the stall point of the fluid coupling. In the Saxomet transmission a centrifugal clutch was fitted and vacuum servo units used to disengage the clutch for a gear change and also to control the carburettor throttle position during gear changing. Ferlec made by French Ferodo used an electromagnetic clutch fitted with a metallic friction material and was offered as an optional feature on Renault and Panhard cars. A Ferlec clutch mounted behind a three-element torque converter was also incorporated in the Renault Transfluid transmission, the stator being mounted on a free wheel to enable the unit to become a fluid coupling when torque multiplication ceased. Drive was taken from the clutch to a three-speed gearbox that included a locking device for parking purposes. An electromagnetic powder clutch was the key component used in Selectroshift transmission developed by Smith Ltd which was used as optional equipment for the BMW Isetta 300. Various operational problems were experienced with these devices and all perished under the onslaught of American developments in automatic transmissions.

Automatic transmissions

Developments in the USA. Automatic transmissions had been proved by General Motors and their success caused other American manufacturers to follow suit. These transmissions consisted of some form of hydrokinetic coupling connected to a planetary gearbox. One type had a fluid coupling coupled to a three- or four-speed gearbox. Other units consisted of either a multistage torque converter giving variable torque ratios or a three-element torque converter coupled to a two- or three-speed gearbox.

After the war the fluid coupling Hydramatic transmission was basically the same design as that of 1939 although reverse gear was now engaged by hydraulic operation of a cone clutch to give an easier and quicker reverse application. The parking lock was still obtained by means of a mechanical pawl operating when the engine was stopped. A later development in 1956 was the addition of another fluid coupling to give improved performance.

In 1948, General Motors introduced the Buick Dynaflow which was the first successful torque converter drive used on passenger cars. The five-element torque converter gave a torque multiplication of 2.25 in the stalled position and gave the car reasonable acceleration and smooth shifts. An auxiliary hydraulically controlled compound planetary gear was provided to give neutral, reverse and extreme low gear operation.

Other multielement torque converters soon followed the Dynaflow. In 1949 Packard introduced the Ultramatic transmission which had a mechanical direct drive clutch together with the torque converter. A lock-up clutch was also used in the Borg-Warner automatic transmission introduced by the Studebaker Corporation in 1950. This transmission was different from others in that the torque converter action was accompanied by a gear reduction during acceleration and the unit shifted into direct drive which locked out the torque converter. 1951 saw the introduction of the Fordomatic and Mercomatic transmissions which were manufactured by both Ford and Borg-Warner. Chrysler also introduced a limited number of Torqueflite transmissions in 1956. Most units used a three-element torque converter coupled to a three-speed power shifting planetary gearbox. The drive was taken through the converter, which had a low ratio of approximately 2.4 to 1 with an intermediate ratio of about 1.45 to 1.

General Motors also introduced the Powerglide transmission for the Chevrolet car but as these were pretty inefficient with six-cylinder engines, in 1953 a standard three-element converter was incorporated. Other manufacturers also used similar two-speed transmissions in their lower power range of vehicles.

Improved performance was obtained in the Dynaflow by the introduction in 1955 of a variable blade angle stator with either 'high' or 'low' angle to give greater efficiencies. A further change occurred in 1956 with the use of a five-element converter having a fixed blade stator between first and second turbines of a twin unit and a variable angle stator between the second turbine and pump. In this way higher starting torque ratios were obtained to avoid the need for more direct gear ratios.

Extensive work also went into the development of the friction elements controlling the planetary trains and to modifications of the hydraulic controls to give smoother gear changing. Manufacturing costs were reduced by simplifying production techniques such as in the fabrication of the vane rotors of torque converters by Chevrolet and also in the casting of the hydraulic control circuitry and general weight reduction. Earlier operational problems were soon overcome and progress was such that automatics dominated the USA industry by the end of 1959.

Developments in Europe. Meanwhile attempts were made in Europe to establish alternative automatic gearboxes for light cars. Smiths Autoselective transmission used two magnetic powder clutches in conjunction with a conventional three speed box and in the form of Easidrive was used as optional equipment by the Rootes Group. Another transmission that avoided the need for either a torque converter or fluid coupling was the Hobbs Mecha-Matic. This used a four-speed epicyclic gear train with the forward and reverse positions being obtained by hydraulic operation of clutches and brakes. This transmission was used on the limited numbers of Lanchester Sprite cars manufactured. These developments made little headway compared with that of manufacturers who preferred to adapt American transmissions to their vehicles. One of these was Rolls-Royce who in 1952 fitted a modified version of General Motors Hydramatic transmission to export models and later manufactured their own version under licence. The modified transmission included the extra drive position to hold third speed for better acceleration and for braking on modest slopes. The Rolls-Royce friction brake servo and the oil pump for ride control of the rear shock absorbers were also arranged in the transmission. Similarly other larger and expensive cars began to fit automatic transmissions and the experience gained with these units enabled other more compact designs to be developed for smaller European cars.

One of the most popular units was that of Borg Warner, who started by manufacturing established designs in the UK but who later developed their own design that became standardised, or widely used as optional equipment on a large number of cars. Initially these used a three-element torque converter coupled to two epicyclic gear units that incorporated free wheels.

The stator of the converter was connected to the casing through a free wheel so that after torque multiplication ceased the converter became a fluid coupling. The locking of the stator rotor to the stationary casing during the converter phase was achieved by a fully phased, one-way clutch.

Three forward and one reverse gear were provided by the planetary transmission. One forward ratio was a direct drive obtained automatically by the single-plate clutch. On the other gear ratios power flowed through the torque converter, which had a maximum multiplication of approximately 2:1, so matching the requirements of smaller engines than those used in the USA. One variant of this transmission used in the Mercedes-Benz 300 in 1958 was the provision for a start in intermediate gear at light throttle openings with automatic starting in low gear for wider openings.

In 1958 one of the most interesting developments was the introduction of the DAF Variomatic infinitely variable gear transmission that gave automatic gear ratio changes to meet varying speed and torque conditions. The drive from the engine was taken through a centrifugal clutch and propeller shaft, and transmitted to a transverse cross shaft through bevel gears. Attached to the cross shaft were two divided inboard V pulleys fitted with belts each driving a separate rear wheel. The change in effective diameter of the pulleys changed the gear ratio as desired, and the rear V pulleys were spring loaded to maintain belt tension. The front pulley's sideways movement was controlled by centrifugal weights revolving with each pulley supplemented by the controlling action of engine vacuum servos. Unlike the previous automatic transmissions discussed this was truly stepless and ideally suited for small, low-powered vehicles. Continuously variable transmissions were later, in the eighties, developed for more powerful cars by suitably reinforcing the belts with metal.

Developments in automatics were such that the reluctance to use them on small cars in Europe was gradually being overcome in the late fifties.

Developments From 1960

Gearboxes

Despite the worldwide trend to automatic gearboxes, manual gearboxes continued to be developed and improved to give greater efficiencies and adaptability. Synchromesh was now fitted on all forward ratios and very effective baulk-type synchronisers have been introduced to make gear shifts smooth and rapid.

Transmission layouts have become very variable. Some manufacturers mounted the gearbox ahead of the engine on their front-wheel drive cars, others mounted it behind the engine, and there are also rear transverse engines and transmission arrangements. Detailed refinements of the various four-speed gearboxes continued. For example, in the front-wheel drive Triumph 1300 the engine was above the final drive and the gearbox below the clutch, flywheel and transfer gear. This layout made it possible to remove the clutch without disturbing the engine or gearbox. An extension of the hypoid final drive pinion acted as the third motion shaft and carried all the main gears and synchromesh devices in addition to the constant-mesh pinions.

In some designs, for example, the Ford Anglia and later in other models, the third motion shafts and rear end plate were made a complete subassembly. This subassembly could be inserted in the rear extension housing and facilitated fitting. This was also a feature of many American and European designs. Later, in the Escort, Ford introduced a remote gear change mechanism that consisted of a novel single-cranked selection shaft (figure 13.10).

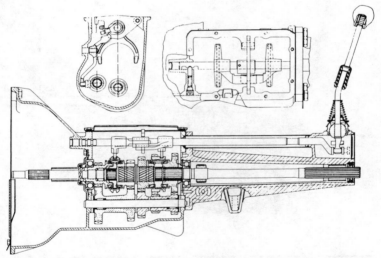

Figure 13.10 Ford Escort gearbox—the actuation lug is pinned to the single selector shaft on which the forks slide. An interlock plate keeps one held while the other is moved to obtain a gear. By permission of Reed Business Publishing Ltd (from *The Automobile Engineer*).

Gears, shafts and sleeves were made of carbonised steel and suitably hardened, and running faces were ground to a 10 microinch surface finish. Traditional helical grooves cut around the peripheries of shaft and sleeve continued to be used to lubricate the gears and needle roller bearings mounted on the layshaft. Attention to such details enabled four-speed gearboxes to cope with the higher engine torques and powers being developed.

The four-speed gearbox was traditional for all but the most expensive cars. In the seventies, with the need for fuel economy, most car manufacturers began to incorporate an overdrive gear ratio in the gearbox, with fourth gear usually direct drive (figure 13.11). The overdrive fifth gear permitted lower engine speeds at high road speeds, thereby using less fuel and causing less mechanical wear. All forward gears were provided with synchromesh and reverse also in some boxes. The use of five-speed gearboxes meant longer shafts which were more difficult to accommodate in the vehicle and the shafts were more liable to deflect under load, which increased noise and vibration. Some boxes overcame this problem by using taper roller bearings to take both axial and radial loads and yet be of smaller diameter and have better life than equivalent ball-type bearings. Other arrangements included the use of a live layshaft to give adequate stiffness as the gears and shafts can be made larger in diameter. Another feature of the designs is the use of an oil pump to lubricate effectively all gears moving

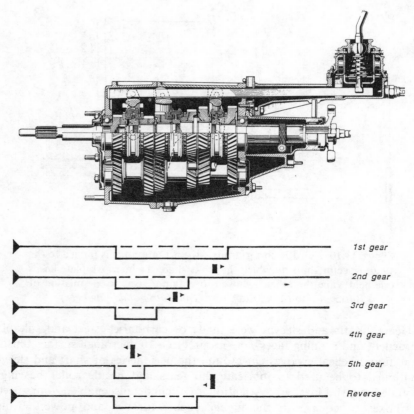

Figure 13.11 Five-speed gearbox and power flow diagram of passenger car transmission (ZF). By permission of ZF Gears (Great Britain).

relative to the motion shaft. Various configurations are possible to match different drive systems with either central shift controls or remote controls for rear-engined cars. The use of heat-treated alloyed stainless steel and helically cut gears with ground or shaved tooth faces has reduced noise. Aluminium housings are now common and these have been made lighter but the gearboxes themselves are more compact and stiff to carry the increasing demands made from higher powered engines. Production methods to make the product cheaper with improved quality are also constantly being made by the manufacturers.

Automatic transmissions

For a short period in the sixties there was a vigorous re-emergence of semiautomatic transmissions in Europe. These included the Ferodo Verto and various Fichtel and Sachs designs which used a torque converter and servo-actuated diaphragm spring clutch and synchromesh gearbox and they found applications in Simca, NSU Ro80, Porsche and Volkswagen 1500 cars.

The main emphasis, however, was on automatics with a smaller Hydramatic 61-05 model being introduced which was used as an option by Vauxhall on their Velox and Cresta from 1960–65. The fluid coupling included a vaned element to multiply the torque and boost low-gear get-away. Torque converters were introduced in 1964 as these offered greater potential for development and they were then used by all General Motors' car divisions. The units became known as Turbo Hydramatic transmissions. The torque converter consisted of three elements and some units consisted of a variable-angle stator blade but later designs settled on stator blades of fixed pitch. Units were to provide initially three forward speeds and reverse. Other US manufacturers introduced new three-speed transmissions that were to give better starting torques and response for overtaking compared with earlier units. Two-speed transmissions were to cease production.

In Europe Mercedes-Benz introduced a four-speed transmission with a gearbox which was very flexible indeed and used most of the engine's full power in all gears. ZF also introduced an automatic transmission with a Ravigneaux compound planetary gear train, which had planetary gears which meshed not only with the ring and sun gear but also with each other. The feature of this four-speed box was the use of plate clutches, disc brakes and spring clutches thus avoiding the need for band brakes. Borg-Warner also continued to develop transmissions to give a wider range of selector characteristics, units being developed for small engines in the 1–2 litre class using multidisc brakes instead of the band brakes as in the earlier model 35 and its heavy-duty derivatives.

One natural development with automatic transmissions was the move to four forward speeds with top gear being an overdrive ratio (figure 13.12) as in manual gearboxes. An innovation included the fitting of a lock-up clutch that prevented any slippage in the converter and cut down power losses, reduced fuel consumption and improved efficiency. This application of the lock-up clutch is made possible by redirecting the converter feed fluid to force the pressure plate against the converter-housing cover and lock the torque converter. The engine is then driving the transmission mechanically.

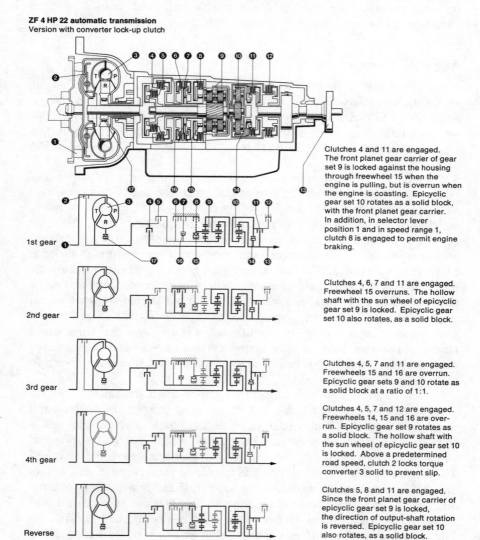

Figure 13.12 ZF 4 HP 22 automatic transmission with converter lock-up clutch. By permission of ZF Gears (Great Britain).

Automatic transmissions have also been adapted for cars with front-wheel drive and with transaxle layouts. Other detailed refinements continue to be made and the basic model of an automatic transmission can usually be tuned to match a particular engine and vehicle. Renault in the late sixties introduced electronic controls on the automatic transmission of the Renault 16TA. More recent developments include the replacement of hydraulic operation of gear shifts by electric solenoids that are controlled by a microprocessor system. Sensors measure engine speed, load speed and temperature, and this information is transmitted to the microprocessor that signals when the gears are to be shifted. Most gear shifting is at present still done by a hydraulic control system.

Chapter 14

Transmission and Axles

Transmission

Benz connected engine and transverse countershaft by a system of belts and pulleys which was based on the belt drive from overhead shafting to machine tools used in contemporary workshops. The countershaft was split by a differential gear, and the drive from the countershaft to the rear wheels was via sprockets and chains. Most cars had chains until well past the turn of the century. Unsprung weight was low, and the chain and sprockets were a convenient means of gearing down engine speeds. On the other hand, chains wore and became noisy, as well as requiring cleaning and lubrication so that by 1910 they were obsolete on private cars, though still used on some heavy limousines. Chain drive lasted longer on commercial vehicles and of course is still used on motorcycles.

The other way to drive the back axle was through a longitudinal propeller shaft with a worm and worm wheel, or pinion and crown wheel, driving half-shafts to the rear wheels through a differential gear.

The suspension springs allowed the axle to move up and down, so universal joints had to be fitted in the drive line to accommodate such movements; in addition the shafts were generally splined to allow for back and forward movement of the axle when the springs deflected. If two shafts, connected by a universal joint, are not in line the second shaft will not rotate at constant velocity, but if three shafts and two universal joints are used the third shaft will rotate at constant velocity provided first and third shafts are parallel. As this could generally be arranged, at least approximately, two universal joints were generally fitted, one at each end of the propeller shaft.

The early pot-type universal joints (figures 14.1 and 14.2) could give trouble; they were difficult to lubricate so that they wore and became noisy, and snatched. By 1914 ring joints and spider joints were widely used; the shafts terminated in forks which carried trunnion bearings with all four bearings being supported by a common ring (figure 14.3) or by a cruciform spider. There was provision for lubricating the bearings; indeed the whole joint could be surrounded by two contacting hemispherical shells, one attached to a flange on one shaft, the other to the other shaft and the

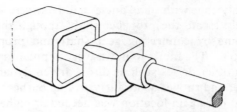

Figure 14.1 Early pot-type universal joint. By permission of Reed Business Publishing Ltd (from *The Automobile Engineer*).

Figure 14.2 de Dion universal joint. By permission of Reed Business Publishing Ltd (from *The Automobile Engineer*).

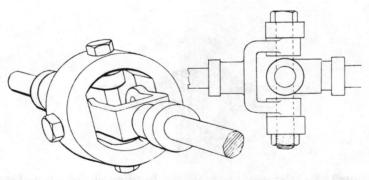

Figure 14.3 Ring-type universal joint. By permission of Haymarket Publishing Co (from *Autocar* and *Autocar Handbook*).

internal space filled with grease. In the thirties the journal bearings were replaced by needle bearings and this type of universal joint (figure 14.4) is essentially that used on propeller shafts today.

In the twenties the fabric joint challenged the metal universal joint on smaller cars. The two shafts were bolted to leather—later rubberised-fabric—discs by two- or three-armed spiders or forks (figure 14.5). Fabric joints did not require lubrication, were silent and gave torsional damping, but they could only cope with small angles and relatively low speeds as they did not accurately locate the propeller shaft, though the latter difficulty could be overcome by connecting gear shaft and propeller shaft by a ball-and-socket joint. The fabric joint was very popular in the early twenties and was used on other drives such as direct driven dynamos. Later joints had a metal spider and rubber quadrants carrying bushes to which the shaft forks were bolted but again location was needed at higher speeds.

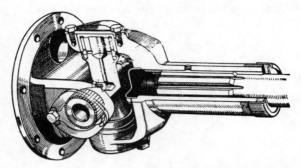

Figure 14.4 Spider-type universal joint with needle bearings. By permission of Haymarket Publishing Co (from *Autocar* and *Autocar Handbook*).

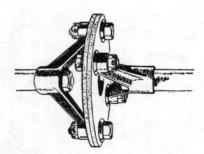

Figure 14.5 Fabric-type universal joint. By permission of Haymarket Publishing Co (from *Autocar* and *Autocar Handbook*).

Shafts could give whirling problems if their natural frequency in bending approached their rotational speed and so on long chassis the propeller shaft was divided into two, and two sets of universal joints used with a central bearing mounted on a cross member. The centre bearing had to be carefully rubber mounted to avoid vibrations, for one problem of the thirties was that, as engine exhaust and tyre noise were progressively reduced, other vibrations hitherto unnoticed became apparent and had to be dealt with.

Differential Gear

The designer had to allow for the difference in speed of the two driven wheels when the vehicle went round a corner, and this was achieved by including a differential gear between the two wheels.

On some small, light cars the differential was dispensed with and the owner had to put up with tyre scrub and high tyre bills. When chain drive was used one wheel could be driven and the other free (an arrangement which apparently worked quite well on good road surfaces), or operated by a clutch. Attempts were also made to use ratchets on some early cars.

The differential itself has a long history; the Greeks may have used differentials in some mechanisms, and in more recent times they have been used on orreries and clocks. In medieval times the Chinese had a chariot that carried a pointer that always pointed to the North. The pointer was not a magnet and George Lanchester suggested that it was worked by a suitably geared differential. When both road wheels rotated at the same speed the pointer stayed put, but when the vehicle turned the pinion the crown wheel moved the pointer through such an angle as to keep it pointing in the same

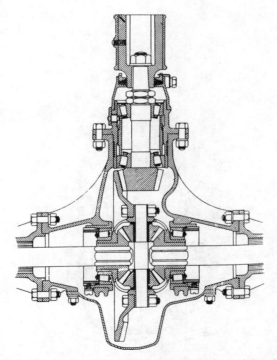

Figure 14.6 Bevel gear differential. By permission of Reed Business Publishing Ltd (from *The Automobile Engineer*).

direction. O Pecquer in 1827 patented a differential gear to enable the wheels of a self-propelled vehicle to turn at different speeds when cornering. Pecquer apparently hit on the idea while making a model for a museum of Cugnot's three-wheeled vehicle and pondering on its shortcomings, rather like Watt conceiving the idea of a separate condenser while refurbishing a model atmospheric steam engine. R Roberts patented a similar device which was used by F Hill on his steam carriage of 1840, and from then on differential gears were used on most road carriages. Consequently the differential gear, the 'balance gear', or 'jack in the box', was a reasonably familiar piece of machinery when Benz and the other early car designers adapted it for the motor car.

There were two types of differential; in one (figure 14.6) pinions carried on a spider or cage on the crown wheel engaged bevel gears on the ends of the half-shafts; in the other, pairs of spur gears were used instead of bevel pinions, the spur gears meshed over part of their length and also engaged corresponding gears on the half-shafts (figure 14.7). The spur gear differential was cheaper and could transmit larger forces then the bevel gear version, but was bulky and heavy and gave way to the latter.

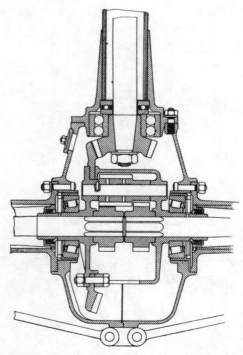

Figure 14.7 Spur gear differential. By permission of Reed Business Publishing Ltd (from *The Automobile Engineer*).

The differential has one big drawback. If one wheel is on a surface of low adhesion, such as an icy patch, it can take all the drive and spin madly whilst the other wheel does not even rotate. Attempts were made to design 'limited slip' differentials in which torque was transmitted to the wheel that was not slipping. This was achieved by providing resistance to motion between differential cage and its housing by cone or disc clutches, or wedge and pawl cams, which were actuated by the relative motion of the cage with respect to the rest of the mechanism.

The wheel carrying the differential cage and gear was driven by a belt and pulleys as on the Benz cars, or by a chain, or if a propeller shaft was used, by a worm gear or a crown wheel and bevel pinion.

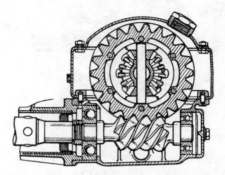

Figure 14.8 Worm wheel and worm wheel drive. By permission of Haymarket Publishing Co (from *Autocar* and *Autocar Handbook*).

Worm and worm wheel (figure 14.8) were quiet even when worn, and were therefore used on a number of quality cars; the gears had to be accurately cut and mounted which made them expensive. Lanchester used Hindley's hour glass worm which was even more expensive to manufacture and fit. The worm had the disadvantage that if it was above the worm wheel lubrication could be a problem, if it was below ground clearance was reduced so that engine and gearbox had to be canted. A few firms continued to use worms for many years and Peugeot did so into the sixties. Bevel gears (figures 14.6, 14.7) were cheaper than worm gears and soon became the usual arrangement, but they could become very noisy when the teeth wore. Spiral bevels (figure 14.9) were quieter, for a larger surface was in contact so that the load was shared between more than two teeth, and they gradually displaced straight bevels. Bollée used spiral bevels in 1896 whereas some manufacturers still used straight bevels into the twenties. The Gleeson Company built the machines for cutting spiral bevels. Citroën used a double spiral, or herring bone, bevel; he had made spiral gears under licence for many years before he started making motor cars.

In 1927 Packard adopted the hypoid gear (figure 14.10) in which the line of the pinion was offset from the centre of the crown wheel. The floor could thereby be lowered, or the transmission tunnel dispensed with. The teeth in the hypoid gear, however, slid as well as rolled on one another which made lubrication a problem, but it was eventually solved by developing special extreme-pressure lubricants. Again Gleeson built the gear cutting equipment. The hypoid was slowly accepted and after the Second World War became almost standard.

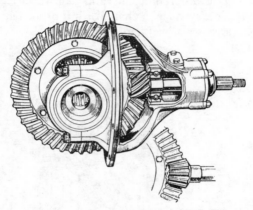

Figure 14.9 Spiral bevel crown wheel and pinion drive. By permission of Haymarket Publishing Co (from *Autocar* and *Autocar Handbook*).

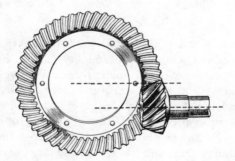

Figure 14.10 Hypoid gear. By permission of Haymarket Publishing Co (from *Autocar* and *Autocar Handbook*).

Back Axle

The axle casing was built up from tubes and castings and this type of construction was used by some manufacturers for many years. Banjo back axles were made well before the First World War and eventually became popular for axles with bevel gears. The steel pressings, welded or riveted together, were shaped to expand from the wheel bearings to a central more or less spherical chamber, one side of which was open to admit the crown wheel and differential assembly. The other side was sometimes open and covered by a shaped inspection plate. The casing therefore looked rather like a double-ended banjo.

The input (or pinion) shaft was mounted on ball and later taper roller bearings in a housing attached to, or cast with, a circular disc, the other side of which carried brackets for the crown wheel and differential ball bearings. The disc was bolted to the banjo casing. The carrier axle was used in the USA in the thirties but not in the UK until after the fifties. This was a development of the built-up axle in that the crown wheel and pinion were mounted in a malleable cast iron housing and the half-shafts carried in steel tubes pressed and welded into position.

The fully floating axle in which the axle casing alone carried the weight on the corresponding wheel was used on most cars before about 1910 but it was progressively replaced by the cheaper, lighter three-quarter and semifloating axles. On a fully floating axle (figure 14.11) the wheel was carried by two bearings, rotated by splines or dogs on the hub end cap. In the three-quarter floating axle (figure 14.12) the wheel was carried by a single bearing in the plane of the wheel centre line and so carried bending moments caused by side thrusts only. However, in the semifloating (more logically non-floating) axle (figure 14.13) the wheel was again carried by only a single bearing but as the wheel completely overhung the bearing the weight of the car fell on the half-shaft which was therefore subjected to bending as well as torsional stresses. Straightforward ball bearings were soon replaced by bearings that would take thrust, and by taper roller bearings. The live back axle has not

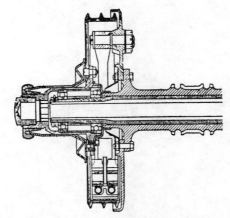

Figure 14.11 Fully floating axle. By permission of Haymarket Publishing Co (from *Autocar* and *Autocar Handbook*).

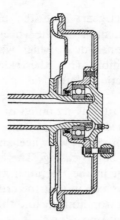

Figure 14.12 Three-quarter floating axle. By permission of Haymarket Publishing Co (from *Autocar* and *Autocar Handbook*).

changed in principle over the years; it has had to carry larger and larger torques at higher and higher speeds and still not increase in weight and bulk; this has been achieved by using improved materials, bearings and seals, and by improvements in production methods which have permitted manufacture to closer tolerances. Live axles on FWD cars are generally very simple beam axles.

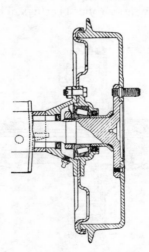

Figure 14.13 Semi-floating axle. By permission of Haymarket Publishing Co (from *Autocar* and *Autocar Handbook*).

Locating the back axle

The springs should constrain the back axle to move in a vertical plane. It should not twist about a vertical axis through its centre, if, for example, one wheel hits a bump, and so cause the rear wheels to have a steering effect. It should also not twist about the axis through the rear axle itself, because of torque reaction, when the vehicle is accelerating or braking for this again may affect suspension geometry.

The simplest way to locate the back axle is with the leaf springs alone, the so-called Hotchkiss drive (which was first used on the 1894 Renault) which worked reasonably well with small, light vehicles. The torque reaction, however, tended to twist the springs into an elongated S form so an early modification was to fit torque rods. These were anchored to top and bottom of the differential housing and shackled at their front ends to a point on a cross member of the frame near the front universal joint. Another approach was to use a torque tube (figure 14.14); the propeller shaft was carried in a large-diameter tube rigidly attached to the differential housing at one end and hinged to a cross member of the frame at the other. The hinge was later replaced by a ball on the end of the tube and the ball engaged a steel cup. The propeller shaft had only one universal joint at the pivot end and in the plane of the hinge, but because the difference in angle between gear shaft and propeller shaft was small and did not change much this was not normally a problem.

As well as torque rods or tube the car could be fitted with radius rods to keep the axle housing perpendicular to the centreline of the vehicle. These extended from the ends of the housing to a point on the cross member near

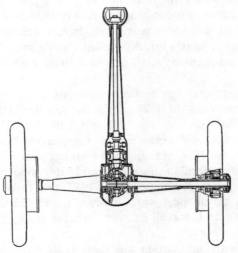

Figure 14.14 Torque tube of about 1910.

the gearbox universal joint. The arcs of movement permitted by the various rods had to be as consistent as possible with that of the springs themselves. Cars that do not have IRS still locate the rear axle with various arrangements of rods.

Semitorque tubes were used on some cars in the sixties. The short tube pivoted at its front end in a rubber bearing rather like a propeller shaft centre bearing, and was connected to the gearbox tail shaft by a shaft carrying universal joints and a muff spline joint. The tube was lighter in weight than a full torque tube and the rubber bearing transmitted less noise to the vehicle structure.

Front Wheel Drive

A number of attempts were made to drive the front, instead of the rear, wheels and no less than a hundred patents were described in a list of 1908. However, the complications of driving a steered wheel defeated most inventors and the obvious advantages of FWD—no heavy drive shaft and consequently a lower floor and lower centre of gravity (CG)—did not appear great enough to justify the effort and expense of developing FWD so that the initial impetus came from racing.

J W Christie built some large racing cars with FWD in the US before the First World War and in 1925 Alvis started building FWD racing cars in the UK, the attraction of FWD again being a saving in weight and a lowering of the CG but also, because the driving wheels were in front of the CG and the weight of the car concentrated over the driving wheels, more stability under power. These racing cars (the later ones of which had IFS) were sufficiently successful for the firm to produce sports cars in 1928 based on them but the sports cars were intended for experienced drivers interested in competition work, and not many were sold. As related elsewhere (p 294) after one car was involved in an accident Alvis returned to making more conventional cars.

Other manufacturers, however, became interested in FWD about this time, Cord in the USA in 1929, for example, and DKW ('Das kleine Wunder'—'The little Wonder') in 1931 and Adler in 1932 in Germany; but the car that made the most impact was the Citroën, the 'Traction Avant' car of 1934. The Citroen 7CV, not only had FWD but also unitary construction and independent suspension. The 7CV and later models were eventually accepted, remaining in production until 1957, and FWD was so successful that it was used on all post-war Citroëns. Other French manufacturers followed Citroën's lead, and all Renault cars, for example, had FWD by the end of the fifties.

One problem with the Citroën was that, as its gearbox and differential were ahead of the engine, the car was long with a large turning circle, and it was difficult to arrange a satisfactory linkage between gearbox and gear change lever. These drawbacks were overcome in the BMC Mini of 1959 by mounting the engine transversely. The Mini was a landmark. It was so capacious for its size, with such excellent handling and cornering, that the 1100 and 1800 models were designed on similar lines and FWD, often in conjunction with a transverse engine, became the norm for small cars.

Drivers had to adapt themselves to the rather different handling behaviour of the FWD car. For example, a FWD car tends to understeer if the throttle is opened when the car is going into a corner and not oversteer like a conventional car.

Once the mechanical problems of transmitting torque to steered wheels had been solved FWD was progressively adopted on larger cars. It had, however, previously been used on powerful cars; the engine of the pre-war Cord developed 190 bhp and the 7.4 litre engine on the 1966 Oldsmobile Toronado 375 bhp. The Toronado did not remain in production long, and except for compacts based on European cars FWD did not reappear on US domestic cars until the eighties.

Universal joints for front wheel drive

The main problem, after the packaging of differential, gearbox and engine, was the universal joint. As mentioned above, constant velocity is obtained with ordinary joints when first and third shafts are parallel but with FWD, the shafts no longer remain parallel when any steering lock is applied.

Alvis did not attempt to obtain constant velocity and used pot joints which, however, did allow for fore-and-aft movement, that is, they could 'plunge'. Citroën used a double Hooke's joint, the two spiders being almost side by side in the one housing. The Bendix-Tracta joint had an ingenious arrangement of components that slid in grooves in one another instead of simple spiders, but was again two joints in the one assembly.

A universal joint will transmit constant velocity if driving contact is restricted to the plane which bisects the angle between the two shafts and is perpendicular to the plane of the shafts. In the Zreppa joint, invented in the mid-twenties, one shaft terminated in a spherical member, and the other in a spherical cup, and balls running in grooves cut in their surfaces transmitted the drive between sphere and cup. A cage and control link (figure 14.15) held the balls in the bisecting plane, the link causing the cage to rotate through an angle half the deflection of the shaft. The Bendix-Weiss joint similarly had balls to transmit the drive between the two surfaces with both the latter being grooved in directions such that their intersections and therefore the corresponding balls lay in the bisecting plane.

The pre-war CV joints gave trouble at higher speeds and a lot of development work was needed to make suitable joints for the post-war FWD cars.

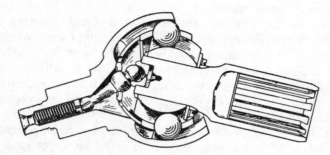

Figure 14.15 Zreppa constant-velocity joint. By permission of Chapman and Hall Ltd (from A W Judge *Motor Manuals*).

In the Birfield joint (figure 14.16) the outer member was cylindrical, the inner spherical and balls running in grooves cut in their surfaces transmitted the torque. The balls were held in a third member or cage, shaped to roll between the other two members but its centre of curvature was slightly offset to keep the plane of the balls in the median plane of the inner and outer members. The whole inner assembly could roll along the grooves in the bore of the outer member and so there was no need for splines. The outboard joint of the half-shaft was one of the constant-velocity type. The inboard joint, because it had to cope with suspension deflections only, was at first a simple splined Cardan joint, but to overcome the limitations of the latter and, in particular, the friction of the splines, a number of plunging joints were developed. The de Dion pot joint was modified and roller bearings fitted to reduce friction. The bipod joint did not give constant velocity but the tripod arrangement with a three-armed spider gave a more

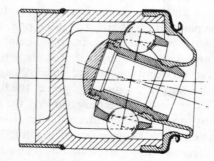

Figure 14.16 Birfield constant-velocity joint. By permission of Butterworth Scientific Ltd.

uniform velocity. Similarly the rubber doughnut joint in which a hexagonal rubber block was sandwiched between two triangular flanges or spiders, one upside down with respect to the other, gave approximately constant velocity and also permitted plunging. In another joint using rubber the arms of the spider were mounted in rubber sleeves with one pair of sleeves clamped to a flange on the input shaft and the other pair to a flange on the output shaft.

Rear-engined Cars

Another way to dispense with the propeller shaft and obtain more body space was to revert to the rear-engined layout used on most cars before the adoption of the Système Panhard. A few oddities like the Trojan and the GKW had rear engines, and rear-engined cars were produced by Mercedes (the 130) in Germany and Tatra in Czechoslovakia in the thirties. The most famous of all rear-engined cars was also designed in Germany in the thirties. Porsche had designed a Volksauto for Zundapp and later modified the design for NSU, but although prototypes were built neither firm had risked production. Hitler, however, became interested in the concept of a cheap mass-produced German car and on hearing what Porsche had done he wanted Porsche and the German automobile industry to collaborate in designing and producing a Volkswagen—a people's car. Progress was, however, slow and held up further by the war but when it did appear the Volkswagen, developed from the NSU design, was phenomenally successful. More than twenty million Beetles were eventually built (and the car was still being made in Mexico in 1989) but this was because of clever marketing combined with the detail design and workmanship of the cars, not because of their general design or rear engine. The Beetle was tail heavy and its handling was poor and, indeed, could be dangerous. The centre of pressure on the side of the car was ahead of its centre of gravity and so in gusty winds the vehicle tended to turn away from the wind, making the steering unstable.

The success of the Beetle caused a number of small rear-engined cars to be made by Renault, Simca, Fiat and other manufacturers. In the UK Hillman made the Imp. The only US rear-engined car was the much bigger GM Corvair which was condemned because of its handling. It was 'unsafe at any speed' according to Ralph Nader.

Four-wheel Drive

Four-wheel drive (4WD) had been used on off-road vehicles and some CVs for many years in order to obtain maximum traction, and had also been tried on racing cars, for example the 1931 Bugatti Type 35. During the

Second World War the 4WD Willys-Overland Jeep was very successful and it inspired a number of four-wheel drive vehicles. The front wheels, driven by a transfer box, were often engaged only when extra traction was required on rough ground, or for stability on icy roads. On ordinary roads only the rear wheels were driven. After the war attempts were made to design cars in which all four wheels were permanently driven. The easiest way to do this is to interconnect front and rear wheels by a differential as on the Land-Rover, but if one pair of wheels loses adhesion it takes all the torque. This can be overcome by giving the driver the means of locking the differential, but Ferguson on their experimental cars of the fifties connected input and output shafts for each axle by two gears which were higher geared than the differential and which incorporated a free-wheel mechanism in the gear on the output shaft. If either output shaft ran at the same speed as the gear revolving it the free wheel locked and prevented the shaft from rotating any faster. To prevent wheel lock during braking a mechanical Maxaret antilock system was fitted (p 353) but as this operated on the select-low principle, stopping distances could be increased. The 1967 Jensen FF (Ferguson Formula) had a similar system but with an electronic antibraking system.

Later systems dealt with the locked-wheel problem by connecting front and rear wheels by a viscous coupling in parallel with the differential; when the speed differential is low the torque transmitted by the coupling is very small but it increases rapidly as the speed differential increases.

Interest in four-wheel drive for passenger cars was stimulated by the 1980 Audi Quattro and the Subari and 4WD is now available on a range of cars from the Fiat Panda to the Porsche 959. On many of the vehicles the second axle is engaged as required by the driver, as on the early systems, on a number the engagement is automatic, and on some it is permanent.

Chapter 15

Chassis Frames

Most frames at the turn of the century were made cycle-fashion of round tubes brazed together, brackets being brazed on to the tubes to allow for the attachment of other parts.

Larger, heavier, cars followed coach construction; a wooden frame reinforced with flitch plates was used and later the frame was sometimes armoured with channel steel. The earliest steel channel frames were made of cold-rolled parallel-section steel, and the side channel/cross member junctions were reinforced with gussets (figure 15.1). By about 1903 large enough presses were available for channel steel to be made by pressing, with the result that soon most frames were of this form.

Engine and gearbox were usually carried on a separate subframe, a practice continued by some manufacturers for many years (figure 15.2).

As the wheel base of the car was increased frames became longer and wider. Some attempt was made to stiffen the frame against racking and lozenging and to keep parts in alignment, but many designers did not aim at much torsional stiffness. Lanchester, however, made his frame very deep and stiff, considering that only the springs should deflect, and not the frame, for the best ride. He used the petrol tank as a cross member to increase stiffness further still. Frames also had to be stiffened when side doors were introduced, otherwise the body sagged after a time causing the doors to jam.

By 1910 the frame was typically constructed from pressed channel steel members which generally swept inwards at the front end to increase the steering lock, and sometimes arched at the rear to accommodate deflection of the rear axle. The front axle was cranked (dropped) so that no arch was needed at the front. Sometimes the only cross members were at the front and rear of the side members. The engine was still often mounted on a subframe which was carried by cross members and the gear box carried by another one or two cross members. Cross members were of channel section though a tubular section was sometimes used. There was a tendency, however, to attach engine and gearbox to the lower flange of the side members and to locate cross members at the spring anchorages and so make them more effective. Little further change occurred until after the 1914–18 war, except for increasing the stiffness a little by using deeper section and heavier gauge steel, in addition to more cross members.

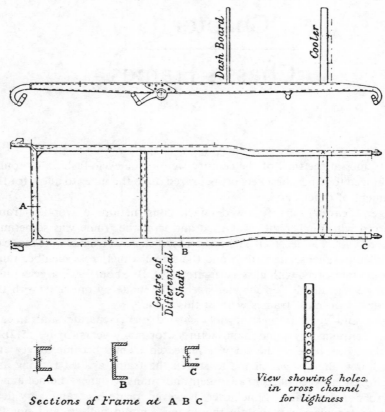

Figure 15.1 Channel frame on 1902 40 hp Mercedes. By permission of The Council of the Institution of Mechanical Engineers (from *Institution Proceedings*).

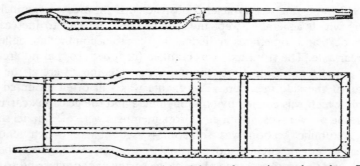

Figure 15.2 Channel frame with subframe for engine and gearbox. By permission of The Council of the Institution of Mechanical Engineers (from *Institution Proceedings*).

In the early twenties the general adoption of front wheel brakes made it necessary for the front end to be stiffened to cope with the braking torque. The more general use of the unit engine/gearbox helped here. There was a tendency to avoid bends in the frame by using plan forms like a truncated letter A and to increase stiffness by using deeper, more complicated, side members and by using tubular cross members and diagonal bracing.

In 1922 Lancia introduced their revolutionary 13.9 hp car, which, instead of having a conventional frame, was made up of steel pressings to which the body was attached to give a rigid structure. A tubular triangulated structure at the front end carried the independently sprung wheels, and there was a tubular subframe for the engine.

By the end of the twenties the engine subframe was obsolete; the engine and gearbox were mounted at that time on brackets attached to the side members. Next, the mountings were further modified by the insertion of rubber pads or bushes. As the engine then no longer contributed towards the stiffness of the frame more cross members were required and the cruciform bracing which had made a brief appearance many years before was used to some extent, particularly on the Continent. Frames could be heavily stressed which resulted in some fractures so nickel steel was widely used, but as designs improved and stresses were reduced low-carbon steel could again be employed. Frames tended to be widened at the rear with the springs mounted under, and not outside, the side members. There was a demand for lower vehicles so that the frame was lowered by further dropping the front axle and using more upsweep over the rear axle.

The 1927 Austro-Daimler had a tubular backbone frame (figure 15.3). The frame was Y shaped in plan with engine and gearbox mounted in the fork of the Y. Such a tubular frame is very stiff in torsion. The Austro-Daimler had IFS and for the wheels to track properly in the absence of a solid axle the frame has to be very rigid indeed. IFS suddenly became very popular in the USA and on the Continent about 1933 and this led eventually to a further increase in the rigidity of the front end.

There was also a trend to weld rather than rivet the assembly together, for, properly done, welded joints were stronger and cheaper, and of course could not work loose. Improvements in welding also made it possible to use box instead of channel section members, with a consequent stiffening of the frame. Cruciform bracing (figure 15.4) was also widely used and could extend from near the engine to the rear spring anchorages. Some cruciform frames were also made (figure 15.5); the frame was X-shaped with longitudinal extensions at the four ends of the cross, and outriggers were welded on to the frame to carry the door posts and other fixtures.

At the outbreak of the Second World War the usual frame consisted of side members of partly or wholly box section braced with a cruciform structure usually of channel section plus further cross members. There were still some ladder frames (figure 15.6) which generally had box cross

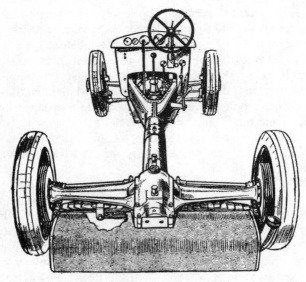

Figure 15.3 Tubular backbone frame of 1927 Austro-Daimler. By permission of Haymarket Publishing Co (from *Autocar* and *Autocar Handbook*).

members. However, instead of making chassis and body separately and then joining them together other manufacturers besides Lancia started to make cars in which the conventional frame was modified or dispensed with.

A logical development of the all-steel body was to combine frame and body so that the latter contributed to the strength and torsional stiffness of the whole vehicle. This could be done by welding body members together and joining them directly to the frame. The Chrysler Airflow of 1934 had such an interlocking body and frame, though curiously the two were bolted and not welded together; the resulting increase in strength permitted much lighter rails to be used instead of the conventional frame.

Frames and bodies were combined on a number of other cars with the bodies carrying a proportion of the load. One way to combine the two was

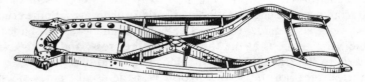

Figure 15.4 Frame with cruciform bracing. By permission of The Council of the Institution of Mechanical Engineers (from *Proceedings of the Institution of Automobile Engineers*).

to use a channel frame with the opening outwards, continue the body panels downwards and weld them to the frame, making the latter effectively a box girder.

The next stage in the development of the body was to do away with the frame and design the body to carry all the load, the vestigial frame and body forming the one indivisible unit.

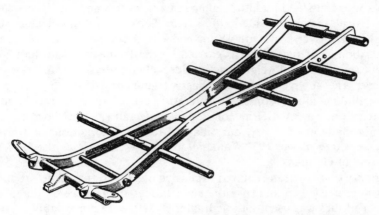

Figure 15.5 Cruciform frame. By permission of The Council of the Institution of Mechanical Engineers (from *Proceedings of the Institution of Automobile Engineers*).

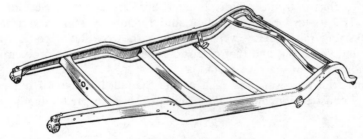

Figure 15.6 Simple ladder frame of the early thirties. By permission of Haymarket Publishing Co (from *Autocar* and *Autocar Handbook*).

Unitary Construction

J Ledwinka of the Budd company (p 401) was interested in unitary construction, and a demonstration body that Budd had had made to try to arouse Ford's interest was seen by André Citroën. Citroën saw the potential of the new construction method and got André Lefebvre, who had been involved in aircraft construction, to design the 1934 Citroën 7CV. The body of the 7CV was built up by welding steel pressings together; two horns projected from the fire wall, and a subassembly carrying the engine and front wheels was bolted to the horns. A simple dead axle carried the rear wheels. Springing was by fore-and-aft torsion bars at the front, and transverse torsion bars at the rear.

The 1937 Lincoln Zephyr employed a structure designed and developed by J Tjaarda for the Briggs Manufacturing Co. of Detroit. A tray located the various units and longitudinal channel members and outriggers carried the vertical members, side buffers and floor, everything being welded together. Transverse strength was provided by a bulkhead and a brace at the rear. About half the load was carried by the framework and half by the panels, so that overall weight was considerably reduced. Within a few years, however, the Zephyr reverted to chassis frame construction.

The Budd company stimulated interest in other manufacturers in unitary construction. Earl A McPherson designed a small car for Chevrolet which was not made in the USA, but, adapted for local conditions and production, appeared as the 1937 Vauxhall 10 in the UK and as the 1935 Opel Olympia in Germany.

Unitary construction required expensive presses so that large production runs were needed to amortise the cost. Model changes were limited while one set of dies was used, although later stylists were remarkably ingenious in altering the exterior trim, and the front and back ends of a car, to make it look different even although the body shell was unchanged.

The advantages of unitary construction, however, were so great that the separate chassis frame rapidly became obsolete after the Second World War. The unitary body was light and torsionally stiff, and free from squeaks and rattles. However, experience showed that it could transmit noise and vibration, and that corrosion could be a serious problem. More cars now rust out than wear out.

Some true post-war cars such as the Standard Vanguard did not use unitary construction but by 1959 the latter was so predominant that the

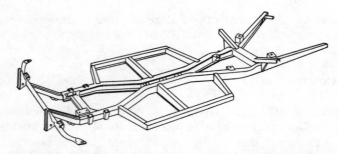

Figure 15.7 Frame of 1959 Triumph Herald. By permission of The Council of the Institution of Mechanical Engineers (from *The Structure of the Automobile*).

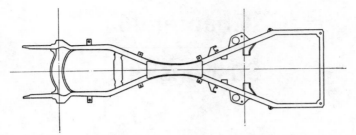

Figure 15.8 Cruciform frame on larger Chevrolet in the sixties. By permission of The Council of the Institution of Mechanical Engineers (from *The Structure of the Automobile*).

Triumph Herald caused quite a stir when it appeared with a separate frame (figure 15.7). Triumph made a point of advertising the virtues of a chassis frame. Even Rolls-Royce adopted unitary construction in 1965. The separate frame persisted for longer in the USA, where the manufacturers wanted to turn out a new model every year; also unitary construction could make the bigger cars heavier because of the weight of stiffening members and sound-deadening material. Frames could be cruciform (figure 15.8), or, later, of the perimeter type (figure 15.9) which gave occupants some protection from side collisions though they might have to step over the side rails to enter the car. Some frames had short, transverse torque boxes to give flexibility to the ends of the frame. The compacts of 1959, and the later compacts and subcompacts were of unitary construction.

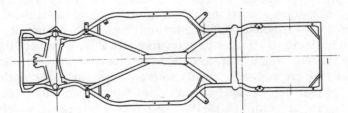

Figure 15.9 Perimeter frame on Oldsmobile in the sixties. By permission of The Council of the Institution of Mechanical Engineers (from *The Structure of the Automobile*).

One development has been the reintroduction of the subframe on a number of cars. The subframe forms a rigid mount for the steering mechanism and the suspension linkages and can be insulated from the body proper to reduce the transmission of noise and vibration to the latter. It also simplifies matters on the production line. The subframe can also carry the engine as well as the suspension and steering—this was the arrangement used on the Citroën Avant Traction in 1934.

Chapter 16

Suspensions

Developments to 1914

Quite a lot was known about springs and suspensions in the nineteenth century as carriage builders had amassed a great deal of empirical knowledge and locomotive engineers also had a considerable amount of knowledge. Carriage makers used leaf springs with a tendency to use fully elliptic springs for pleasure vehicles and semi-elliptics for business vehicles, but more complicated springs were also used. The early automobile engineer followed suit and generally used leaf springs which could be assembled as quarter-elliptic or elbow; half-elliptic or grass hopper, nipper or double elbow; three-quarter elliptic or fully elliptic springs. These various types of spring are illustrated in figure 16.1.

Most of the cars of the nineties had fully elliptic springs front and rear but these did not survive very long into the new century because the body swayed too much at speed and the springs took up a lot of space. However they did persist for quite a while on the Franklin and the Stanley Steamer. Usually, the front elliptic was replaced by a semi-elliptic and the side frame was carried forward as a dumb (imitation) iron with an upward curve similar to the spring it replaced to allow for vertical movement of the axle. Similarly the side frames were arched at the rear to clear the rear axle. Incidentally for quite a while the rear springs, etc, were known as the hind springs. The rear spring could be a semi-elliptic or three-quarter elliptic with the latter persisting until about 1920. Another arrangement widely used until about 1910 was the platform suspension. This had two semi-elliptics carrying the rear axle but their rear ends were shackled not to the chassis but to a third semi-elliptic transverse to the chassis and secured to the latter at its midpoint. It was realised, however, that a well designed conventional suspension was just as effective and much cheaper than the platform suspension. There was also mild enthusiasm for cantilevers just before the First World War. These were semi-elliptics pivoted at their centre and held between pins at their inboard end. Unsprung weight was somewhat reduced, but radius rods or links of some sort were required to constrain the axle. Lanchester had pioneered the use of cantilever springs at the turn of the century. Some vehicles had transverse springs on the front axle, but they

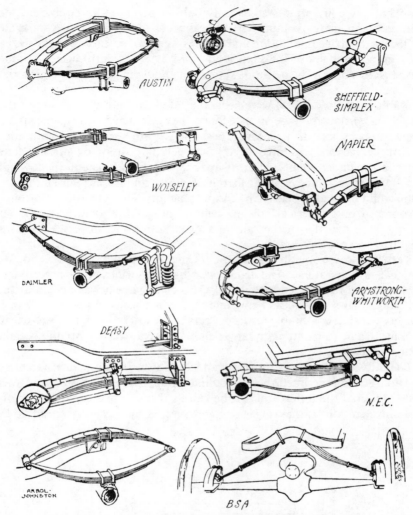

Figure 16.1 Different types of leaf springs: elliptical, Austin, Arrol-Johnson; three-quarter elliptical, Wolseley, Armstrong-Whitworth; half-elliptical, Daimler (plus auxiliary springs), Sheffield-Simplex; cantilever, Deasy, NEC; transverse, BSA; platform, Napier. By permission of Haymarket Publishing Co (from *Autocar* and *Autocar Handbook*).

were rarer at the rear. However, the Ford Model T had transverse springing front and rear.

Coil springs were used on some early cars. These were mounted between the rear underframe and the chassis of the Constatt-Daimlers of the late nineties. In America the Brush car had coil springs and so did the 1903

Vauxhall. A coil spring weighs only half as much as an equivalent leaf spring (on an energy/volume basis) but it needs radius rods, sway bars, etc, to locate the axle. In addition coil springs were not so easy to make as leaf springs so coil springs disappeared until IFS was introduced. Rubber suspension and air suspension were both tried but never really got into production.

Radius rods (see p 271) were generally fitted to chain-driven cars so that the chain tension could be adjusted, and they also acted as torque arms and kept the back axle in position. The springs were shackled at both ends. When the live axle was introduced it was held in position by a triangulated arm or a member connected at its apex to the chassis. The propeller shaft was sometimes protected by a surrounding tube and it was realised that this tube could be made to carry the torque reaction. The tube was accordingly terminated in a spherical housing behind the gearbox (figure 14.14). The torque tube had the further advantage that only one universal joint was required when it was used.

Very light cars sometimes took the torque drive and braking reaction on the rear springs alone. The front ends of the springs were accordingly anchored and the rear ends shackled. This arrangement was known as the Hotchkiss drive, Hotchkiss being one of the first firms to use it. At first it was not considered suitable for larger cars but it worked well and was cheap so that it came to be fitted on larger and larger cars and eventually became widely used.

The differential and transmission added a lot to the unsprung weight so de Dion Bouton mounted the differential rigidly to the chassis (figure 16.2) and connected the differential to the wheels through telescopic half-shafts and universal joints, the wheels being connected by a cranked 'dead' axle.

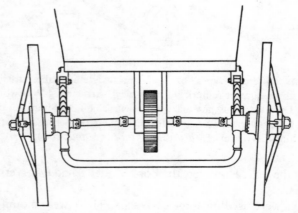

Figure 16.2 de Dion axle. By permission of Seeley Services and Co (from C C H Davis 1965 *Memories of Men and Motor Cars*).

Only the wheels and part of the half-shafts were unsprung. Later de Dion reverted to the conventional Hotchkiss drive. The de Dion suspension was revived many years later but today is mainly used on racing and sports cars.

Leaf springs

The finest quality steel was required for springs and so their manufacture was generally left to specialist firms. Indeed, for a number of years springs had to be imported from the Continent as very few firms had sufficient knowledge and skill to make them in the UK. At first the best Swedish steel was preferred but in the 1900s chrome vanadium and silicon manganese alloy steels were used. They were less likely to snap and had better fatigue properties.

The leaves were cut to length and eyes rolled into the ends of the top leaf and the tips of the other leaves tapered in plan form. The leaves were rolled to radius and the shorter leaves given a smaller radius so that when the spring was assembled the ends of the leaves 'nipped' one another. This meant that the upper leaves were prestressed in such a way that the normal stress developed during use acted in the opposite sense thereby reducing the overall stress in the metal. In addition, the blades did not separate when the spring rebounded past the no-load position. Nibs and slits were cut in the leaves, the nib in one half sliding in the slit in its neighbour so that the leaves could not be displaced sideways from one another. The most critical stage during manufacture was the heat treating and tempering of the finished leaves.

The springs were held together by clips and more complicated clips secured the axles to the springs. Hard rubber buffers were mounted above the centre of the springs to act as chock blocks when the spring bottomed.

Design was by rule of thumb and trial and error, and there was not a great deal of agreement about what size and shape the springs should have. Cars of more or less the same shape and size could have springs of very different lengths and number of leaves, though there was a general tendency to make the springs very flexible. At the turn of the century it was considered that the springs should be at least a metre long, and should deflect 15–30 mm/100 kg weight for light cars and 12–13 mm/100 kg for heavy cars.

Lanchester gave the suspension a lot of thought when designing his cars. He considered that the chassis should be absolutely rigid and torsionally stiff, and that the suspension should be 'soft' with a frequency equal to the normal adult walking pace (that is 1.33–1.5 Hz). The wheels should move vertically up and down, and not in the arc of a circle.

He achieved these requirements by using very deep chassis girders, very long cantilever springs front and rear, and parallel-motion radius and torque links. His vehicles gave an excellent ride which was far better than that of their contemporaries and is said to have been better than that of many cars designed fifty years later.

A major difficulty was that the chassis designer often did not know what weight of body the chassis had eventually to carry; it might carry a light touring body or it might groan under limousine coach work. He would compromise for a medium weight and so could not please everybody. There were also troubles with very light cars. With passengers and load aboard the suspension might be quite satisfactory but with the driver only on board the car might, to use Lanchester's phrase, 'dance all over the road'. Not that any car carrying Dr Fred was likely to dance—he weighed 18 stone in his prime! The weight of the passengers was, of course, much smaller in relation to the weight of the car on big cars and so this problem did not arise. Designers were therefore not very keen to reduce the weight of smaller vehicles to a minimum even though this improved their performance.

The general adoption of pneumatic tyres in the early years of the century reduced the demands on the suspension and little further development occurred, and, despite Lanchester's pioneering work, many years were to elapse before designers in general had a sound knowledge of the suspension and how it affected ride and handling.

The springs took the shock and absorbed some of the bumps but to damp out the subsequent vibration of the spring the designer had to rely on the friction between the leaves. The leaves were generally lubricated, but this was to try to keep the friction constant rather than to reduce it, and to minimise squeaks and corrosion. Sometimes soft metal strips were placed between the leaves to increase friction.

Shock absorbers

Coaches with very soft suspension had used leather-covered friction discs on lazy tongs assemblies which acted between coach body and axle to damp out the vibrations of the springs. When dampers were used on cars they were known as shock absorbers and apparently their mode of action was a mystery to a lot of people. Commandant Krebs of Panhard Levassor patented a damper but their first recorded use was on the 70 hp Mors in the Paris–Madrid race of 1903. Few pre-1914 cars had rebound dampers. They were generally of the friction type and consisted of two arms hinged together with leather and, later, friction (Ferodo) discs at the hinge, one arm being secured to the frame and the other to the axle. The Hartford (figure 16.3) was an example; it had several discs and on later models the clamping pressure could be adjusted by the driver via a Bowden cable. However, because static friction is greater than dynamic friction, friction dampers gave most damping when it was least required.

Auxiliary springs were sometimes fitted which were also known as shock absorbers (figure 16.1 Daimler). They were small coiled springs which replaced the rigid spring shackle. The idea was that these would absorb small bumps without the main leaf springs coming into action as some appreciable force was required to overcome the interleaf friction of the latter.

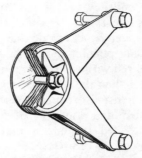

Figure 16.3 Cutaway view of friction-type shock absorber. By permission of Haymarket Publishing Co (from *Autocar* and *Autocar Handbook*).

Developments Between the Wars

Even after the 1914–18 war there was no general agreement on how to design the suspension to give a good ride. Very little development had taken place since the early days and the suspension was probably the least understood part of the car. In a few cases designers had worked out the best frequencies for the springs and designed light and efficient springs, but on the majority of cars the springs were heavy and inefficient. A great variety of types of springing was used and the most complicated arrangements did not necessarily give the best rides. Indeed, a driver could often improve things on an expensive car by fitting a shock absorber, or by simply reducing the pressure in his tyres.

Matters tended to get worse rather than better from the early twenties, particularly with light cars. Four-wheel brakes, balloon tyres and heavier axles increased the unsprung weight, speeds were higher, and the springs had to carry larger torques.

Accepted 'good practice' in the twenties was for the front springs to have a frequency of about 150 cpm (2.5 Hz) and the rear springs a frequency of 90 cpm (1.5 Hz). It was known that a frequency of above about 2.3 Hz jarred the passengers and much below 1.5 Hz could make some people car sick; it could also reduce adhesion as the wheels would be lightly loaded during rebound. The lower frequency was also limited by spring clearances, and typically the front clearance was 2.5 to 3 times the static deflection and the rear clearance 1.75 to 2 times the static deflection.

The front springs were made harder than the rear springs, for they had to be relatively stiff as they had to control the deflection of the front axle and wheels relative to the frame. If the vehicle rolled badly during cornering the resulting deflection of the axle could alter the steering geometry. Moreover, the front springs were closer together than the rear springs to increase the steering lock and so any tendency to roll was correspondingly increased. The front springs also had to carry the weight of the engine. The softer rear springs gave the rear passengers a better ride than the driver but they could be subjected to an unpleasant pitching motion because of the way the front and back vibrations interfered with each other. Pitching was reduced if the front suspension was softer than the rear suspension.

For lighter cars the difference in frequency with the vehicles loaded and unloaded was still a problem and a number of suspensions were invented whose frequency was intended to remain more or less constant and independent of load. One way was for the loaded spring to come progressively into contact with specially shaped chocks at each end. This was tried on a Turner car in 1923 and was used for some time on some buses. The big drawback of this arrangement was that it clattered badly when it was just on the verge of acting.

Though spring frequency certainly affected comfort it was found that the location of anchorages and constraints such as radius arms also had a considerable effect.

The springing was quite hard and the frames of the vehicles, particularly many English ones, were fairly flexible and speeds were increasing, so it was a major problem to prevent transmitted road shocks, and movement of the body with respect to the frame, from eventually wrecking the body. There was a move, therefore, to stiffer frames and more flexible suspensions, but the pressure for further developments was eased by the general adoption of balloon tyres in the mid-twenties. Cars with badly designed or badly constructed suspensions could be made to give a tolerably comfortable ride and so the conventional suspension received a new lease of life. Indeed, in 1926 it was said that you could pay £2000 for a motor car and yet still get a suspension identical to that on your great grandfather's dog cart. However, even with balloon tyres the springs had to be made stiffer and the shock absorbers harder so that the road holding of small cars at high speeds was poor, and indeed became even worse in the thirties.

Shock absorbers were used more on Continental than on UK or American cars. Besides the friction disc type there was the Gabriel snubber—a spring took up slack in a coil of belting during bump, and friction within the belting caused the belt to move out slowly during rebound.

Types of springs

At the beginning of the twenties most cars had semi-elliptics at both front and rear. The larger and more expensive cars tended to have semis on the front and cantilevers on the rear, and smaller, cheaper cars had quarter-elliptics front and rear. Cantilevers and quarter-elliptics would take drive but not torque without further constraints. Quarter-elliptics were cheap and were used on small cars but when they deflected they could have an anti-castoring effect on steering. The average designer preferred to use semi-elliptics and expected them to act not only as springs but as radius rods, torsion bars and anti-roll devices and anything else he could think of.

There were, however, a few vehicles with more enterprising systems. The Short-Ashby of about 1921 had an IFS like the early Sizaire, and Lancia had similar arrangements with enclosed vertical springs and dashpots. The 1923 Marlborough-Thomas used torsional suspension. Webb had duplex springs, one quarter-elliptic above the other so that deflection of the springs did not twist the front axle and thereby interfere with the steering. Vermerol had a similar duplex arrangement using cantilevers. The Bignan car of 1922 had hydraulic suspension using pistons and cylinders but with spiral springs as well to make the system fail safe. Some of the small cycle cars, particularly the French, had curious suspensions; one vehicle had two longitudinal springs supporting the axles at both front and rear. The Leyland 8 was fitted with an anti-roll torsion tube which passed through the front dumb irons, and torque arms at each end of the tube were coupled to the springs.

By the mid-twenties front wheel brakes were more or less standard equipment and as a result quarter-elliptics and cantilevers gradually disappeared over the next few years for it was cheaper to fit semi-elliptics and dispense with radius rods, etc. More leaves were used in the springs which were made more solid and more clips were used.

In 1923 only 53 % of the cars exhibited at the Motor Show had semi-elliptic springs both front and rear, and in all 14 different types of suspension were shown. By 1929 83 % of the cars had semi-elliptics all round and in 1930 91 %. Only ten systems were shown in 1929, six of which had persisted from 1923, including the Lancia system. The four new types were represented only on single makes, for example, the cantilever front, semi-elliptic rears on the Trojan, and full ellipsics on the Franklin. The second most popular suspension was semi-ellipsics front and cantilevers rear (25 % in 1923, 5 % in 1929).

Since the inception of the car some twenty different types of suspension had been used with sufficient success to remain in production for considerable periods, and many others had been tried but for various reasons did not go into successful or large-scale production. All, however, were eventually displaced by the semi-elliptic, so that by the early thirties practically every make had semi-ellipsics, the exceptions being a few with independent suspensions, and some strays like the Ford and the Austin 7.

Rebound plates attached above the main plate had a vogue for some years, first in the USA before the twenties, then in the UK and finally on the Continent, but it was eventually concluded that rebound clips to hold the plates together were more effective and cheaper. The blades were bolted together by a centre bolt made of high-class steel so that the bolt was as small as possible and therefore did not appreciably weaken the leaves. On more expensive cars U clips were used which did not weaken the spring, but did reduce its effective length slightly and could make the spring difficult to dismantle. Clips along the blades held them in position laterally and also prevented separation on rebound. The more expensive the cars the greater the number of clips, although American cars also used a number of clips. It was almost universal practice to roll eyes on the ends of the top plate and insert brass phosphor bronze or gunmetal bushes in the eyes. On expensive cars the plate ends were made thicker and eyes machined out of the solid metal; the spring was stronger but much more expensive.

In the UK studs and slits were largely used to prevent side movement—a stud pressed in one blade sliding in a slit cut in the blade below, the slit being near the end of the blade. Alternatives such as ribs and grooves were developed but not adopted to any extent.

The springs were anchored at the front and shackled at the rear. Instead of using shackles the top plate could move in a slot in a phosphor bronze cylinder which could rotate. It became increasingly popular to protect shackles and clips with leather gaiters.

Later there was a trend to shackle the front end of the spring and pivot the rear end; this helped the steering somewhat. Spring eyebushes were troublesome to lubricate and the 'Silentbloc' rubber bushing introduced in the late twenties was a considerable improvement over the metal bush. Rubber bushes had been used on horse vehicles but in this situation the rubber had been lightly stressed. In the 'Silentbloc' bush the rubber was contained between two concentric metal shells under a very high compressive stress. It deformed little under load but twisted easily in shear; it did not squeak and it did not need lubrication. The inner shell was later made eccentric so that it became central under load.

Though there had been great improvements in production methods, the leaf spring manufacturers could not, even as late as 1939, predict the spring rate of their products when they were mounted in a car.

Hydraulic shock absorbers had been made in the early years of the century but they did not become popular until the mid-twenties. At first they were single acting only but later acted on rebound as well as bump. They were, in effect, dashpots in which oil was forced through a small aperture within the body of the absorber by a vane or plunger connected to the spring (figure 16.4). The setting of a number of shock absorbers could be controlled from the dashboard. About 1937 another type of hydraulic absorber became available—piston and cylinder formed a single telescopic strut and fluid passed through the piston via disc valves, these could operate on bump, rebound or both. The cylinder had a separate recuperator chamber with a spring-loaded valve to act as a reservoir. When they did have to use shock absorbers some manufacturers tended to economise and fit the cheapest available which resulted in shock absorbers acquiring a poor reputation.

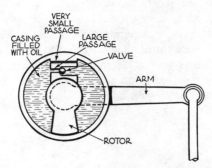

Figure 16.4 Principle of hydraulic shock absorber—vane type. By permission of Haymarket Publishing Co (from *Autocar* and *Autocar Handbook*).

Independent suspensions

The extra weight of the front brakes and low-pressure tyres were in the long run to cause a great deal of trouble for they brought in wheel wobble or shimmy, axle tramp, and other evils. Vibrations caused by slackness in the steering gear had been known for many years, but the violent wobble that came in at high speeds was new; it could twist the whole front end and was not only uncomfortable but very alarming to the driver. It was soon found to be due to gyroscopic effects. The axle vibrated about a horizontal longitudinal axis through its centre and as the wheels rotated about an axis more or less at right angles to this a vibration was set up about a third axis perpendicular to both the first two and as a result the wheels could oscillate about their kingpins in phase with one another. Mathematical analyses were made but the mathematicians could not suggest a simple cure, or, for that matter, how, except within broad limits, to design a suspension that would not 'shimmy'. Shimmy was a tremulous movement made in the foxtrot and both the foxtrot and wheel shimmy had their hey-day in the twenties.

One obvious solution to these problems was to dispense with the front axle and make the front wheels independent of one another. De Lavaud suggested this about 1928 and designed an IFS accordingly, but other engineers had also realised that IFS should be a certain cure for wheel wobble and a trickle of cars with IFS had begun some years earlier. De Lavaud considered that the front wheels should be independent of one another, even to the extent of almost duplicating his steering, which was completely reversible, and he considered that the wheels should be displaced by translation not rotation. He used a double-parallelogram arrangement of rods. He disliked IRS as it gave excessive heeling over on corners, rolling and pitching, and increased tyre wear. He recommended instead what appears to have been a de Dion Bouton suspension though he realised this could transmit noise to the bodywork. He said that the usual design of front axle was 'unpardonable' and must soon disappear in favour of IFS. For a number of years, however, independent suspensions were used on only a few cars in the UK, for example, the Morgan and Alvis, (figure 16.5). Alvis used front wheel drive as well as IFS in 1928 and were unfortunate in that one of their cars was involved in an accident, and the insurance companies looked askance at a car with so many new ideas.

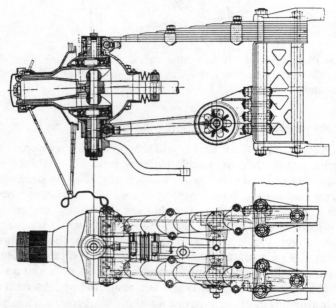

Figure 16.5 Alvis independent front suspension (1926). By permission of The Council of the Institution of Mechanical Engineers (from *Proceedings of the Institution of Automobile Engineers*).

On the Continent independent suspensions were more widely used. A big breakthrough occurred in the USA in 1933, when they were introduced to improve ride. Ride could be improved by softening the front suspension and so reducing pitching, but this aggravated shimmy and steering problems which could only be solved by using IFS. IFS was used as a gimmick to help sales; the publicity people called it 'knee-action' and knee action certainly appealed to the public.

Rolls-Royce in the twenties had become interested in the problem of ride. They found as a result of their abortive attempt to manufacture in the USA that a car that behaved perfectly in the UK might not give such a good performance on roads overseas. They had a very simple ride meter—a container was filled with water and the amount that slopped over its edges in a given distance was measured. Rolls-Royce also built a bump rig—the wheels of the vehicle were mounted on large rotating drums and 'bumps' of the required height and shape could be attached to the drums. The inertias about various axes, and spring and frame stiffnesses, were also measured.

Maurice Olley went from Rolls-Royce to General Motors in 1930 and worked on ride for the latter. He modified a seven-passenger limousine so that its inertias and spring stiffnesses, etc, could be quickly altered, and subjectively assessed the ride for different values of these parameters. He found that if the front springs were softer than the rears he got a great improvement in ride but at the expense of handling when using a conventional front axle. He therefore converted two experimental Cadillacs to IFS using the Dubonnet system on one and wishbones on the other. Ride and handling were improved enormously though there were some teething troubles with the steering, but by March 1933 he had sorted things out to such an extent that he demonstrated the two cars and a standard Buick for comparison to the GM top brass, including Sloan and Kettering. They were very impressed and despite the country being in the depths of the Depression they had the new suspension introduced very rapidly; indeed Kettering said that it seemed to him the Corporation could not afford not to have it!

Designers in various countries became very interested in IFS and reacted in characteristic manner. Manufacturers in the UK were conservative and the advantages of IFS were not very obvious as road surfaces were good and speeds relatively low. As a result the change to IFS was slow, and indeed no Austin, Morris or UK Ford car had such a suspension before the Second World War.

On the Continent, however, most manufacturers had adopted IFS by 1939 and a wide range of systems was in use, many of the designs being novel and ingenious. In the USA simple systems were required which could be easily mass produced, and in some ways it was more important that the driver could say his car had 'knee-action' than that the suspension should be particularly effective.

Independent suspension already had quite a long history. One of the first vehicles with IFS was the Bollée steam car of 1878 which had two transverse laminated springs, one above the other and connected by the wheel pivot. The Stephen car of 1898, the Decauville of 1899, and the Sizaire-Nadin (figure 16.6) of 1905 used a transverse leaf spring the ends of which engaged extensions of the steering knuckles which slid up and down in vertical cylinders. The Sizaire-Nadin car was a very simple low-priced vehicle and it is likely that the independent suspension was used for its cheapness and with no other end in mind.

Independent suspensions continued to crop up from time to time but the vehicle that drew attention to the advantages of IFS was the 1922 Lancia.

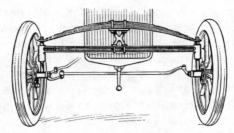

Figure 16.6 Sizaire-Nadin independent front suspension. By permission of Haymarket Publishing Co (from *Autocar* and *Autocar Handbook*).

Again the steering knuckles could slide up and down vertical tubular guides (figure 16.7) but these also carried enclosed coil springs together with a hydraulic shock absorber. Unsprung weight was reduced and the riding qualities of the car were very good but the manufacture of the springs was apparently a difficult job. The Lancia suspension, with minor modifications, remained in production for many years.

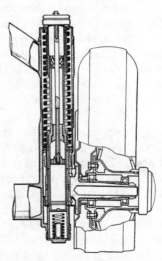

Figure 16.7 Lancia independent front suspension. By permission of Haymarket Publishing Co (from *Autocar* and *Autocar Handbook*).

By the end of the twenties possibly 20 different makes were using independent suspensions of one kind or another. Some simply used swinging axles (Rumpler, Buccaili). Others (Sizaire, Steyr—rears only) leaf springs and radius arms to keep the displacement of the wheel parallel to itself. On some vehicles the wheel was mounted on suspension arms whose movement was controlled by springs which could be helical (Harris Leone Laisne) or rubber (Adams system). Clement-Rochelle, De Lavaud, G.H.W. and Morgan used systems similar to the Lancia arrangement. Alvis and Cottin-Desgouttes employed laminated springs in a box-like formation. Very often independent suspension was used on the rear as well as the front wheels and, curiously, Steyr, Austro-Daimler and Tatra, for example, used independent suspension at the rear only. Practically all of the front wheel drive vehicles had IFS; although this simplified design, the front end even so became very complicated.

The Rohr eight-cylinder car of 1928 and the Mercedes 170 of 1931 appear to have been the first with all four wheels independently sprung. A little later Mercedes used transverse links with coil springs on the front wheels. The links were of equal length.

The first US cars with IFS used the Dubonnet system (Pontiac, Chevrolet) or wishbones (Buick, Cadillac, Dodge, Olds, Plymouth). In the Dubonnet 'knee' (figure 16.8) a hollow casing was attached to the steering pivot at one end and at its other end a horizontal shaft carried short trailing arms to which the stub axle was attached. Springs in the casing acted on a cam attached to the horizontal shaft in both take-up and rebound. The casing

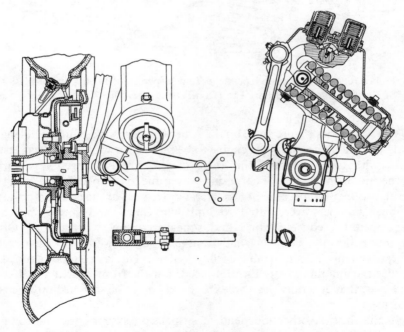

Figure 16.8 Dubonnet independent front suspension (Pontiac). By permission of Reed Business Publishing Ltd (from *The Automobile Engineer*).

also contained a shock absorber. The pivot was very heavily loaded. The whole unit was difficult to produce and gave trouble in service so that it was replaced on the Chevrolet by a wishbone system.

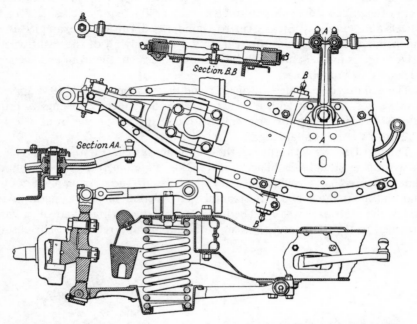

Figure 16.9 Wishbone independent front suspension (Buick). By permission of Reed Business Publishing Ltd (from *The Automobile Engineer*).

The wishbones (figure 16.9) on the American cars were of unequal length. This was merely to allow the engine to be moved forward, but it was found that, with unequal arms, the angle of the wheel changed less during cornering, with less loss of cornering power due to camber change, and tyre scrub was reduced because of less lateral movement of the tyre contact area. Rubber bumpers were fitted to limit movement and special bushings developed to reduce friction and squeaks at the many pivots. Other arrangements of coil spring and transverse leaves were also used in the Voran system and the André Girling system. The latter was intended to reduce the amount of modification needed at the front end. It was found, however, that generally the whole front end had to be stiffened up when IFS was used.

In the thirties, other independent suspension systems were introduced including a number with longitudinal, not transverse, links. The Gordon Armstrong system (figure 16.10) incorporated a spring within the linkage, and bump pads on the central rod limited movement. Dubonnet showed his

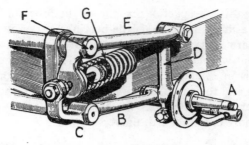

Figure 16.10 Gordon Armstrong independent front suspension; the pivots F and C are mounted on the frame and deformation of the parallelogram is resisted by the spring G.

system in 1932. He was the son of the aperitif manufacturer and he built a complete car to demonstrate his ideas which he exhibited at the Paris Motor Show. Besides Pontiac and Chevrolet, Vauxhall in the UK used the Dubonnet suspension system. Vauxhall modified it by using a short torque tube instead of coil springs but used subsidiary coil springs to take up small deflections. At this time Dr Porsche was working on torsion bars and these were fitted to the Rohr Olympier and a Mathis car. In the original Porsche suspension the torsion bars extended through a tubular cross member. Each

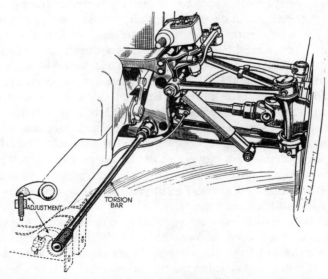

Figure 16.11 Torsion bar suspension on the front wheel drive Citroën; the bars are connected to the lower link of the wishbone suspension. By permission of Haymarket Publishing Co (from *Autocar* and *Autocar Handbook*).

bar was anchored in a bracket at one end of the tube and splined into the head of a lever arm mounted in needle bearings at the other end, the free end of the lever arm carrying the wheel assembly. Torsion bars were used on the VW, the front wheel drive Citroën (figure 16.11), and on Lagonda and Vauxhall models in England. In most designs the torsion bars extended along the frame and their free ends were pin-jointed to the steering heads. Independent rear suspensions were of the swinging axle type, that is, each half-shaft had only one (inboard) universal joint.

Roll

The springs on a car fitted with IFS were softer than those on a car with a solid axle and, although the car rolled about the contact patches of the tyres and not the much narrower spring base of the solid axle, the roll axis was lower. Consequently the penalties of IFS and a soft ride could be greater roll during cornering, and squat and dive when accelerating and braking, respectively. With solid axles, the camber angles of the wheels were fixed but this was not so when independent suspensions were used; the change in angle with body roll depending on the geometry of the suspension. Changes in camber angle changed the cornering power of the wheel; for example, if the wheel leaned inwards in a bend its cornering power was increased, so that the handling properties of the vehicle changed.

When a sideways force is applied through the centre of gravity (CG) of the body the latter rolls about its roll centre (which can change in position as the body rolls). The roll centre is obviously in the plane of the spring shackles in the case of a beam axle, half way between the wheels, and in the vertical plane through the tyre contact areas. For a parallel equal-length wishbone the roll centre is at ground level, and the wheels remain parallel to the body when the latter rolls. The distance between GG and roll centre can be changed by altering the suspension geometry; thus by making the upper wishbone shorter than the lower one the roll centre is raised. In this way changes in camber angle can be controlled and minimised though this can be at the expense of sideways movement of the wheels ('tyre scrub').

The designer had as usual, to compromise and try to give as good a ride, with as little rolling and as good positive predictable steering as he could manage, and still keep costs down.

Anti-roll bars (stabilisers) were a palliative so far as rolling was concerned and had first been used before the First World War. In one early arrangement the mid-point anchorages of the rear springs were mounted at each end of a transverse axle running in bearings on the chassis frame (figure 16.12); if both wheels went over a bump the springs acted independently, but during cornering the outer spring tended to deflect the inner one, so reducing rolling. Later angled torsion rods were used which were revived and used on a number of cars just before the Second World War.

They consisted of steel rods mounted on the body in rubber bushings so that they could rotate, the angled ends of the rods being connected to the steering heads or suspension arms. They had their disadvantages; when only one wheel went over a bump the deflection of its spring was resisted by the other spring, and the ride and handling of the car could be adversely affected.

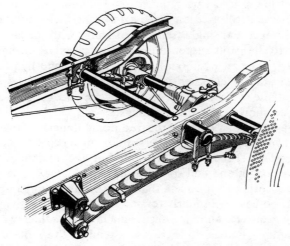

Figure 16.12 Early (about 1917) anti-roll bar interconnecting cantilever springs. By permission of Chapman and Hall Ltd (from A W Judge *Motor Manuals*).

Developments Since 1946

IFS did not become standard in the UK until the first new post-war designs had appeared. The broader principles which determined the handling of the car were understood—the front and rear roll stiffnesses, the heights of the roll axes, the weight distribution, the slip angle of the tyres (see p 327), and the way these and the suspension and steering geometry changed when the tyre passed over a bump or during cornering. However the designer had to unravel a very complex situation if he wanted to study the problems in any depth. Indeed detailed studies of ride and handling were not made until the sixties when computers were used. Progress also depended upon obtaining more information about cornering forces, slip angles and the self-aligning torques of tyres.

A number of rear-engined cars were designed during the fifties. Since these could behave oddly, if not positively dangerously at times, it was soon realised that the suspension systems of such cars had to be designed very carefully; this problem gave further impetus to the study of the suspension.

Except for expensive cars, and rear-engined cars, IRS did not make much progress during the late forties and fifties. IRS reduced unsprung weight which improved ride and handling. However, it was more expensive and required more space than a beam axle and its advantages showed up more on bad roads than on good ones. It also transmitted more road noise and vibration to the body structure. Consequently UK designers of medium- and low-priced cars retained semi-elliptics and Hotchkiss drive. Torque tubes were still used on some cars in the fifties and later the Ford Anglia and Prefect had single transverse springs with torque tubes.

Returning to the front suspension the most popular arrangement in the fifties was two unequal wishbones plus coil spring and hydraulic damper, but there were at least ten other types in production. However, as time went on some of these were dropped as they were too expensive or because experience had shown their shortcomings.

The Lancia Appia of 1952 used the traditional Lancia arrangement of coil springs and hydraulic shock absorbers sliding in vertical cylinders mounted at the ends of an I-section axle beam, but the Flaminia of 1958 had transverse wishbones and coil spring whereas the later Flavia used a single transverse leaf spring with no less than eleven leaves on the upper member with a wishbone below. Similarly torsion bars were expensive; they had to clear steering gear, etc, and have secure anchorages, which was not easy to arrange on unitary bodies, so there was a tendency to replace torsion bars by coil springs, though they are still used on a number of cars.

In general there was a move to tidying up and simplifying designs and to obtain the required performance by modifying geometrical, spring and damping, parameters.

The general adoption of IFS meant that shock absorbers had to cope with much greater deflections than with beam axles, and coil or torsion springs gave very little damping compared with laminated types. Development of shock absorbers lagged behind that of the suspension on the whole and it was not until ten years or more after the war that telescopic shock absorbers (figure 16.13) were available which were reliable, required little maintenance and were free from fade. Lever arm shock absorbers were used into the fifties. They could double as transverse links, and were popular in the UK. The lever arm worked a rocker arm, each end of which operated a piston in a cylinder and the two parallel cylinders were connected by a channel containing the damper valve. The telescopic absorber generally had two valves, one in the piston and the other at the base of the cylinder, the latter communicating with an annular reservoir or recuperator surrounding the cylinder. Instead of a reservoir, later shock absorbers could have only one valve in the piston and a space at the bottom of the cylinder separated from the rest of the cylinder by a floating piston and filled with nitrogen under pressure. Entry of the piston rod in the hydraulic cylinder compressed the nitrogen instead of causing fluid to flow into the reservoir. There was no aeration of the fluid in this type of absorber and it was lighter and ran cooler than the other type.

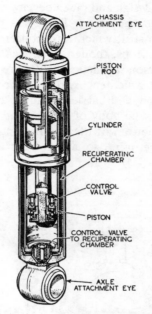

Figure 16.13 Telescopic hydraulic shock absorber. By permission of Haymarket Publishing Co (from *Autocar* and *Autocar Handbook*).

Cars, particularly the large American ones, could roll badly when cornering and the front end dive during braking. Stabiliser bars reduced rolling to some extent but brake dive remained a problem for a number of years. If a driver misjudged his distance at traffic lights his bumper could go under the rear bumper of the car ahead and when the front end came up the two cars would become locked in an embrace that was difficult to break. Over-riders, which were short vertical bars mounted on the bumpers, solved this particular problem.

Brake dive could be offset by using fore and aft suspension arms and arranging for the reaction on the brake caliper to act in the opposite direction to the inertial forces. It was found that this could be done when using transverse double wishbones by suitably inclining the two wishbone axes to one another instead of having them parallel and horizontal.

Transmission of road noise was troublesome; this could be reduced by using rubber pivots wherever possible, but another way was to mount the suspension on a subframe insulating the body from the subframe by rubber mounts. This also improved the steering. Subframes also made production easier as they could be assembled on subsidiary lines and fed into the main production line.

Most of the systems had been used before the war but there were improvements in detail design and much development work and testing had gone into improving their geometry. There was, however, one major new design—the McPherson strut which was pioneered by the European Ford companies in the early fifties, though developed originally by McPherson for GM. The McPherson arrangement (figure 16.14) consisted of an inclined telescopic strut which enclosed a telescopic damper and carried a coaxial spring. Its upper end carried a thrust bearing which fitted in a compliant anchorage in the body of the car and its lower end carried the steering knuckle. The bottom of the steering knuckle engaged a wishbone through a ball and socket joint. An anti-roll bar was fitted for fore-and-aft location.

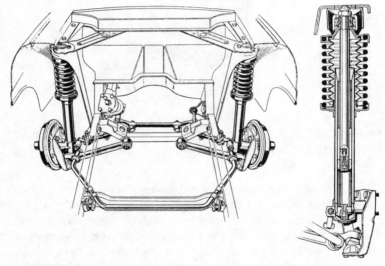

Figure 16.14 McPherson strut on Ford Consul. By permission of Haymarket Publishing Co (from *Autocar* and *Autocar Handbook*).

Unconventional systems

The curious Citroën of 1949 had a novel suspension system. Four bell crank levers were mounted at the ends of two tubular cross members. The horizontal arms, the front leading, the rear trailing, carried the road wheels, and the vertical arms were interconnected by rods and springs so that when one wheel went over a bump the end of the arm carrying the other wheel on the same side of the car was pulled down thereby tending to lift its pivot and therefore the corner of the car. Pitching was thereby reduced, thus improving the ride. The wheel arms were trailing so that the braking torque tended to lower the rear end of the car offsetting the otherwise excessive nose dive. The rear of the car was so high when the car had no passengers that the headlamps were made adjustable so that the beam could be kept horizontal.

Pneumatic springs have the advantages that their frequency does not vary with load, the spring rate can be made low, it is easy to arrange for the vehicle to be kept level and at a constant height, and the springs act as sound and vibration insulators. The disadvantages, of course, are their extra cost and complexity. Citroën, however, used hydropneumatic suspension on their DS19 of 1955. The suspension system consisted of four metal cylinders terminated by hollow hemispherical chambers which were divided by rubber diaphragms. The upper part of each chamber (figure 16.15) contained gas under pressure and the lower part hydraulic fluid. Movement of the road wheel about a pivot moved a piston up and down the cylinder. The Citroën system also automatically kept the vehicle at a constant height. An actuating arm from the suspension controlled a valve which passed extra fluid under pressure or removed fluid from the hydraulic cylinder, depending upon whether the ground clearance was lower or greater than the selected value. Citroën were helped by the experience they had gained in developing hydropneumatic devices for aircraft.

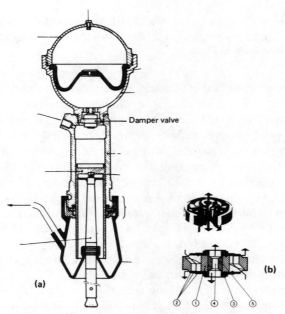

Figure 16.15 Citroën gas spring suspension unit. By permission of Pentech Press.

306 SUSPENSIONS

In America air suspensions were available on a number of Ford and GM models in 1958 but they were dropped after a year or so.

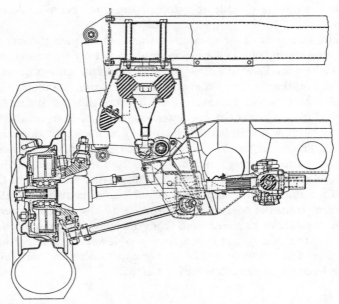

Figure 16.16 Suspension on the BMC Mini showing rubber cones (shaded areas). By permission of Reed Business Publishing Ltd (from *The Automobile Engineer*).

Rubber suspensions had often been considered and design and development work had been undertaken on a number of occasions, but it was not until the Mini of 1959 that the first successful rubber suspension was used on a volume-produced car. Rubber springs were used because they could be designed to give a variable rate, which is very desirable for very light cars, and because they were more compact. The springs (figure 16.16) consisted of inner and outer conical members with an annulus of rubber bonded between them, the varying thickness of rubber giving the required spring rate variation. The front springs were used in conjunction with double transverse links with a semitrailing radius rod at the front wheel making the lower link into a wide-based wishbone. The rear springs were almost horizontal in order to save space, and used a simple trailing arm. Both front and rear suspensions were mounted on subframes.

A few years later the Hydralastic system (figure 16.17) was used on the BMC 1100. The suspension unit (figure 16.18) contained a rubber spring like the Mini but also a lower compartment filled with fluid. Movement of the wheel acted on a rolling diaphragm in the lower compartment and

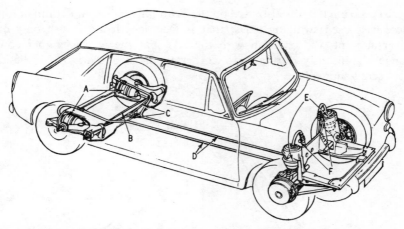

Figure 16.17 Hydralastic system on BMC 1100: A, rear Hydrolastic units; B, anti-roll bar; C, torsion bars; D, interconnecting pipes; E, front Hydrolastic units; F, charging valves. By permission of Reed Business Publishing Ltd (from *The Automobile Engineer*).

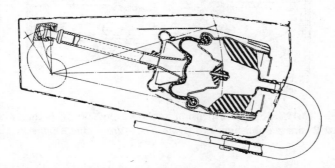

Figure 16.18 Hydralastic unit on BMC 1100. By permission of Reed Business Publishing Ltd (from *The Automobile Engineer*).

forced fluid through a valve in a port plate to compress the rubber spring with some fluid passing under pressure to the suspension unit on the other wheel on the same side of the car, tending to lift that corner of the car and so reduce pitching. Valves in the port plate acted as dampers in both bounce and rebound.

Rear suspensions

IRS made some headway on more expensive cars, on Jaguars for example in 1961, and the de Dion suspension was used on some sports cars such as the Aston Martin and on the 1964 Rover 2000 (figure 16.19). At the other end of the spectrum IRS was also used on the Triumph Herald and Hillman Imp. These cars used swinging axles, that is fixed-length axles with only one universal joint; the problem with these is that the wheel camber could change significantly during hard cornering in such a way that the car oversteered suddenly.

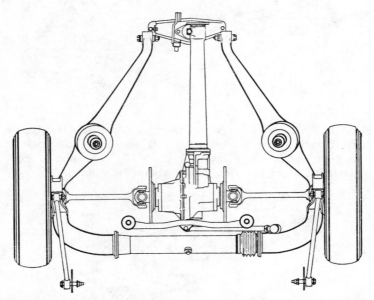

Figure 16.19 de Dion suspension on the Rover 2000. By permission of Reed Business Publishing Ltd (from *The Automobile Engineer*).

This phenomenon plus Mr Nader killed the GM Corvair in the sixties. A better arrangement was to use two universal joints, one at each end of the half-shaft, and a splined joint to allow for changes in the length of the half-shaft. Suspensions similar to those used on the front wheels could then be used, though generally speaking there was less space available.

One independent rear system which became widely used was to mount the wheel carrier on a 'semitrailing' arm with a coil spring and damper over the wheel carrier. The axes of the pivots of the arm were inclined backwards to the centreline of the car to minimise variations in the length of the half-shaft, as well as changes in the attitude of the wheel (figure 16.20).

The live rear axle suspension did not change much. As suspensions tended to become more flexible to give a softer ride there was a trend to use linkages of one sort or another, particularly on the more expensive cars, to locate the axle more precisely.

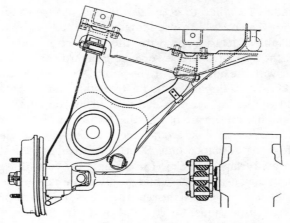

Figure 16.20 Independent rear suspension on Hillman Imp. By permission of Reed Business Publishing Ltd (from *The Automobile Engineer*).

Sixties and later

About 1960 the wishbone was the most popular form of front suspension. There were many variations on one theme; the wishbones could be solid stampings or forgings, or levers, and the spring a coil spring, or in a few instances a torsion bar. Small BMC cars used rubber springs and the Citroën DS19 hydropneumatic springs. Sometimes upper or lower wishbones were replaced by a transverse leaf spring. The main alternative to wishbones were fore-and-aft parallel arms, again in conjunction with coil springs or torsion bars. Rubber was used in increasing amounts in pivots to isolate road noise.

Very simple arrangements were used on some cheaper cars, for example, a single lever for the lower wishbone located by a radius rod, plus a McPherson strut.

The Hotchkiss drive was still used at the rear of many cars, the live axle being supported and located by semi-elliptic leaf springs, rubber being used in the spring eyes and between axle and spring to reduce road noise. Improvements in steels enabled fewer leaf springs to be used. The Chevy II of 1961 had a single leaf spring which decreased in thickness but increased in width towards the ends to give torsional stiffness. The upper surface was shot-peened while in tension to increase its fatigue life. By the end of the sixties most rear springs, if not single leaf, had only a few leaves with the extra leaves being quite short. Leaf springs, however, were progressively replaced by coil springs. Leaf springs could wind up, causing wheel hop and patter, particularly with softer springs, and were not as efficient as coil springs or torsion rods.

Links or radius rods had to be used with coil springs to locate the rear axle. One arrangement was to use longitudinal rods below the axle, and shorter rods above it to cope with the torque; to prevent sideways movement of the axle a transverse rod, the Panhard rod, could be fitted to link one wheel with the bodywork on the opposite side of the car. Instead of a Panhard rod a Watts linkage could be used which had the advantage that the wheel end of the linkage was constrained by the latter to move in a straight line and not on the arc of a circle. Another way to control the axle was to incline the shorter links so they met the axle at an angle or even for the links to form an A-shaped unit. Short torque tubes were still used and again were located transversely by, for example, a Watt's linkage.

Very simple rear suspensions could be used on cars with front wheel drive, and solid axles were so light that there was not much point in using independent rear suspensions. IRS could be of the trailing arm type, but the semitrailing type became the most popular as it was light and compact and camber change was reduced during cornering. The half-shafts had to be splined or plunging universal joints fitted. Double wishbones and McPherson struts were also used.

Returning to the front end the McPherson strut became more and more popular in the sixties and was no longer used only on Ford cars. It was cheap and simple but had its drawbacks—space had to be found for the vertical member (which needed some type of bearing at its top end), it was not as flexible as double wishbones in that the designer could arrange for all sorts of subtle geometrical interactions with the latter, and it did not insulate road noise as well as some other designs.

By 1970 the great majority of cars used double wishbones or McPherson struts. Coil springs were the most popular form of springing, some torsion bars were used, and rubber springs on some makes. Pneumatic systems were used on a few cars.

In 1973 the BL Hydralastic system was modified and the rubber spring replaced by a gas chamber. A conical piston sealed by a rolling diaphragm acted on fluid in a lower chamber which interconnected with the upper gas chamber through a damping valve and the fluid in the upper chamber compressed the gas (nitrogen). Again fluid in the lower chamber also passed through to the corresponding chamber in the suspension unit on the rear wheel on the same side of the car, and the body lifted accordingly.

The designer still has to compromise between ride and handling, and in general an improvement in ride is made at the expense of handling. Most contemporary cars are firmly sprung to give good handling characteristics and this is particularly marked on small cars on which the ride tends to be harsh.

The suspension systems so far described have been passive; reacting to the forces acting on the system without the system parameters changing (though they may be nonlinear) and the latter are chosen as the best compromise for the vehicle concerned.

A quite different approach is to use active suspensions in which sensors detect the position and attitude of the body with respect to the wheels, and actuators automatically maintain the body in the desired position. Thus roll can be minimised during cornering (or the body even tilted in the opposite direction), the height of the body kept constant despite changes in load, and its attitude maintained unchanged during accelerating and braking.

One way to do this is to supplement the springs by hydraulic struts. Fluid passes from an engine-driven pump to the struts via rapidly acting valves which are controlled by a pendulum supported by a spring and mass system, and the struts also keep the distance between wheel and body constant. Later developments have been to sense the relative position of wheels and body and use a microprocessor to control the hydraulic struts and so keep the body on an even keel at all times.

Suspension systems that automatically cope with changes of load and loading have been used for many years, notably by Citroën, but the more sophisticated active systems are not yet in use on production cars.

Chapter 17
Steering

Developments to 1914

Benz experienced difficulty in designing a satisfactory steering mechanism for a four-wheeled vehicle and so his first cars were tricycles. The front wheel was mounted in a vertical fork (completely unsprung in his first car) which could be turned by a short hand lever or tiller through a rack and pinion. The 1888 Model III was similar but the wheel was sprung, and steering was via a double rack and pinion. The tricycle arrangement was unstable so Benz persevered with the development of steering gears suitable for four-wheeled cars. He had developed a successful design by 1892 and patented it the following year. A solid front axle carried stub axles at each end and rods from the axles were connected with rack bars which engaged a pinion at the base of the vertical steering column. The top of the column carried a handle.

Daimler's first car—the first four-wheeled car—was merely a modified American horse-drawn carriage. The front axle swung on a central pivot and the car was steered by revolving a handle on the top of a vertical pillar which rotated the whole fore carriage through a pinion engaging a circular toothed track. The centre-pivoted axle worked well with horses because of the leverage of the long shafts, but was not satisfactory with the horseless carriage even at low speeds. Bumps and pot holes in the road—and there were many—wrenched the axle round. The Daimler car of 1889 had Ackermann steering operated by a tiller, a cross tube connected the knuckle arms and was pivoted on a horizontal pin so that it could tilt laterally within guides on the main frame. The Daimler-Maybach vehicle of 1895 reverted to a pivoted axle steered by a pinion and toothed sector. The 1898 Canstatt-Daimler car used a pinion and chain, rather like the old-fashioned steam roller.

Even after 1900 there was at least one vehicle still being made with centre pivot steering. With such steering the front wheels had to be small in diameter to enable the fore carriage to rotate under the front of the vehicle; front wheels still tended to be smaller than the rear wheels even when pivoted stub axles were in general use.

Ackermann steering

Ackermann steering did not become general until about 1895 though it had been used on the Bollée 'L'Obéissante' of 1873 and Jeantaud later used it on his electric vehicles. Two stub axles were pivoted at either end of the Ackermann axle and were rotated by levers or knuckle arms (figure 17.1). The ends of the knuckle arms were connected and rotated together by a tie rod, but the length of the tie rod was a little less than the distance between the king pins. The effect of this was that when the vehicle rounded a curve the outside wheel turned through a smaller angle than the inside wheel. The ideal geometrical arrangement is for lines drawn through the axes of all four wheels to intersect at a common point; there should then be no scrubbing of the tyres or interaction between the wheels. The Ackermann geometry gave perfect steering for only one turning radius, but by suitable design the error could be made small over a wide range of radii. The linkage was invented by George Leckensburger of Munich; he communicated his invention to Rudolph Ackermann, publisher and print seller of the Strand, London, who had it patented in 1818. Nothing came of the invention until it was taken up nearly 80 years later. Ackermann incidentally was born in Saxony and when a young man joined his family coach building business as a designer, working in Paris and London. He married an English woman and to give his family more financial security set up a print shop in the Strand and published books which are collectors' items today.

With the Ackermann linkage one stub axle carried a second lever, the drag link arm, in addition to the steering knuckle, which was moved by the

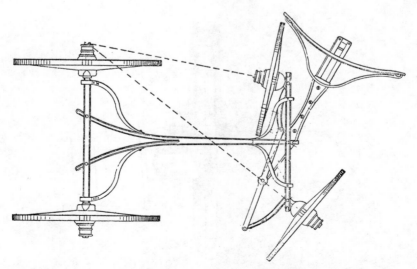

Figure 17.1 Ackermann steering as illustrated on original patent drawing.

steering mechanism. Steering was controlled by a tiller, a handle or a wheel mounted on a vertical pillar or column, and rotary motion of the column was converted into translatory movement of the drag link arm by the steering gear via the drag link.

Stub axles and steering heads

For many years carriages had used Collinge or Lemoine axles (p 364), the former in England, the latter in France. These were used on the early cars both as stub axles for the front wheels (figures 17.2 and 17.3) and for the rear wheels. These axles ran in journal bearings but the latter were soon replaced by ball and roller bearings.

The stub axles swivelled in Elliot or Lemoine heads at each end of the

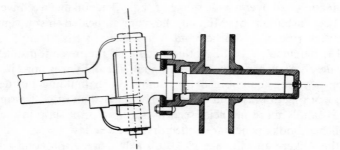

Figure 17.2 Collinge stub axle.

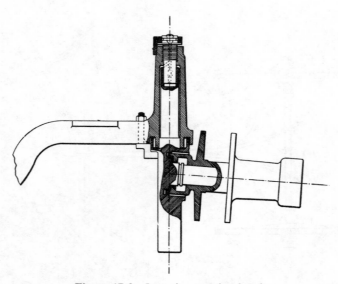

Figure 17.3 Lemoine steering head.

front axle beam. Axles were forged out of I-section beams and carried pads where the beams and springs were bolted together. Later when the engine was moved forward the centre part of the beam was dropped. Still later, when the engine had to be lowered to increase stability at higher speeds, the whole centre portion including the spring pads was dropped.

In the Elliot head (figures 17.2, 17.4) the front axle terminated in a Y-shaped steering knuckle with the stub axle attached to a vertical member which fitted in the opening of the Y and at first a pin ran through knuckle and vertical member to hold everything in place. The steering arm was attached to the latter.

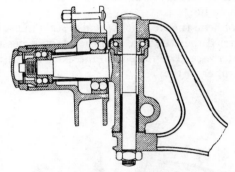

Figure 17.4 Elliot head as on a de Dion Bouton car of about 1912. By permission of Reed Business Publishing Ltd (from *The Automobile Engineer*).

In the Lemoine or French-type (figures 17.3, 17.5) the head had the stub axle attached to a long vertical shaft which was held in a corresponding long bearing at the end of the axle. The Lemoine head was simpler as it required

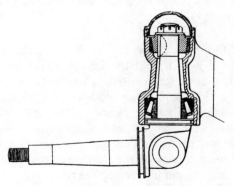

Figure 17.5 Lemoine head as on the Winton car of about 1912. By permission of Reed Business Publishing Ltd (from *The Automobile Engineer*).

fewer parts than the Elliot head and was easier to lubricate but was considerably more bulky. The Lemoine head raised the front axle somewhat, so a reversed head was developed with the stub axle above the vertical member with the thrust loads being taken by a thrust bearing at the bottom of the vertical member.

The next stage with regard to the Elliot head was to do away with the central pin and to make the vertical member, or kingpin, rotate in bushings at top and bottom of the steering knuckle. The upper fork had to carry the whole wheel load so some form of thrust bearing was later fitted. To prevent the kingpin falling through the steering knuckle, a hole was drilled through stub axle and kingpin so that half the hole was in each and a cotter pin passed through the hole.

The reversed Elliot head (figure 17.6) was also soon in use. The Y-shaped steering knuckle carried the stub axle and the kingpin passed through the knuckle and a cylinder forged on the end of the axle. The thrust bearing was above the lower member of the Y.

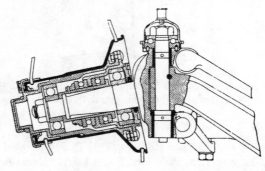

Figure 17.6 Reversed Elliot head as on the Sheffield Simplex car of about 1911. By permission of Reed Business Publishing Ltd (from *The Automobile Engineer*).

Before long the Elliot head was by far the most popular and by 1910 85 % of vehicles were fitted with it, but the Lemoine type persisted for many years, mainly on light cars. It was used as late as the mid-twenties on the 15 hp Wolseley. The 1920 Armstrong Siddeley and Angus Sanderson cars had T-shaped stub axles with a hub bearing either side of a vertical pivot.

At first king pins were vertical and stub axles horizontal. Bending moments were large and roads bad, so that the steering head suffered, and bumps on the road reacted back on the steering control. It was realised that if the line through the king pin intersected the road tyre contact area (centre point steering) these troubles would be overcome. This could be done in principle by tilting the wheel, that is, giving it a camber angle as well as by inclining the king pin axis, but even so centre point steering was difficult to arrange, and made steering heavier because the front end of the car had to be lifted slightly when making a turn, though conversely the car tended to straighten up automatically after the turn. About the turn of the century a straightforward castor effect was obtained by slanting the king pin axis to make its axis intersect the road slightly ahead of the tyre contact patch. Increasing the stability of the car in these ways made it easier and less tiring to drive.

An alternative to centre point steering was to insulate the driver from road shocks by using irreversible steering mechanisms. The steering could be 'locked' or 'direct' (or 'free'). If locked the steering remained set—in theory the driver could take his hands off the control and the vehicle would continue to turn. The difficulty was that the steering gear then had to withstand all the road shocks, though expedients like springs in the rods were used to take some of the shock. If the steering was direct shocks were lessened and could be accommodated by small movements of the steering wheel but this could be fatiguing to the driver, and dangerous on fast, heavy vehicles. Consequently, direct steering could only be used on light cars of moderate performance as a large measure of irreversibility was necessary for safety in fast, heavy vehicles. There could be no castor action or 'feel' if the steering was completely irreversible and argument about how direct or indirect the steering should be lasted for decades. Rack and pinion, bevel pinion and sector or epicyclic gearing (figure 17.7) gave direct steering; worm and worm wheel (or wheel sector) or worm and nut gave indirect steering, though some degree of reversibility could be given by increasing the angle of the worm.

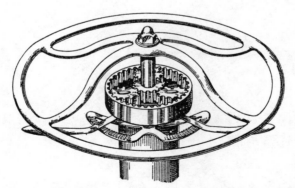

Figure 17.7 Epicyclic steering gear on the column of the Model T Ford.

Tillers ('cow tails') for steering worked after a fashion but they were very dangerous at higher speeds if the steering gear was reversible, so steering wheels began to displace tillers about 1897. Lanchester, however, retained his version of the tiller as long as he could. He criticised the conventional tiller very strongly because when the driver made a turn the centrifugal force on his body tended to throw him on to the tiller making him move the tiller even more and so tighten the turn. Lanchester put the steering column behind the driver and the tiller passed under the driver's arm to his right hand. The tiller was lightly weighted. This arrangement was inherently much safer.

The steering column of the 1901 Mercedes was steeply raked and this eventually became the conventional arrangement. In a few cars of this period the steering column was rather like a joystick; pulling it back braked the car, pushing it forward sped it on its way. Steering controls were on the right-hand side of the car on Continental, American and British cars. About 1910 the steering wheel was shifted from the right to the left-hand side of the car in the USA; the change took place much later on the Continent, only becoming general by 1929 although some more expensive cars used right-hand steering into the fifties.

The 1901 Humberette had a single-spoked steering wheel, the Citroën DS19 used a similar wheel nearly sixty years later.

Steering gears

To return to the other end of the steering column. The worm and nut steering gear (figure 17.8) consisted of a multi-start square-threaded screw and corresponding nut with trunnions on its outside carrying square trunnion blocks located in slots in the arms of a forked lever which was keyed or otherwise secured to the steering arm shaft or integral with it. The

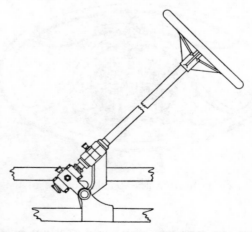

Figure 17.8 Early worm and nut gear. By permission of The Council of the Institution of Mechanical Engineers (from *Institution Proceedings*).

worm and nut gear wore less than the worm and sector and gave less backlash, but was more expensive and more difficult to adjust.

In another type of worm and nut gear the nut carried a rack which engaged a pinion on the drop arm (figure 17.9).

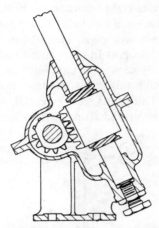

Figure 17.9 Worm and nut gear as used on early de Dietrich cars. By permission of The Council of the Institution of Mechanical Engineers (from *Institution Proceedings*).

In the worm and sector gear (figure 17.10) the worm was keyed to the steering wheel shaft and the sector bolted to a spindle which operated the drop arm. A drag link connected the drop arm to the steering arm on the stub joint by means of cup and ball joints to allow for the slight rotation that accompanied the translatory movement. Indeed at one time it looked as if universal joints would replace the ball and cups. As one ball was on the end of the drop arm the drag link could fall away if wear was excessive. The two steering arms were connected by a track or tie rod usually in front of the axle and the tie rod was held in place by simple jaw and pin joints. The latter wore rapidly because of dirt from the roads and the tie rod was easily damaged. Adjustment was not possible and parts had to be replaced when they wore.

Sometimes the steering gear was not enclosed, which aggravated wear. It was not until after 1900 that stops were fitted on the steering knuckles to limit wheel lock—previously the driver could turn the wheels until they were rubbing against the frame.

Gearing was low on the average Veteran and Edwardian car with one revolution of the wheel turning the wheels from lock to lock so although steering could be on the heavy side the driver had very good control over the car.

The worm and sector gear was soon improved and by about 1910 was by far the most popular type. Instead of a sector only, a full wheel was sometimes used and it was arranged that this could be rotated to fresh areas when wear occurred; the drop arm had to be reset at the same time but this was easy if the shaft was splined. To reduce the risk of the drag link falling off the ball was mounted at right angles to the bottom of the drop arm. Ball joints were spring loaded, each ball being held between a spring-loaded cup or sometimes between two such cups. Tie rods, however, usually continued to be held in place by plain pins in plain unbushed holes. These, however, did not hold grease nearly so well as ball and cup joints, so that they were eventually replaced though this was sometimes not until the thirties. With inclined steering pivots ball joints were essential, and only cheap cars used pin joints. Joints were protected from dust by leather covers.

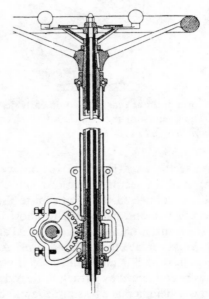

Figure 17.10 Worm and sector gear. By permission of Reed Business Publishing Ltd (from *The Automobile Engineer*).

Although there were improvements in detail design of the components, the steering linkage itself remained virtually the same as long as the front axle persisted (figure 17.11). There were some minor variations, however; the steering was sometimes turned round so that the drag link ran parallel to the tie rod instead of more or less perpendicular to it and if the drag link was transverse the tie rod could be divided and both parts connected to a common arm pivoted on the axle.

Although overall design was fairly static for a decade or more many improvements in detail were introduced. The value of castor action was appreciated, the geometry of the steering linkage improved and some attempts were made to compensate for wear. But to find the most satisfactory gear reductions, king pin angles, camber, etc, the designer only did what he had had to do for many years and progressed by trial and error and evolved rules of thumb.

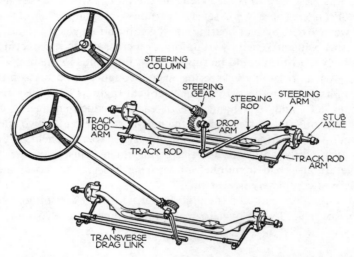

Figure 17.11 Steering linkages. By permission of Haymarket Publishing Co (from *Autocar* and *Autocar Handbook*).

Developments Between the Wars

There were no notable changes in Europe during the war years but in the early twenties four-wheel brakes were introduced and became universal by the end of the decade. However, these reacted on the steering and space also had to be found for the brake and brake drum. One way to do this was to reverse the normal Elliot head; the knuckle forging could then easily be provided with a large flange for the attachment of the brake cover. The reversed head also coped better with braking torque, and by the end of the decade it was practically standard. The axle itself remained of I section between the spring pads but an elliptical or circular section could be used between pads and steering head to stiffen the axle. Four-wheel brakes introduced other problems—steering, braking and suspension could interact so that designers were compelled to study the whole front end. Steering linkages, movement of springs, stiffnesses of axles and other factors had to be improved to reduce these interactions, but some problems were never really solved. This led eventually to independent front suspensions.

Throughout the twenties the worm and wheel, and worm and sector continued to be the most popular types of steering gear. The wheel was always a bit of a problem because it was bulky and heavy but wear could be allowed for merely by rotating the wheel round to use a fresh segment, as mentioned above. The gears were generally made of case-hardened steel, though one member was often still made of phosphor bronze. The worm and nut was also used to some extent on the more expensive cars and was liked in France. There were several variants. In one a floating-column spindle was carried by ball bearings at the top and engaged the nut at the bottom, this being attached by trunnions or a cross pin to the steering arm. Alternatively the worm could be supported at both ends in the box with the nut engaging a fork on the steering arm through slotted discs fitting in recesses in its sides. In the 1916 Lanchester gear (figure 17.12) the screw had intersecting left- and right-hand threads each of which engaged a corresponding half-nut. The lower ends of the half-nuts bore on the two ends of a short arm on the steering spindle in the box. There was no backlash and the steering was easily adjusted but the arrangement must have been very expensive. It also had a number of wearing faces. The worm and nut assembly was completely irreversible.

There were a few cars like the 9/20 Rover with rack and pinion steering (but the Rover 9/20 went over to worm and segment in 1925), and a few

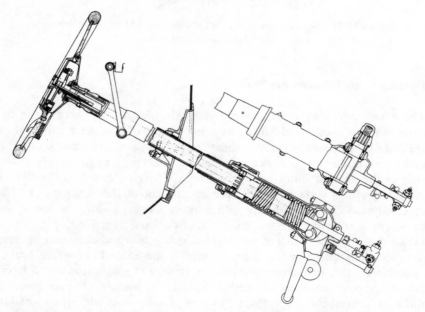

Figure 17.12 Lanchester worm and nut gear of about 1916. By permission of Reed Business Publishing Ltd (from *The Automobile Engineer*).

with epicyclic gears in which a pinion engaged an internally-toothed ring were still in use.

Marles invented a revolutionary new steering gear during the First World War and introduced it about 1919. It consisted originally of two spiral cams fixed to the lower end of the steering column which engaged a pair of rollers mounted on a crank connected to the drop arm. The rollers were mounted on an axis parallel to the column and the cams were in opposite phase so that when the wheel was turned one roller approached the pillar and the other receded, but both remained in contact with the cams, so preventing backlash. By substituting rolling for sliding contact, friction and wear were reduced.

In the early twenties Marles brought out a cheaper gear in which the cams became a very coarse helix of hour glass shape and a single roller mounted on a short forked arm engaged the helix (figure 17.13). Later the roller was replaced by a double roller, rather like a pulley, to increase the forked

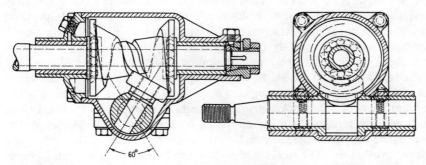

Figure 17.13 Early Marles gear; a roller engaged a helix. By permission of Reed Business Publishing Ltd (from *The Automobile Engineer*).

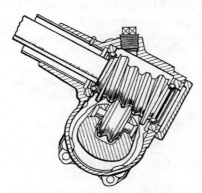

Figure 17.14 Marles steering gear; a roller mounted on a forked arm engaged a helix. By permission of Haymarket Publishing Co (from *Autocar* and *Autocar Handbook*).

bearing area (figure 17.14). Other designers followed the Marles' lead and a number of other proprietary gears were developed in which rollers or pins engaged worms or cams.

About 1925 high-pressure beaded-edge tyres gave way to low-pressure ones of larger cross section. The area of contact between tyre and road was much larger making steering much heavier, particularly at low speeds. This was overcome by lowering the steering ratio so that four or five turns of the steering wheel were required to go from lock to lock. Sensitive control of steering became impossible and it was difficult, for example, to control a skid, small deviations could not be corrected, so to restore vehicle stability and make the car safe at high speeds considerable castor action was needed.

Proprietary steering gears

In the thirties many types of gear were in use. The worm and wheel, or worm and sector types were used on small and medium cars, and the worm and nut on larger cars. The cam and roller or cam and peg types generally required special production equipment and so were made by specialist firms. They gave good control and little friction but were only slowly adopted, though eventually they displaced the simpler gears. It could be that the friction in the other gears helped damp shimmy. The Marles type has already been described. In the Bishop gear (figure 17.15) a hardened conical peg on a rocker arm engaged a specially generated groove cut in the column member. The rocker arm was mounted at one end of a cross shaft and the drop arm at the other end. The rocker arm of the Marles-Weller gear carried a parallel-sided peg free to rotate in a bush at the end of the rocker arm so that the peg could follow the helical groove. Hemispheres were recessed in each face of the peg. The Burman-Douglas gear (figure 17.16) was of the nut and worm type; a ball peg on the rocker arm engaged in a socket in the nut which could rotate slightly, as well as moving up and down, and so the arcuate movement of the rocker arm was allowed for.

Some very simple steering gears continued to be used, however. Jowett

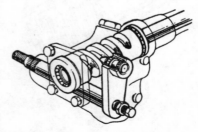

Figure 17.15 Bishop steering gear; a peg on a rocker arm moved in a specially generated groove. By permission of Haymarket Publishing Co (from *Autocar* and *Autocar Handbook*).

had a pinion on the end of the steering column which engaged an internally toothed ring the centre of which was above the pinion. Later BMW had a rather similar arrangement in which a bevelled pinion engaged a semi-circular rack on which bevelled links were cut. These were reversible, and as very light reversible steering was tiring dampers were sometimes fitted to make the steering feel safer and also prevent wheel wobble.

Some attempts were made to make the steering column adjustable for length and angle but without much success.

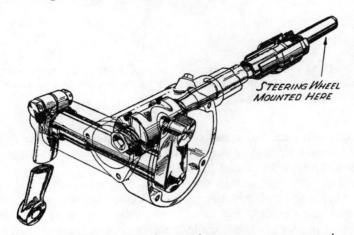

Figure 17.16 Burman-Douglas steering gear, a worm moved a nut which in turn rotated the rocker arm. By permission of Haymarket Publishing Co (from *Autocar* and *Autocar Handbook*).

Linkages

In the thirties independent front suspensions were widely used in the USA and on the Continent and to some extent in the UK and this had a considerable influence on steering. For one thing the conventional tie rod could not be used, for if one wheel and not the other went over a bump the wheels would have been pulled inwards. The tie rod was therefore divided into two and the joint connected to the chassis by a short pivoted arm and the other ends of the rods connected to the steering arms (figure 17.17). The pivoted arm was operated by a transverse drag link. Each wheel was in effect positively steered.

Another arrangement was to use a transverse rod connected to short arms at each end; the arms were pivoted on the frame and their other ends connected to the steering arms. One of the subsidiary arms was lengthened and connected to the steering gear drag link or operated directly (figure 17.18). Just before the Second World War several manufacturers reverted to the rack and pinion gear which had been out of fashion for some twenty

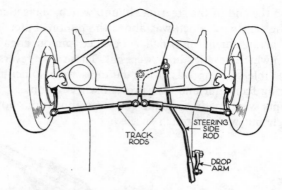

Figure 17.17 Arrangement of track rods for independent front suspension. By permission of Haymarket Publishing Co (from *Autocar* and *Autocar Handbook*).

years. The steering head also had to be modified and given some form of bearing at top and bottom so that it could, hopefully, retain its orientation irrespective of any deflection of the suspension.

It was quite difficult getting the steering geometry right. Unless the linkages were very carefully designed, deflection of the springs could interact with the steering, and to minimise this some quite complicated linkages were produced. These could incorporate a number of ball joints and pivots but, even if the steering was satisfactory to begin with, these bearings could wear and the steering become sloppy and insensitive.

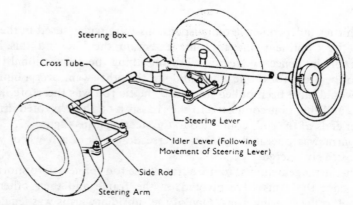

Figure 17.18 Arrangement of track rods for independent front suspension. By permission of Penguin Books Ltd.

Handling

In the thirties much was being learnt about the handling of vehicles. It was realised that during cornering the wheels did not move in the direction they were pointing, and that the cornering force developed depended on the angle between the plane of the wheel and the instantaneous direction of motion—the slip angle—and the load carried by the wheel. Ackermann steering had assumed the slip angle was zero, which is only approximately the case at low speeds. The cornering force depends also on the camber of the tyre which in general changed with spring deflection.

Differences in the cornering forces developed by front and rear wheels could cause the vehicle to understeer or oversteer, terms coined by Olley. If a neutral steering vehicle travelling in a straight line is subjected to a sideways force it moves in a straight line at an angle to the original direction of motion. An understeering vehicle develops a smaller cornering force at the rear and so the vehicle follows a curved path veering away from the neutral line, whereas on an oversteering vehicle the rear wheels develop a larger cornering force and so the vehicle veers in a curve which eventually points in the direction of the disturbing force. Similarly an understeering vehicle tends to come out of a curve and an oversteering vehicle to turn faster into a curve, consequently a driver has to wind on the steering wheel in the one case and unwind it in the other to keep the car on a curve of constant radius.

The suspension could not only alter the cornering forces but some suspensions allowed the rear axles to rotate slightly about a vertical axis when the car rolled, introducing a roll–steer effect. Radius rods did not always give the positive location the designer intended.

Eventually the designer had to reconsider what the various angles and what the camber, castor and toe-in should be, and examine how these changed with the suspension deflection. He had also to consider the effect on the latter of load transfer from inner to outer wheels during cornering, and try to work out how to give the car optimum handling characteristics, and minimise tyre scrub and adhesion problems.

Developments Since 1945

The worm and sector type of steering gear disappeared but the different proprietary gears continued to be used for many years. There were a number of designs; cam and roller (Marles), worm or cam and peg (Bishop) and worm and nut (Burman). Changes in detail design made them more efficient and easier to adjust.

In the Bishop gear, in which a peg at the end of the rocker arm engaged in a thread cut in such a way as to allow arcuate movement of the peg, the peg was mounted in bearings to reduce friction. The efficiency of the worm and nut gear was improved by using a half-nut that engaged the worm through steel balls which moved between nut and thread. When these balls left the nut they passed through a semicircular transfer tube back to the other end of the nut and so recirculated. The back of the nut carried a conical projection which engaged the inner faces of a forked arm on the rocker shaft.

Another type of gear (figure 17.19) had a full nut with recirculating balls and the back of the nut carried teeth which meshed with a toothed sector on the rocker shaft. This gear became widely used on Japanese cars.

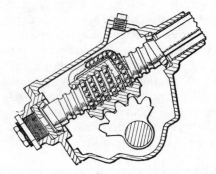

Figure 17.19 Recirculating ball type of worm and nut gear. By permission of Haymarket Publishing Co (from *Autocar* and *Autocar Handbook*).

An advantage of the cam-type worms was that they could be cut to give an automatic variation in gear ratio, so that a relatively large movement of the steering wheel in the straight ahead position gave a small deflection of the road wheels but a small movement gave a large deflection near full lock, making parking easier.

Rack and pinion steering had been discarded in the early years because it was too reversible but it was used on the Traction Avant Citroën (figure 17.20) and on some light cars before the Second World War. It was light, precise and well suited to IFS, in addition there were fewer bearings and pivots to lubricate and wear than on conventional linkages. The rack and pinion type became increasingly popular until by the mid-seventies it was used on all but a handful of cars in the UK. The teeth were at first perpendicular to the rack but later helical teeth were used to reduce tooth pressures and to give some irreversibility. Damping was also provided by a spring-loaded plunger which forced rack and pinion together.

In the mid-sixties collapsible steering columns appeared to minimise injury in accidents. GM used a lattice mesh in the column as an energy absorber, but later replaced the mesh by an arrangement of steel balls in a plastic cage; the balls ploughed into steel surfaces during axial collapse. Other arrangements were also used.

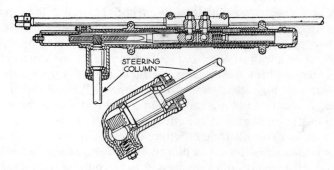

Figure 17.20   Rack and pinion gear used on Citroën Traction Avant. By permission of Haymarket Publishing Co (from *Autocar* and *Autocar Handbook*).

Simplification of the steering head, begun in the thirties, continued. Simpler, cheaper, ball joints replaced the bearings between head and upper and lower wishbones. Some vehicles retained ball bearings between stub axle and steering head but eventually these were omitted and the steering knuckle or bracket, carrying stub axle and brake caliper, was connected to upper and lower suspension arms by ball joints which thus coped with deflection of the suspension as well as rotation of the stub axle. On some light cars the steering knuckles (and suspensions) were remarkably simple.

The McPherson strut (p 304) became increasingly popular in the sixties and seventies. Suspension spring and damper were contained in the strut the upper end of which terminated in a thrust bearing in a compliant anchorage on the underside of the wing valance. The lower end of the strut was mounted on a bracket which carried the stub axle (and brake caliper), and the lower end of this bracket carried a steel ball that engaged a socket on the lower suspension arm.

In the sixties impregnated nylon ball joints were used more and more on the steering linkages as well as suspensions to decrease the amount of maintenance needed and also to reduce steering effort. Intrinsic castor action and parking torques tended to be high with radial tyres; this too was an incentive to reduce friction in the linkages.

The designer tried to arrange that the tyres developed the required cornering power despite deflections in the suspension linkages, and that there were no sudden changes in cornering power in particular manoeuvres. Often the greatest difficulties were in finding space for the steering gear and associated equipment, and ensuring that they did not foul the brakes on extreme locks and deflections. Indeed the designer could have quite a problem in finding any space at all for the steering gear, particularly if the car was made for export with left- and right-hand drive models. The geometry approximated Ackermann conditions to a certain extent but, as mentioned above, work on vehicle handling had shown that the slip angle of the tyre as well as the angle of the wheel should be considered. Indeed, with high lateral decelerations the slip angle could cause the instantaneous centre of rotation to lie ahead of the front axle, and not on the extension of the rear axle as required for Ackermann steering. Lack of Ackermann steering does not matter at high speeds when the steer angle is small but at very low speeds the resultant scrubbing can loosen a gravel drive, or make the car difficult to push in a turn, but the effect on tyre life is negligible.

If the king pin axis falls inside the tyre contact patch uneven braking of the front wheels causes the car to pull to one side or other during braking. In 1973 Audi arranged for the king pin axis to fall outside the contact patch, uneven braking in these circumstances tends to turn the wheels in the opposite direction and so correct for uneven braking.

Even in the seventies steering on medium-sized cars was not altogether satisfactory; the minimum steering circle could be large, and the steering not as light and responsive as on small cars. Larger cars required power steering. Some people considered that the steering had been better on cars made more than fifty years ago than on a number of contemporary cars.

Power steering

A power steering system devised by F W Davis was demonstrated on a Pierce Arrow car in 1926 and was considered for the 1933 Cadillacs but rejected because of the Depression. The War interfered with further plans, though power steering was used on some military vehicles, and it was not until 1951 that it became available as an optional extra on Chryslers and de Sotos. It was badly needed on the big American cars. The 1929 Chrysler (2672 lb weight) had a 9.5 to 1 steering gear ratio and 2.5 turns took the wheels from lock to lock. The 1951 Cadillac weighed 4290 lb and for the same wheel effort the ratio was 21.3 to 1 and no less than 5.3 turns were required to go from lock to lock. A driver really sweated it out trying to park in an awkward spot. Such low gearing did not improve directional control at high speed either. Chrysler's power steering enabled them to reduce full lock to 3.5 turns and to reduce the effort at the wheel rim to 10 lb, a quarter of what was required without power assistance; if the wheel effort was small operation remained manual, as it did when the car was moving straight ahead. Two systems were used in the USA. One system could be used in conjunction with a normal manual steering box; the drop arm, for example, operated a spring-loaded spool valve, movement of which admitted fluid under pressure to one or other side of the piston in a hydraulic cylinder and the piston in turn was connected to the steering linkage (figure 17.21). By the mid-sixties, however, integral gears had largely replaced these booster systems.

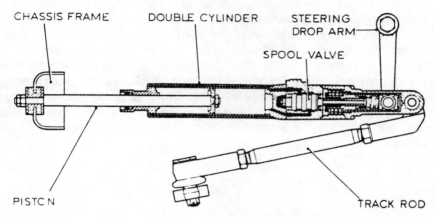

Figure 17.21 Booster-type powered steering. By permission of Reed Business Publishing Ltd (from *The Automobile Engineer*).

In the integral powered steering gear, control valve and power assist were all combined in one mechanism. Originally such integral boxes were bulky and expensive, and so were used mainly on bigger, more expensive cars. The steering column was connected to a torsion shaft and torque applied to the steering wheel twisted the shaft slightly (which gave a measure of feel) and thereby opened the appropriate fluid passage to admit fluid under pressure to one side or other of a piston in a hydraulic cylinder.

The hydraulic cylinder took up a lot of space and later systems in effect incorporated the cylinder in the column and made the nut of the ball and nut the piston—an ingenious solution. In some systems pressure increased as the wheel lock was increased, so reducing parking effort. On the 1970 Citroën SM a separate self-centering servo unit controlled by a centrifugal device was fitted, so that power assistance increased with speed as well as when the steering wheel turned to full lock.

Acceptance of power steering was slow at first but by 1966 it was used on about a third of American cars and by 1978 on more than 90 %. It is still used only on the more expensive cars in the UK.

The advent of powered steering renewed the old arguments about what feel the driver should be given. Fortunately the human being is very adaptable. Rolls-Royce demonstrated this years ago when they modified the steering of a car so that when the wheel was turned one way the car went the other way. Test drivers very rapidly learned to cope and could drive the car with no trouble. One poor man, however, after a session on the modified car, had adapted so well that when he went to go home that evening in his own car he created havoc in the works' car park!

Rear wheel (and four wheel) steering had been tried on some early vehicles but had been found to be unstable and the driver could have parking troubles. However, on some recent Japanese cars all four wheels are steered. In the Honda system the steering wheel operates on the front wheels directly with the steering input transmitted to the rear wheels through a centre shaft. For small and moderate steering inputs the rear wheels turn in the same direction as the front wheels, making lane changing easier and handling more responsive, but for greater deflections the rear wheels turn in the opposite direction to the front wheels, dramatically reducing the turning circle and making the vehicle more manoeuvreable and easier to park.

Chapter 18

Brakes and Braking Systems

Developments to 1914

The brakes on the first motor cars were based on contemporary bicycle or carriage brakes although Benz braked his first car by applying a small brake block to a countershaft in the transmission. Simple hand-operated spoon, or block, brakes operating directly on the solid tyres were often used but were not very effective, especially in wet weather and could damage the tyres, so they became almost extinct by 1900. Later as the car developed and became faster and heavier, brakes were designed specifically for it, but there was no obvious best solution to the problem of stopping so that many more or less experimental systems were tried.

Successful designs like that on the 1891 Panhard-Levassor tended to be copied. The Panhard-Levassor car had contracting band brakes on the hubs operated by a hand lever and a pedal-operated brake on the countershaft. Band brakes increase the torque because of their self-servo action—the capstan effect.

About the turn of the century the use of pedal-operated external contracting brakes fitted to drums on the sprockets of the rear wheel drive was fairly general. At first the band brakes were effective only when the vehicle was moving forward and so many cars were fitted with a sprag, a pointed iron bar hinged at its rear end to the chassis and held clear of the ground by a cord. On approaching a hill the driver released the sprag which would hopefully dig into the road and hold the car if the driver stalled the engine or missed a gear change. The sprag was less necessary when band brakes were introduced that applied a direct force to both ends of the band making them equally effective regardless of which direction the wheels rotated, but its use persisted for some time. Operation of the brake pedal declutched the engine, for it was not realised then that the engine could be an effective brake when it was in the appropriate gear.

Lanchester in 1900 combined cone clutch and brake in the one assembly; the sliding part of the cone when released past the neutral position engaged a fixed cast iron ring and so braked the vehicle. Daimler, in 1902, used an external band brake fitted to a shaft at the rear of the gearbox together with band brakes on each rear wheel (figure 18.1). A pedal operated the transmission brake and a hand lever the two wheel brakes through a balance bar to give compensation, and so ensure equal pull on each brake. In the following year Mercedes introduced the expanding internal brake; two shoes lined with friction material were forced outwards against the inner surface of the brake drum by means of an expanding ring actuated by levers and, as the assembly was enclosed, it was not affected by dirt and water. The cam-operated internal expanding brake was used by Renault in 1903. Rotation of a cam was used to turn the shoes about a pivot and force them against the drum (figure 18.2). The frictional force on one shoe, the leading shoe, produces an increase in the effective force applied by the cam, a self-servo action, and the ratio of drum drag to shoe tip load is called the shoe factor. On the other shoe, the trailing shoe, the frictional force opposes the applied load and tends to move the shoe away from the drum and reduce the torque. Other manufacturers soon followed suit and designed various mechanically operated leading–trailing shoe brakes.

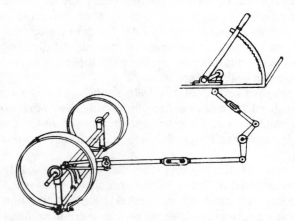

Figure 18.1 General arrangement of hand brake on a 1904 Daimler. By permission of The Council of the Institution of Mechanical Engineers (from *Stopping Revolutions*).

By 1905 shaft transmissions instead of the earlier chain or belt transmissions were in general use, with most vehicles having a pedal-operated transmission brake and band brakes on the hubs which were operated by a hand lever. The band brakes on the wheels gradually gave way to internal expanding brakes. The 1903 Mercedes used two separate transmission brakes; they were operated by separate pedals, the idea being that they should be used alternately to prevent either brake overheating, that is, reaching such high temperatures that its performance was impaired or its components damaged. The Mercedes drums incidentally could be cooled by dripping water over them. The 1912 Hispano-Suiza even went to the extent of using hollow aluminium brake shoes connected to the cooling system. Some transmission brakes were of the locomotive type in which two rigid shoes clasped the external surface of a drum, but later transmission brakes were of the external contracting band type.

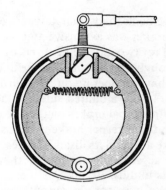

Figure 18.2 Principle of internal expanding drum brake. By permission of the Haymarket Publishing Co (from *Autocar* and *Autocar Handbook*).

Some cars had quite unconventional braking systems. The 1904 Hutton Co. car for example, used a pressurised hydraulic system; a pump driven by the engine delivered fluid to one of two sets of brakes on the rear wheels at a pressure determined by the driver. A similar system was used in the twenties and in a developed form is used today on some Citroën models and on the Rolls-Royce. On the 1904 Rover the brake pedal shifted the engine camshaft, altering the valve timing and converting the engine into a compressor. Rover found this brake so successful that it was used on larger cars in later years. Lanchester used oil-immersed multidisc brakes in 1906.

Braking arrangements

As brakes on the front wheels would have to be fairly complicated if they were not to interfere with the steering, or be themselves affected by the steering mechanism, all vehicles for many years were braked on the rear wheels only, either directly or through the transmission. Legislation also played a very important part in the development of brakes. For example, a motor car should have, according to a law enacted in 1904, two independent brakes, either of which should be capable of locking the wheels on one axle. The law did not state whether one or both axles should be braked, or whether primary and secondary brakes should operate on different axles.

A number of different braking arrangements were used to meet these requirements. On some vehicles two independent sets of brakes and brake drums were fitted on the rear axle. On others the brakes were independent but mounted side by side to share a common drum, or two concentric brakes engaged a double-flanged drum. Sometimes an external contracting and an internal expanding brake both operated on the one drum. Another popular arrangement was to brake the rear wheels and fit a third brake on the transmission shaft.

These various arrangements had their advantages and disadvantages. Merely duplicating the brakes was expensive and the brakes took up space. The use of a single drum for two independent sets of shoes meant that if the primary brake failed through overheating (a quite likely state of affairs), the secondary brake would have to operate on a hot drum and might also fail. Furthermore, if the shoes were mounted side by side the inner shoes were inaccessible. The internal/external brakes on the one drum were particularly prone to overheating, as the external brake acted as a good insulator.

The transmission brake also had its disadvantages. Braking torques could be at least double the accelerating torque so these brakes stressed the transmission components unduly. Because of their intrinsic high self-servo action, band brakes gave high torque outputs and could be vicious in action. Because they were so powerful the designer was tempted to make the transmission brakes small and light, again aggravating overheating troubles.

Whichever brake arrangement was used, a pedal operated one system and a hand lever the other, but for many years there was no uniformity as to which control should operate which system. Some designers thought that the more powerful brake (for example, the outer of two concentric brakes) should be the emergency brake, presumably because, by definition, a driver would want the maximum deceleration in an emergency. Others thought that the primary brake should be worked by the hand lever so that the driver could keep it applied by a ratchet when descending hills. On cars fitted with transmission brakes it became almost universal practice for the foot brake to operate the transmission brake, and the hand lever the wheel brakes. Because the foot brake could strain the transmission it was often recommended that it should only be used for gentle applications such as descending hills, and that in traffic the driver should use the hand brake even though he had to steer the car and operate the hand throttle at the same time! On the other hand, it was thought that the hand brake should be the emergency brake as it operated directly on the wheel brakes and so should be more reliable than the transmission brake, and the latter should therefore be the primary brake. The driver had to work out for himself what suited him best.

'Sideslips' and skidding

The maximum deceleration of a car can be increased by making the brakes more and more powerful, but a limit is ultimately reached which is set by the adhesion, or friction, between tyre and road surface. Too powerful brakes stop the wheels from turning, and the vehicle then slides along and skids on the locked wheels. Any further increase in pedal effort then has of course no effect whatever. Skidding can increase the stopping distance, but even more important, directional control of the vehicle is lost once the wheels are locked.

The braking force at a wheel should be proportional to the load carried by the wheel but when a vehicle decelerates it tends to rotate about the contact patches on the front tyres and weight is transferred from the rear to the front wheels, and consequently the higher the deceleration the easier it is to overbrake the rear wheels. Once the latter lock they tend to over-run the front wheels and a rear-end skid or 'side-slip' occurs.

Obviously all four wheels, not only the rear wheels, should be braked, and the braking should be divided between front and rear wheels in the proper ratio so doubling the deceleration and halving the stopping distance. Furthermore, the front wheels should preferably lock before the rear wheels for when these lock, the car, though unsteerable, moves ahead in a straight line.

All this was well understood by the early years of the century, indeed adhesion and its importance with regard to braking had been investigated on the railways in the 1870s by Galton.

It took many years, however, for four-wheel brakes to be universally adopted. The major difficulty was, as mentioned earlier, to arrange for braking not to interfere with steering and vice versa. Other difficulties were that the front axle and suspension had to be stiffened up and designed to withstand the braking torque, with a consequent increase in weight and cost.

A number of attempts were made to introduce four-wheel braking before the First World War. The three-wheeler Phoenix had its front wheels braked in 1903. P L Renouf patented a system for front wheel brakes in 1904 but it created little interest. In 1910 at least half a dozen makes of car appeared with four-wheel brakes. The Allen-Liversidge design was used on the Arrol-Johnston, Crossley and some other makes but in this system the main brake control rod passed through the steering pivot and under certain circumstances torsion of the pivot locked the steering when the brakes were applied so that this system was rapidly abandoned. The front brakes on the Argyll car used the system developed by Henri Perrot; these brakes were more successful than the firm making them, which got into financial difficulties and production stopped in 1914. Isotta-Franschini in Italy produced vehicles with front wheel brakes in 1911 as optional extras. Peugeot used the Isotta-Franschini system and Delage the Perrot system in the 1914 French Grand Prix; although Mercedes won the race the Peugeot and Delage cars had such better braking performances that their brakes made a considerable impression.

Developments Between the Wars

Before the First World War speeds were low and there was little traffic, so the main use of the brakes was to check speeds downhill. The driver therefore required only a hand lever with ratchet to operate the brakes. On a particularly long descent he would use hand and foot brakes alternately to prevent overheating either set of linings. With rear wheel brakes only, he would be wary about braking too hard for fear of sending his vehicle into a skid or 'side slip'. He preferred to steer his way out of trouble rather than use the brakes. But after the war, as vehicle speeds and performance increased, the driver wanted powerful brakes which would stop him quickly without skidding. He wanted a foot brake so that he could slow down for corners and avoid banging into other vehicles and at the same time keep his hands on the steering wheel. He further required a hand brake with ratchet for hill descents and parking.

Four-wheel brakes

Because the driver wanted powerful brakes, and powerful brakes on the rear wheels only sent him into a skid, there was great pressure on the designer to develop satisfactory front wheel brakes. As a result the use of four-wheel brakes became almost universal in the ten years after the First World War. Front wheel brakes first appeared after the war on the more expensive and larger cars like the 37.2 hp Hispano-Suiza and the 18/22 Hotchkiss of 1919, and then on cheaper and lighter vehicles. There were exceptions to the rule; the Austin 7 had four-wheel brakes in 1922 but Rolls-Royce persisted with two-wheel brakes until 1924. In many cases front wheel brakes were at first optional extras. Some UK manufacturers considered them to be dangerous but despite their adverse reactions, the changeover was so rapid that although 70 % of all cars available in the UK in 1923 were braked at the rear only, by 1929 all British models but one (the Trojan) had four-wheel brakes. Adoption of four-wheel brakes was even more rapid on the Continent but the Americans lagged behind both the Continent and the United Kingdom. Four-wheel brakes created much interest and controversy in the USA, as effective brakes were not so essential as in Europe and initially there was great reluctance from manufacturers to redesign the front suspension to accommodate front brakes and make other changes that increased costs. Nevertheless most American cars fitted them by 1927.

In the early twenties many different front wheel brake mechanisms were tried; some were so complicated that their complexity tended to defeat their object. However, after the middle of the decade only a few systems were in general use. The Perrot system was the most popular. The Rubery system was also used on a number of makes; hollow kingpins, and cable and pulley

DEVELOPMENTS BETWEEN THE WARS

arrangements were used to some extent, while hydraulic brakes made their first serious appearance.

In the Perrot system (figure 18.3) the brakes were operated by cams with the camshaft connected to the horizontal operating shaft by a universal joint located on the axis of the steering pivot. The brakes could therefore be applied with the wheels in any position. The operating shaft was telescopic and fitted with two universal joints so that it could extend and deflect when the springs deflected. The camshaft was above the centre of the wheel. Unless protected by suitable boots, the joints and splines could be affected by mud and corrosion.

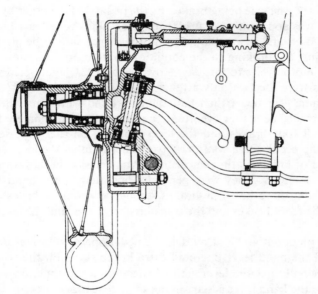

Figure 18.3 Perrot front wheel brake. By permission of The Council of the Institution of Mechanical Engineers (from *Stopping Revolutions*).

The Rubery mechanism was very similar but the operating shafts were mounted on and below the axle beam so one joint less was needed. Arms on the operating shafts connected with the brake pedal.

Another arrangement was to use a hollow kingpin: a shaft passed through the kingpin and the upper end of the shaft engaged the shoe-expanding cam or toggle which actuated the shoes while the lower end engaged a cam on a fixed horizontal shaft mounted on the axle beam. Changing the direction of the wheel only rotated the shaft about its axis on the lower cam. Rotation of the horizontal camshaft lifted the shaft in the kingpin and applied the brakes.

A number of manufacturers (often makers of sports cars) simply ran a cable round a pulley above the steering head to the brake camshaft lever.

A simple way of applying the front wheel brakes was to use a hydraulic system. Depression of the brake pedal moved the piston in a master cylinder and forced fluid from the master cylinder through pipelines and flexible hoses to wheel cylinders in the brake. Pistons in the wheel cylinders were moved accordingly and the pistons forced the shoes against the drums. Because the hosing was flexible, movement of the wheels had no effect on the brakes. Hydraulic brakes, however, did not appear to be as safe and reliable as mechanical brakes and for years did not make much headway.

It was thought at first that front and rear brakes should be operated independently, one axle being braked by the pedal and the other by the hand lever. This was apparently because front wheel skids were feared more than rear wheel ones and it was thought that the driver should be alert, and be able to adjust the braking to the conditions obtaining. The same situation exists with motorcycles to this day. However by 1923, uncoupled front and rear brakes were used on only a few models.

In the same year fewer than half the cars fitted with four-wheel brakes used the arrangement that eventually became standard, namely the pedal operating all four brakes and the lever working on the same set of rear brakes as the pedal. On a considerably smaller number the lever acted on a separate set of linings in the rear drums. Transmission brakes were used on a number of makes. They were generally operated by the hand lever, the pedal working the other four brakes, though on some vehicles the pedal operated the front brakes and the transmission brake, and the lever the rear brakes.

By 1929 on about 40 % of models the pedal operated brakes on all four wheels and the hand lever operated extra brake shoes on the rear wheels; about 20 % had the hand lever acting instead on a transmission brake, and the rest had the hand lever acting on the same rear brake shoes as the pedal.

Compensation

Connecting rods linked pedal and lever with levers on cross shafts which were in turn connected by rods or cables, often through relay levers, to levers on the brake assembly. Some problems gave much trouble, including lost motion due to flexure and stretch, and friction which had to be reduced to a minimum (and frequently was not) in what was very often a fairly complicated arrangement of rods and levers. Another problem was to what extent should the brakes be compensated, that is, what allowance should be made for uneven wear in the brake linings?

Without compensation one brake could come on before the other and so affect the stability of the vehicle. Some designers compensated left- and right-hand brakes on only one axle, others compensated front and rear brakes on one side, and some used full compensation on all four brakes. One objection to compensation was that if one component, say a lever pivot or a rod, should fail, all the brakes served by the compensator would also fail and in many completely compensated systems only one lever had to fail for all brakes to be inoperable. A second objection was that many compensating linkages were complicated and unreliable and did not work properly. Complete compensation, therefore, soon died out. A contributing factor was probably that the lack of rigidity in the linkage and chassis gave some degree of automatic compensation.

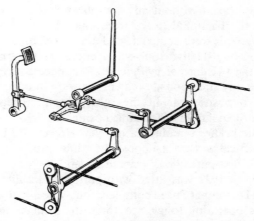

Figure 18.4 Fully compensated four wheel brakes—but if one cable broke all brakes were inoperable. By permission of The Council of the Institution of Mechanical Engineers (from *Proceedings of the Institution of Automobile Engineers*).

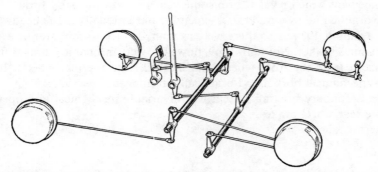

Figure 18.5 Four-wheel braking system of the late twenties. By permission of The Council of the Institution of Mechanical Engineers (from *Proceedings of the Institution of Automobile Engineers*).

By the time relay levers, compensator toggles, wear take-up points and other components were added the actuating system could become very complicated. Every pivot and joint even if properly lubricated contributed to the overall friction and to the overall lost motion. Many systems were grossly inefficient, and some extraordinary and potentially dangerous arrangements were in use (figure 18.4). A logical layout is shown in figure 18.5.

Hydraulic brakes

The advantages of hydraulic systems were realised very early—the front wheels could be braked without any complicated gear, compensation was automatic and the frictional losses were much lower than in mechanical systems; such systems were demonstrated before the First World War. One vehicle exhibited in 1911 had four-wheel brakes and the hydraulic system was split with the two master cylinders interconnected by a whipple tree; one front brake and the opposite rear brake were in one circuit, the other front and the opposite rear in the other circuit. There were doubts, however, about the reliability of flexible hosing; the system was not fail safe and so hydraulic brakes remained a novelty. However, M Loughead later invented a practical system (supposedly while prospecting for gold in California!); his first patent was granted in 1917.

The Bugatti 8 in 1921 was fitted with partial hydraulic braking but the 1920 American Duesenberg used complete four-wheel hydraulic brakes, the wheel cylinders operating toggles to force the linings against the drum. Pressures were of the order of 400 lb in^{-2}, and there were arrangements for topping up the master cylinder and for blocking off individual wheel cylinders should they not operate properly. Hydraulic brakes were used in the United Kingdom in 1924 on the 13/35 hp Triumph which had hydraulic brakes on the front wheels and mechanical brakes on the rear wheels, an arrangement which persisted on some vehicles until the early fifties.

Hydraulic brakes were at first expensive but gradually made headway so that by about 1934 or so there were as many makes with hydraulic as with mechanical brakes. Not until 1949, however, were mechanical brakes finally displaced from the front wheels and in 1952 from the rear wheels. In the USA mechanical brakes had practically disappeared by 1939.

The secondary braking system was invariably mechanical and operated on the rear wheel brakes.

Servos

A considerable pedal effort was required to brake a heavy car to the limits of its adhesion when all four wheels were braked. This led to the introduction of external servos to assist the driver, the pioneers being

Hispano-Suiza who fitted a servo on their 1919 model. For four or five years servos proliferated. They were at first mechanical, operation of the brake pedal loading a band against a rotating drum driven by the gearbox, and the resulting friction on the band operating the brake rods. Later plate clutches were used instead of band brakes; the clutch was driven from the gearbox and operated by the brake pedal and the torque developed used to apply the brakes. A complicated system of levers ensured that the servo was equally effective when the car was reversing. In 1924 Dewandre produced a servo which depended upon the inlet manifold vacuum for its operation; when the brake pedal was depressed, air was exhausted from a cylinder containing a piston; atmospheric pressure acted on the other side of the piston and the resulting force was used to apply the brakes. The vacuum servo displaced the mechanical servo but the Rolls-Royce mechanical servo was so successful that it was used until the sixties.

Within a few years, however, the use of external servos was limited to heavy, expensive cars as Perrot had developed the duo-servo brake in the early twenties, which had such a high shoe factor that servos were not usually necessary.

In the Perrot (originally Perrot-Farman) duo-servo brake the two shoes were linked together to make the brake similar to an internal expanding band brake (figure 18.6). The secondary shoe was pivoted and so the brake was correspondingly ineffective for stopping in reverse. The brake was improved by V Bendix between 1924 and 1927 and made equally effective for stops in either direction. At first the upper ends of the shoes of the Bendix brake were slotted and hung loosely from two pins mounted on the back plate, while their lower ends were connected by a link, springs holding the shoes away from the drum. The brake was applied by forcing the upper ends of the shoes apart by a floating link lever. The front shoe was forced against the drum and friction tended to move it round so that it forced the other secondary shoe against the drum, and the latter came up against its anchor pin. Wear was taken up by expanding the link between the shoes.

In the early thirties the brake was modified with a single anchor stop being used instead of two; the shoes were moved by an internal lever and Bowden cable instead of by a cam and levers.

The external contracting band brake operating on the wheels died out in the later twenties in the United Kingdom. This type of brake had been particularly popular in the USA, but also went out of favour there by 1929. The thirties saw the emergence of the specialist brake manufacturer. Bendix and Girling were the main protagonists of mechanical brakes, Lockheed of hydraulic brakes. Automotive Products Ltd started importing Lockheed brakes into the UK in 1927 and began their manufacture in 1929. Within a few years they had designed and started to make their own hydraulic brakes.

Mechanical systems were further simplified and made more efficient to cope with the increasing competition of the hydraulic brake. Single cross shafts became usual and cables eventually became more popular than rods. No compensation was used between front and rear brakes. In a properly designed system levers were hung in such a position that vertical movement of the wheels did not interfere with braking, whip in the chassis did not tighten the cross shafts, etc, but even so, with so many bearings and pivots present, friction and lost motion in the linkages could still seriously affect brake performance.

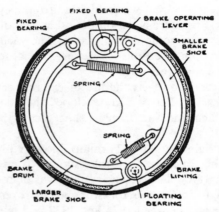

Figure 18.6 Principle of the duo-servo brake. By permission of The Council of the Institution of Mechanical Engineers (from *Proceedings of the Institution of Automobile Engineers*).

In the late twenties Captain A H Girling tackled the problem of braking afresh and invented a system (figure 18.7) which was much simpler and more efficient than most of its predecessors. He did away with cross shafts and reduced the number of rods to single rods to the front and rear axles, each acting through a compensator to rods connected to the individual brakes. All rods worked in tension, there being a small initial tension to take up lost motion. The pedal operated all four brakes and incorporated a fail-safe compensator between front and rear brakes so that if the rod to one pair of brakes failed the brakes on the other axle would still operate. The hand brake engaged a tab on a relay lever to operate the rear brakes. To enable rods in tension to actuate the brakes the brake shoes were forced apart by a hard steel conical plug which moved at right angles to the plane of the wheel (figure 18.8). The plug operated on plungers acting on the shoes and hardened-steel rollers were interposed between cone and plungers to reduce friction. The other ends of the shoes engaged a fixed cone through two other tappets or plungers and this cone could be advanced by a threaded shank to allow for lining wear. Shallow flats were formed on the cone which prevented vibration from loosening it and helped adjustment of the brake.

Girling arranged with the motorcycle firm New Hudson Ltd of Birmingham to manufacture his brakes and in 1933 they were used on Rover and Lagonda cars. The brakes, with minor modifications and improvements, were used by 1939 on a large number of British makes and remained in production for many years. In 1943 the Joseph Lucas Organisation acquired New Hudson and combined it with their own chassis engineering company, Bendix, to form Girling Ltd.

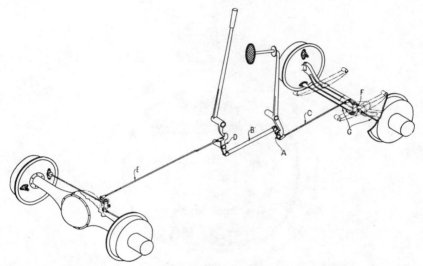

Figure 18.7 Diagrammatic arrangement of the Girling braking system. A, swinging link; B and C, rods; D, pivoted lever; E, rod; F, vertical shaft; G, rods operating brake shoes. By permission of The Council of the Institution of Mechanical Engineers (from *Stopping Revolutions*).

Considerable ingenuity went into making the various braking systems as effective as possible, for example the Bendix Auto Control transmitted braking force to the rear wheels until a certain force was reached, a spring was then sufficiently compressed for a wedge to lock against two balls locking the rear brake rod so that any further increase in braking effort went into the front brakes only, thus allowing for weight transfer. Girling in 1938 obtained the same effect by arranging the geometry of the levers so that the lever ratio changed as their deflection increased. A progressive band brake was developed on rather similar lines; a chain or cable ran round a cam on the hand lever, as the latter was pulled back the radius of the cam decreased. Consequently lost motion was taken up rapidly but once lining and drum were in contact the leverage was very high.

The introduction of synchromesh gears in the thirties made changing down on hills far less of a skilled operation, so drivers changed down more often, using the pedal for the occasional downhill check. The hand brake was relegated to a parking brake, and a standby brake if the primary braking system of the vehicle failed. As the hand brake was little more than a parking brake there was a tendency to make it as unobtrusive as possible, and the pistol grip type mounted below the dashboard became popular. The use of a separate set of linings for the hand brake died out in the early thirties but the transmission brake still persisted on a few makes.

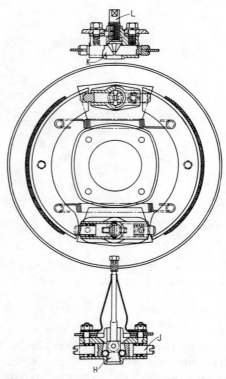

Figure 18.8 Girling brake: H, cone; J, plungers with inclined faces; K, members in contact with cone faces; L, adjuster cone. By permission of The Council of the Institution of Mechanical Engineers (from *Stopping Revolutions*).

During the thirties hydraulic systems gradually displaced mechanical brakes. All Morris cars except the Minor had hydraulic brakes in 1931 and all American cars had such brakes by 1936. The move towards hydraulics was accelerated by the increasing use of independent front suspensions, for these made great demands on any mechanical system. Some UK production models had tandem master cylinders and split systems in 1935, one half of the master cylinder operating the front brakes, the other half the rear brakes. Fiat had used split systems earlier. Stepped wheel cylinders were introduced so that more force could be applied to one lining than the other, either to make the lining wear more equal, or to make the brake more powerful still without further complicating it.

Some curious devices were tried, for example, one valve linked the clutch and the brake, so that if the clutch pedal was depressed while the brake pedal was applied the brakes remained on until the clutch pedal was released, so preventing the car from rolling backwards on an incline.

Lockheed had hitherto used wheel cylinders within the brake assembly but in 1937 they introduced a Bisector expander in which a central plunger attached to the piston of the external wheel cylinder forced two sectors of cams apart which in turn operated tappets which engaged the shoe tips. The Second World War delayed some interesting developments, for example, a valve which isolated the pipeline to the rear brakes when the deceleration reached a certain figure was suggested in 1940. The valve contained a fluted cylindrical plunger in an inclined cylinder; the plunger slid up the cylinder and closed a port to the rear wheel cylinders at a deceleration determined by the inclination of the cylinder.

After 1930 the brake assembly itself did not change for many years, and practically all brakes were of the Perrot or Bendix-Perrot type. The Girling brake, after its introduction in 1933, did not change much until it was eventually superseded by hydraulic brakes by 1948.

The duo-servo brake could be unduly fierce and was very sensitive to fade or water. Two-leading-shoe brakes are less powerful and more stable but as they act as two-trailing-shoe brakes in reverse and become relatively ineffective, they were only used on racing cars. Bendix in 1938, however, introduced a system of bell cranks and struts which made them equally effective in both directions of rotation. These did not become popular on cars but were used on some heavy Girling commercial vehicle brakes. Just before the 1939 War Bendix also developed a floating cam brake which overcame the problem of camshaft wear and also increased the output of the brake.

In 1938 Lockheed introduced sliding instead of pivoted shoes. At first the single pivot pin carried a square washer with the shoes cut to receive the washer and allow the shoe to move slightly up and down the washer. Higher shoe factors could be used without the shoes being so likely to grab or lock.

Automatic wear adjusters which operated when the lining movement exceeded a critical distance were introduced about 1935. However, they could also operate when thermal expansion of the drum exceeded this distance. One way to overcome this was to arrange for the adjuster only to operate when the car was in reverse, when the drum would be less likely to be hot. Automatic adjusters only became widely used when disc/drum combinations became common in the sixties.

Developments Since 1945

Mechanical brakes were not sufficiently flexible to cope with independent front suspensions and so the first few years after the Second World War saw the end of mechanical primary brakes. Production on the Hillman Minx with mechanical brakes on all four wheels ceased in 1948, and on the 10 hp Austin and Daimler DB18 2.5 litre in 1949. The basic Ford Popular retained them until 1959 and the Volkswagen standard model until 1962. The arrangement of hydraulic fronts, mechanical rears favoured in the UK also began to be phased out, for example, it was used until 1951 on the Austin A40, A70, A95 and until 1952 on the Riley 1.5 and 2.5 litre models.

In 1948 Girling introduced their first hydraulic brake and braking system for cars. At first they produced the hydrostatic brake in which there was no clearance between drum and lining, springs being used to keep them in contact and to prevent rattling. Much less fluid displacement was required to operate the brake. Later models had part of the leading shoe raised to reduce the risk of the leading shoe grabbing. Girling also made a range of more conventional hydraulic brakes. In 1946, Lockheed developed two-leading-shoe brakes that were first fitted to the front axle of the Morris 8 and 10.

In 1950 the brakes on all cars made in the UK except Vauxhall were made by either Girling or Lockheed and even the Vauxhall models used Lockheed hydraulic components.

By 1950 leading/trailing shoes on the front brakes were almost universal, exceptions being the Jaguar Mk VII and the Rover 90, 105 and 105E models, which used two-trailing-shoe brakes to obtain maximum stability and a servo to compensate for their very low shoe factor. The Peugeot 404 used two-trailing-shoe brakes until the disc-braked model appeared in 1968. Shoes were of the floating type, Girling preferring inclined abutments, Lockheed parallel abutments. Girling used snail cam adjusters to take up wear, Lockheed used adjusters between cylinder and shoe. Lockheed slotted their shoes to reduce squeal. Linings bonded instead of riveted to the shoes were standard in the USA but it was some little time before bonded shoes were used to any extent in the UK.

In the United States the Bendix duo-servo brake continued to be the most popular type, and generally either one or both shoes were fitted with linings of very low coefficient of friction (μ) to increase the stability of the brake. Drum brakes did not change much in the UK after the fifties; the rear brakes tended to carry less and less of the braking and so there was not much incentive for further development, though in the seventies automatic wear adjusters became mandatory. These were of two basic types, the one-shot adjuster and the incremental adjuster.

Servos which were standard on heavier American cars staged a comeback in the United Kingdom in the mid-fifties, partly because cars were getting larger and heavier. Clayton Dewandre had been manufacturing vacuum servos for many years and in 1954 Lockheed and in 1955 Girling entered the field. The earlier servos were air suspended, that is, air had to be evacuated from one part of the power cylinder to operate the servo. Later models were vacuum suspended, the cylinder on both sides of the power piston was evacuated and air admitted to one side of the piston when the driver depressed the brake pedal. As disc brakes slowly displaced drum brakes the use of servos to reduce pedal efforts became more widespread and they were eventually used on all but the smallest and cheapest cars.

Disc brakes

Lanchester patented a mechanically actuated spot-type disc brake in 1902. Friction pads on a pair of jaws gripped the outer portion of a sheet metal disc riveted to the wheel hub and in 1906 he used oil-immersed multidisc brakes on his 20 and 25 hp models. Disc brakes, however, were not generally adopted on cars, though an AC light car had a disc brake acting on the propeller shaft at the rear of the differential gear in 1920 while in 1928 the Tru-stop transmission brake was introduced in America and was used thereafter on some models for many years. In 1931 a mechanical disc brake operating on a ventilated disc was used on the transmission of a Dodge 4 ton truck. Disc brakes were also used on trams, in industrial applications and on aircraft. The Dunlop Rubber Company, for example, designed disc brakes for aircraft in 1935 and Lockheed disc brakes were fitted to Airspeed Oxfords in 1937. The advantage of such brakes for aircraft was that a number of discs and annuli of friction material could be used in the one assembly, reducing the work done on unit area of rubbing path and improving the cooling rate of the brake. Disc brakes for aircraft were further developed and are used on large aircraft to the present day. To return to vehicles, a Hawley spot-type disc was fitted to the 1937 model Crossley and in the same year disc brakes were used on Captain Eyston's Thunderbolt with which he captured the world speed record. During the war years Girling disc brakes were fitted to Armoured Fighting Vehicles. Chrysler in the USA in 1950 employed self-energising self-adjusting disc brakes on their Crown Imperial Model, but little further development took place in America for more than a decade. This was partly due to the greater weight of the American cars which made very high pipeline pressures necessary; also disc sizes could not be much increased thus making temperatures high so that pad life was short. If the servo failed very high pedal efforts were required to stop the car. Ventilated disc brakes were used on a few high-performance cars.

There was, however, much activity in the UK. Dunlop disc brakes were fitted to a Jaguar racing car in 1952 with such spectacular results that within a very short time disc brakes were used on all Grand Prix cars. In 1954 the D-type Jaguar and the Austin Healey 100s had Dunlop disc brakes. In the following year the Citroën DS19 appeared with floating caliper disc brakes and in 1956 Girling brakes were used on the Triumph TR3 and the Lotus II. In the same year disc brakes were first used on a production saloon car—the Jensen 541—as well as on sports cars.

Drum brakes were satisfactory for most purposes but they could have difficulty in coping with higher energy dissipations and smaller wheel sizes. Disc brakes were the solution to heating problems, and were more stable as they had no inherent self-servo to exaggerate changes in lining μ etc, but they were initially expensive so a number of years elapsed before their use became general.

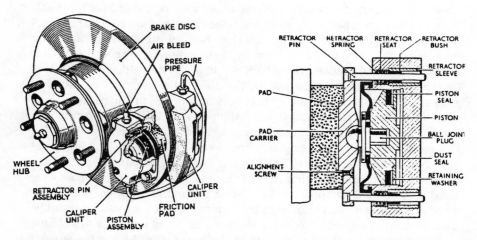

Figure 18.9 Early Dunlop production opposed piston disc brake showing automatic wear takeup mechanism. By permission of Dunlop Ltd.

The first disc brakes were of the opposed piston type (figure 18.9) but they were heavy and bulky. Lighter and more compact brakes were later made by using only one hydraulic cylinder and allowing the caliper to rotate or slide; on applying the brake one pad was forced against the disc and the resulting reaction on the caliper moved it slightly to bring the other pad into contact with the disc. Another arrangement (figure 18.10) consisted of a static cylinder body which carried a sliding yoke. The early single-cylinder calipers had their drawbacks as they could be noisy, and corrosion and seizure could occur, together with unequal pad wear. To overcome these problems pin sliders were introduced. The pads were supported in a frame and the clamping force provided by a 'fist' carried on sealed low-friction pins (figure 18.11).

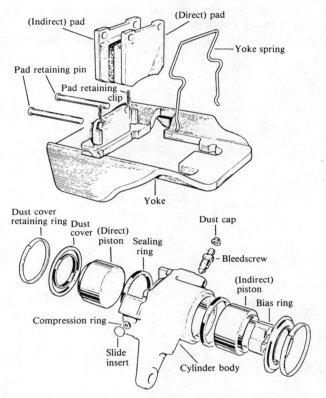

Figure 18.10 Girling A-type single-cylinder caliper. By permission of The Council of the Institution of Mechanical Engineers (from *Stopping Revolutions*).

Although disc brakes were eventually used on the front wheels of practically all UK, Continental and Japanese cars, the difficulty of obtaining sufficient torque from manually operated disc brakes to meet the legal requirements for the parking brake caused the leading/trailing drum brake to be retained on the majority of cars.

Disc brakes are used 'all round' on a number of more expensive cars but the need for high mechanical advantage has made their hand brake mechanisms more complicated than the corresponding mechanisms on drum-braked cars.

A similar trend was to follow in the USA, the short-lived Studebaker Avanti Sports car fitted disc brakes in 1962 and by 1965 the main American manufacturers offered them on the front wheels as optional extras. Afterwards they became standardised and the designs were similar to those outside the USA, the brakes generally being larger versions of those described previously to cater for the higher energies dissipated during braking. Again many vehicles retained duo-servo brakes at the rear, together with devices in the system to avoid excessive rear wheel braking.

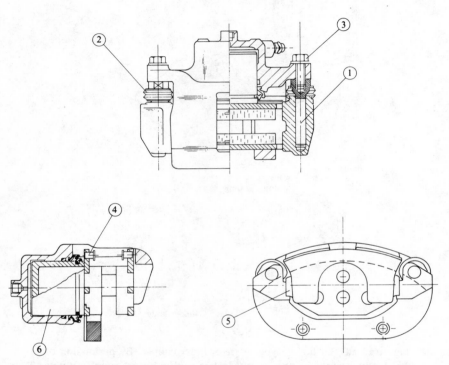

Figure 18.11 Girling Colette caliper: 1, guide pin; 2, dust cover; 3, guide-pin bolt; 4, pad springs; 5, caliper body; 6, piston. By permission of The Council of the Institution of Mechanical Engineers (from *Stopping Revolutions*).

Minimum braking performances which vehicles in the UK had to meet were laid down in the sixties in the Motor Vehicles (Construction and Use) Regulations and these were amended later by EEC regulations. To meet the latter and also, for example, Swedish regulations and the FVMSS 105 in the USA, tandem master cylinders and divided systems were used; if one subsystem failed there was sufficient residual braking in the other subsystem to meet the EEC requirements. A dashboard light warned the driver if a subsystem had failed. The simplest way to divide or split the braking system is to operate the front brakes with one half of the master cylinder and the rear brakes through the other half, but other arrangements were used though these could require two calipers to the one disc; for example, a diagonal split preserves the braking ratio but requires that the line through the kingpin axis should pass through the tyre contact patch in order that braking one front wheel only should not react on the steering wheel; by making the line intersect the road surface inboard of the contact patch, that is, using negative offset, the braking torque tended to steer the car in the direction of the unbraked wheel if one system failed.

Weight transfer on small front wheel drive cars can be so great that the lightly loaded rear wheels lock during severe braking. To overcome this, pressure limiting and pressure reducing valves were developed which isolated the rear wheel brakes at a specified pressure (and therefore deceleration) and any further increase in pressure was then diverted entirely or preferentially to the front wheel brakes. Later these valves were modified to work in divided systems, if one subsystem failed the valve was overridden so that all available adhesion could be utilised on the remaining subsystem.

A completely different type of braking system was pioneered by Citroën and used by Rolls-Royce. A pump driven by the engine develops pressure in the fluid and the brake pedal operates a valve which meters fluid through to the wheel cylinders at a pressure controlled by the driver. To allow the brakes to be applied when the engine is not operating the fluid is stored at high pressures in accumulators (reservoirs).

Antilock systems

Maximum deceleration is obtained when each wheel is braked to the point that it is just about to lock, and antilock braking systems exploit this behaviour. Antilock braking was developed in the first place for aircraft and proved to be very successful. It was adapted for vehicles but was at first so costly that it was used only on a few expensive cars, though its effectiveness eventually justified its use on commercial vehicles like petrol tankers and articulated vehicles. It was introduced for commercial vehicles prematurely in 1975 in the USA and was unsuccessful; for example, the control units could be triggered by electrical interference. Further developments and, in particular, the use of microprocessors in the control units, have now made it a more practical proposition; in Germany over half a million cars have been fitted with antilock braking and it is now standard equipment on Ford Granadas in the UK. The principle of antilock systems is straightforward; as the wheel starts to lock it decelerates very rapidly, which is sensed by an electromagnetic pickup mounted on the wheel, the braking force is then reduced so that the wheel speeds up again, and this cycle is repeated so long as the wheel is on the point of locking. A light warns the driver to reduce his pedal effort. As the wheel does not lock, the driver retains directional control of the vehicle and because the brake operates almost at the limit of adhesion stopping distances can be only a little greater than the minimum set by the tyre–road adhesion.

Drums and Discs

For many years drums made of pressed steel were used. Speeds were low so there were no thermal problems as far as the drum was concerned and the woven asbestos linings did not wear unduly or score the drum. But as speeds increased the higher temperatures caused drum distortion and scoring. Grey cast iron had been used successfully for many years for brake blocks on railway rolling stock, and about 1930 high-quality cast iron alloy containing considerable uncombined graphite (called *Millenite* in the United Kingdom) was introduced and in 1934 *Chromidium* in which, as its name implies, chromium was also incorporated. Cast iron drums and discs are used to the present day but various modifications and refinements have taken place to make the drums less susceptible to heat spotting and crazing.

Bimetallic drums consisting of concentric annuli of cast iron and aluminium were tried in the twenties—the cast iron had good friction and wear properties and the aluminium conducted heat rapidly away from the cast iron. These drums were expensive and there was difficulty in obtaining good thermal contact between the cast iron and aluminium; they are still used, however, from time to time. Another composite drum had a steel wall and the annulus making the frictional contact was made of cast iron.

Drums have changed radically in shape over the years. At first they were large in diameter but narrow in width. In the thirties the introduction of low-pressure tyres and streamlining caused a great reduction in diameter, so to obtain a reasonable area of frictional contact wider drums had to be used.

Discs have similar compositions to drums. Ventilated discs are used on some heavier and high-performance cars, particularly in the USA, to improve cooling; they are in effect two discs joined by radial or spiral ribs.

Friction Materials

In early car brakes, wood, leather, fabric or metal blocks were pressed against the steel rim or rubber tyre of the wheel, or against a steel drum. Later, band brakes were used which were lined internally with blocks of wood. Before long, speeds and weights were sufficiently high for the organic blocks to char and wear rapidly though they could be improved somewhat by suitable treatment, for example, boiling wood blocks in oil. Metal brake blocks withstood the higher temperatures but were noisy and erratic, and could score the opposing surface and they had to be accurately turned and fitted to the drum to give a good performance. They were much used on the locomotive type of transmission brake until after the First World War.

In the nineties, Michelin, before he went on to develop the pneumatic tyre, invented a brake block made of canvas and rubber but it was expensive and wore rapidly. In 1901 Herbert Frood patented a block made from layers of textile material impregnated with rubber if the block was to be used against steel, or with wax if it was to be used against rubber. After some initial prejudice his Ferodo blocks were gradually accepted and in 1904 the London General Omnibus Co. used them on their motor buses. The cotton-based material duly passed the requirements specified by Scotland Yard; for example, the vehicle had to stop in 14 ft from 12 mph (the speed limit), that is the vehicle had to decelerate at 0.35 g. As the duty of the brakes increased the cotton tended to char so in 1908 Frood replaced it with asbestos. The asbestos was woven into a loose fabric and impregnated with resins and varnishes of high-melting point. By 1914 the use of Ferodo brake linings was widespread and in the twenties all friction materials were of the Ferodo type. In the thirties Ferodo and the other friction material manufacturers turned to thermosetting resins and later introduced moulded instead of woven linings. The moulded linings were made by mixing fibre and resin together and polymerising the resin under pressure and temperature; fillers such as mineral and metal particles which modify the friction and wear properties of the lining could be introduced in these linings and polymers could be used which were impracticable with woven linings. As the woven materials were more flexible they were easier to handle, so that garages could keep large rolls of lining and cut off and drill lengths as required.

Pads for disc brakes were developed from the moulded type of lining. Some forms of asbestos cause lung cancer and though the chrysotile asbestos used in friction material is harmless, 'asbestos free' materials have been marketed in recent years.

Friction materials are examples of the many components which are continuously improved without any dramatic changes occurring. Even after 1950 some materials 'faded', that is, their coefficient of friction fell to very low values at high temperatures, whereas present-day materials can operate satisfactorily at very high temperatures with little change in coefficient of friction. Similarly wear rates are much lower and the brakes much less prone to troubles like 'early morning sharpness', grab and squeal.

Chapter 19
Wheels

Solid wheels made of shaped planks held together by dowels were in existence before 3000 BC but they were very heavy and broke along the grain of the wood. This problem was overcome by using spoked wheels, a remarkable invention, which were in use in Asia Minor by about 2000 BC. The rim of the wheel was made by steaming a single piece of wood and bending it into a complete circle and the rim was connected to the hub or nave by the spokes. Later the continuous rim was replaced by shaped wooden segments called felloes which were joined together by dowels, each felloe taking two spokes. A wheel built up in this way was stronger and easier to make than one with a continuous rim. When a knowledge of metal working became more widespread the wheels were strengthened by nailing curved strakes or shoes across the joints of the felloes and by clamping iron bands round the nave, one in front and one behind the spokes. Tyres made of a single hoop of iron instead of a number of strakes were not used in this country until the latter part of the eighteenth century. The centre of the nave was bored out and a metal box wedged in to make a bearing for the axle.

The craft of the wheelwright changed very little over the centuries; wheels were made in much the same way possibly for millennia, and the dimensions were determined by experience and rule of thumb. One important change took place in the sixteenth century when dishing was introduced. The axle was inclined so that the lowest spoke was perpendicular to the road and the face of the rim was inclined outwards. This permitted the upper part of the vehicle to be made wider and the wheel lasted longer as it did not work back and forth along the axle with the motion of the horse. The idea originated in the East; bronze models of Chinese carriages of the second century AD had dished wheels with twelve spokes and the stub axles were possibly cranked.

Artillery Wheels

In 1805 T J Plunkett invented the artillery wheel which was, as its name implies, used on guns. In the 1830s Hancock used such wheels on mechanically propelled vehicles. Instead of the spokes being tongued into a wooden nave the ends of the spokes ended in wedges which nested together and were bolted between two circular metal flanges (figure 19.1). The wheel was therefore made stronger than the carriage wheel and could withstand side thrusts better.

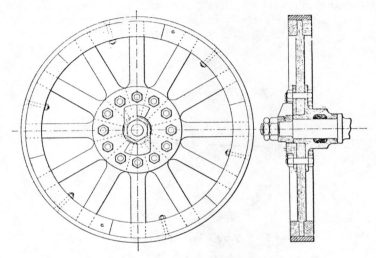

Figure 19.1 Hancock's wheel.

The car designer had to use artillery wheels, or wire wheels, rather than carriage wheels, for though the latter were light and graceful and often beautifully built they were just not strong enough for cars.

The felloes of the artillery wheel were made of ash or sometimes elm, and the spokes of oak, ash or hickory, and tangs on the spokes fitted into holes in the felloes. The wheel was assembled and the steel tyre channel heated and shrunk on or forced on to the wheel hydraulically, the tyre channel first being rolled to shape and its ends welded together. Screws were used in addition to prevent the channel from moving on the wheel. The inner ends of the spokes were held between an inner and outer flange, the inner flange being shrunk on to the box, and the whole assembly was clamped together by bolts passing from back to front. However, the artillery wheel had some disadvantages; it was heavy, and on the earlier wheels the spokes could shrink in hot, dry weather and work loose. The Sankey wheel, introduced about 1908, and its imitators, were similar in appearance to the artillery wheel but were made of two pressed steel halves welded together. In another type of metal wheel the steel rim, spokes and hub were all cast in one piece. The wheel looked very light and was about as heavy as a wooden wheel, but the casting was a difficult process.

Wire Wheels

Some of the very early cars like the Benz, Peugeot and Lanchester, were fitted with wire wheels based on the tricycle wheel, but by the end of the century wire wheels were generally used only on smaller vehicles while larger vehicles used artillery wheels. Even so a small minority of cars used wire wheels until after the First World War when they became much more popular.

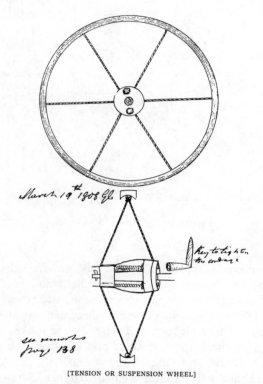

[TENSION OR SUSPENSION WHEEL]

Figure 19.2 Cayley's original sketches of his tension wheel.

The wire, or tension, wheel (figure 19.2) was invented by Sir George Cayley in 1808. Cayley was one of the pioneers of flight and he wanted very light wheels for his aircraft. He was the man who invented a practicable glider and expected his coachman to fly it. The coachman apparently offered his resignation on the grounds he had been engaged to drive, not to fly, but was nevertheless prevailed upon to pilot the contraption, which he did successfully.

In the wire wheel the load is suspended from the top of the wheel and the spokes carry the load in turn and maintain the rim in its circular shape.

Because the spokes are in tension the wire wheel is much lighter than the wooden wheel and so was eventually universally adopted for bicycles. By putting in two sets of spokes inclined to the vertical the wheel could take some side thrust. James Starkey patented tangent spoked wheels in 1874 though they were not widely used on bicycles; inclining one half of the spokes forward from the radial direction and the other half backwards the wheel could be made to absorb accelerating and braking torques. So far as cars were concerned the light weight of the wire wheel was a major advantage; climatic changes did not affect it and it was also flexible, but it was expensive to assemble and the wire spokes were difficult and tedious to clean. It was also thought to detract somewhat from the appearance of the vehicle, which seems strange when one looks at cars like the early Hispano-Suiza. Even so wire wheels lost favour in the USA in 1903/4 because of their appearance and apparent frailty; indeed the tide turned against them so suddenly that thousands of sets had to be scrapped and they were not used again in any numbers until the twenties.

Disc Wheels

A third type of wheel, the disc wheel, was introduced in France about 1903 by Messrs Arbel, who used two dished discs instead of spokes. It did not become popular for many years, possibly because the tyre valve and security bolts (needed to stop the tyre sliding round the rim) tended to be inaccessible and because the wheel was rather ugly.

By about 1914 half the cars in England had wooden wheels, about a third wire wheels and the remainder disc wheels. In France and Germany about three quarters had wooden wheels and the others wire and disc wheels, wire wheels being the more popular in France and disc wheels in Germany. Practically all American cars, however, had wooden wheels, though by 1918 a few of the more expensive cars had wire wheels and disc wheels were used on a small number of makes.

On the very early cars with centre pivot steering the front wheels had to be small to permit a reasonable lock. Rear wheels were large to give a better ride as on the conventional carriage; large wheels taking the irregularities of the road better than small ones. They damaged the road less and caused less dust, and their rolling resistance and bearing friction was also less, but they were heavy and long spokes were easier to bend so that the wheels were less stiff and laterally weaker than smaller wheels. With the adoption of Ackermann steering all four wheels soon became the same in size simplifying the manufacture of wheels and provision of spares, and improving the appearance of the car. Since small wheels were stronger than large ones, wheels tended to diminish in size, a trend which has continued throughout the entire history of the car. The outside diameter, that is, diameter of wheel plus tyre, decreased rapidly over the first couple of decades of the century and then rather more slowly. The inner diameter of the wheel proper was decreased more or less in proportion until the introduction of the low-pressure or balloon tyre during the twenties when it was reduced drastically.

Detachable Rims and Wheels

Punctures plagued the life of the early motorist. To make his life a little easier auxiliary wheels, detachable rims and detachable wheels were introduced.

T M and W Davies invented the Stepney Spare Motor Wheel in 1904. This was a specially strong auxiliary rim carrying an inflated tyre which could be hooked on to the side of the wheel with the flat tyre. It had four hooked clamps fitted on to its inner surface, two being fixed and provision being made for the other two to be tightened on to the wheel.

In the Challiner wheel the wheel rim was of channel section but the outer side was held on by bolts. In the event of a puncture this outer ring was removed, the flat tyre on its rim removed and the inflated tyre slid on and clamped in place by the outer ring. The channel was deep in section to accommodate the security bolts holding the tyre on its own rim, and was chamfered so that the tyre rim automatically centred itself.

Instead of replacing tyre and rim the motorist could remove the whole wheel and substitute a spare wheel with an already inflated tyre. The Rudge-Whitworth detachable wheel was invented by J V Pugh in 1905. The hub of the wheel slid on to a hub running on ball bearings on the axle, flutings on one engaging in flutings on the other. The wheel was secured to its axle by screwing a cap on to the inner hub and locked by a pawl engaging in a notch. It was claimed that a Rudge-Whitworth wheel could be changed in 30 seconds. Rudge-Whitworth wheels (figure 19.3), modified and improved, remained in production for many years.

Detachable wheels (figure 19.4) were in general use in the Old World by 1914 becoming almost universal by 1920, but the detachable rim was preferred in the USA where its use lingered on until 1932; the detachable rim was lighter than the detachable wheel and so easier to replace, that is, unless it became corroded into place.

Immediately after the 1914/18 War disc wheels became much more widely used in the UK and indeed had come into extensive use on vehicles during the war. Wheels could be of the single- or double-disc type. To make the single disc sufficiently rigid it was coned and was also usually reduced in thickness towards the rim, either by turning or rolling. Disc wheels made it easier to find space for the steering head and brake drum. Also wood suitable for rims and felloes was in short supply and expensive, particularly in Europe, so that manufacturers had to turn to metal or wire wheels and by 1925 practically all British makes had such wheels.

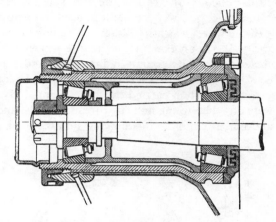

Figure 19.3 Rudge-Whitworth detachable wheel of about 1912. By permission of Haymarket Publishing Co (from *Autocar* and *Autocar Handbook*).

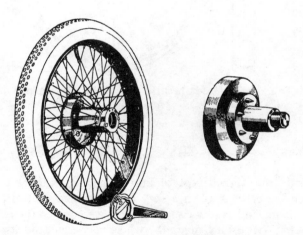

Figure 19.4 Early Riley detachable wheel.

Stronger wheels were required in the twenties because of higher speeds and better braking, including four-wheel braking. Triple lacing enabled wire wheels to be used on heavier and more powerful vehicles, so wire wheels were widely used on big expensive cars as well as the lighter ones. Wire wheels and disc wheels displaced the artillery wheel, but the cheaper disc wheel in turn eventually won out over wire wheels, though the latter have persisted to the present day on sports cars. This is probably a matter of tradition and appearance, for the saving in weight is small and the brakes do not run much cooler.

To overcome the difficulty of cleaning the wire spokes, very light metal discs, or rather shallow cones, were sometimes used to enclose the spokes. The appearance of the wheel was also thought to be improved. The Sankey disc wheel (figure 19.5) was an example of the disc wheel. It consisted of a separate pressed steel rim welded to a dished pressing, and a second smaller disc was welded to the inside of the disc to increase lateral rigidity. A number of tubes round the hub carried the attachment studs.

Disc wheels were attached to the hubs by threaded studs, and wire wheels by coned hub and nut, the hub and axle being splined to transmit torque.

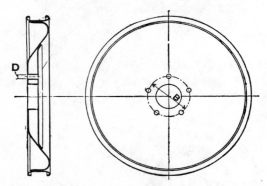

Figure 19.5 Sankey disc wheel. By permission of Chapman and Hall Ltd (from A W Judge *Motor Manuals*).

Magna Type

In the thirties the Magna type of hub attachment (figure 19.6) became popular. The wheel was given a very deeply dished central recess which carried the fixing studs and a circular nave plate covered the recess and hid the studs, etc. The dishing made the wheel very rigid and its rigidity was further increased by reducing the distance between the hub shell and the rim. The Magna hub was originally used on wire wheels but later the hub shell was further extended and welded to the rim so doing away with the wire spokes. The outer part of the wheel between cover plate and rim was generally perforated at intervals to give the appearance of spokes, and to help cool the brakes. All disc wheels were eventually of this type; after the 1939–45 War most cars had disc wheels (figure 19.7) not very different from those used in the thirties, with the same general design persisting to the seventies. The dished centre had a large central hole surrounded by smaller holes for the fixing studs and the holes were shaped so that the wheel spigotted accurately on to the hub. The disc could be corrugated or otherwise shaped to increase its rigidity and again generally carried ventilating holes. The cover plate was sprung into position. On a number of vehicles a light decorative pressing, the rim embellisher, masked the disc between cover plate and rim.

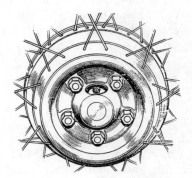

Figure 19.6 Magna wheel hub. By permission of Haymarket Publishing Co (from *Autocar* and *Autocar Handbook*).

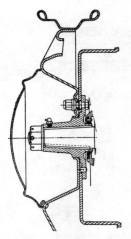

Figure 19.7 Typical wheel of the sixties. By permission of Chapman and Hall Ltd (from A W Judge *Motor Manuals*).

The wheels of cars of different makes tended to be of similar appearance so in the sixties the stylists got to work and rang the changes on the cover plate and visible parts of the wheel. The wheel could be given a different number of chunky 'spokes' and the cover plate decorated in various ways, or even omitted to expose the capped securing units.

Light alloy wheels had been used on racing and sports cars for a long time but, although very light, they were expensive and so came under the category of 'bolt-on goodies' bought by enthusiasts, so far as ordinary cars were concerned. They could, however, be readily cast into the shapes the stylist required and retain their light weight and strength and so became used on the top models of a range.

Rims were standardised in section and dimensions to ensure interchangeability of tyres, the seats of the rims having a 5° taper so that as the tyre was inflated the beads were forced up the taper to give a wedge fit. When tubeless tyres appeared in the fifties the rims had to be welded and not riveted on.

Wheel Bearings

As mentioned in the chapter on steering early motor cars used Collinge or Lemoine axles. In the Collinge axle (figure 19.8) the box or wheel hub was held in place by a dust ring and collar inboard, and a conical ring at the outer end, which also carried the oil cup. The original Collinge patent was dated 1787. The Lemoine axle (figure 19.9) had a collar clamped between

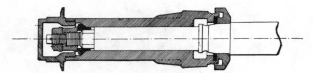

Figure 19.8 Collinge axle.

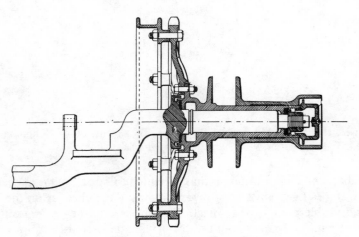

Figure 19.9 Lemoine axle.

leather washers and this held the box on the spindle. The very early Panhard cars had bought-out Lemoine axles. Soon these straightforward journal bearings were replaced by ball and roller bearings.

Roller thrust bearings had been used in classical times and it is even possible that roller bearings had then been used on some vehicles. In modern times ball bearings were successfully used by C Varlo on his postchaise in 1772 and this was followed by the invention of cages, roller bearings, and even self-aligning bearings. The widespread use of ball bearings along with the foundation of a ball bearing industry had, however, to wait until better steels were available and bicycles had established a market; it was found that ball bearings dramatically reduced the effort required to propel a bicycle.

Cars at first used ball races similar to those used on bicycles but improved materials and manufacturing methods had to be developed for these bearings to carry the greater loads involved. Design was straightforward. The stub axle was tapered and ended in a threaded section; the wheel hub had a corresponding tapered central shell and at each end carried a ball bearing race, the outer members of the two races being carried by the wheel and the inner members by the stub axle. Tightening the nut on the threaded part pulled everything together (figure 19.10). The difficulty with a simple arrangement like this was in coping with the side thrust on the front wheels during cornering. Various ways were tried—the balls could roll in grooves, or special thrust bearings could be used with balls rolling between vertical races (figures 18.3, 18.4).

Taper roller bearings, patented by H Timken in the USA in 1898, and intended for horse-drawn vehicles, were soon used on automobiles as they

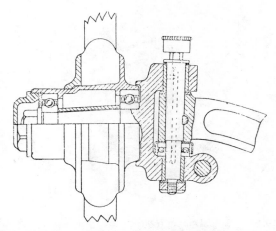

Figure 19.10 Early ball bearing. By permission of The Council of the Institution of Mechanical Engineers (from *Institute Proceedings*).

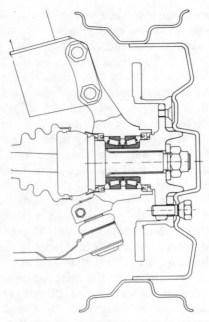

Figure 19.11 Taper bearings on Ford Fiesta. By permission of British Timken.

could take side thrust as well as normal loads (figure 19.11). Ball bearings, however, continued to be used, particularly in the Old World. One development of the thirties was the double-row angular contact bearing which contained two sets of bearings in the one assembly. This type of

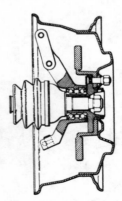

Figure 19.12 Typical single-row bearing system. By permission of The Council of the Institution of Mechanical Engineers (from *The Automotive Engineer*).

bearing is axially compact and has therefore become popular on front wheel drive cars, replacing the more usual two single-row bearings (figure 19.12).

Another development has been the semi-integral bearing. One ball race (the outer) was dispensed with, and the balls run directly on the machined hub. The number of surfaces to be machined and fitted, and therefore costs, were reduced. The next step was to make integral assemblies in which the balls ran directly on the machined surfaces of spindle and hub.

There have been remarkable improvements in materials and manufacturing methods in the lubrication and sealing of ball and roller bearings and in their detail design so that instead of lasting only 5000 miles or so they now last the life of the car.

Chapter 20

Tyres

Developments to 1914

The rubber tyre has quite a long history. In 1845 R W Thomson of Edinburgh patented a pneumatic tyre which consisted of a rubber tube inflated with air and enclosed in a leather case (figure 20.1). His tyres were used successfully on a brougham. They were silent and road resistance was considerably reduced, for example, from 45 lb for conventional tyres to 28 lb on a hard macadamised road for a 10.5 cwt carriage, and they ran for 1200 miles without wear. Later, Thomson replaced the outer leather case by a 'peculiar kind of canvas expressly manufactured for the purpose' and on the load carrying area a band of vulcanised india-rubber was attached. On a 15 cwt carriage these tyres reduced the rolling resistance from 48 to 28 lb on paved streets and from 130 to 40 lb on newly laid broken granite. The tyres did not sink into soft ground and they yielded to irregularities on the road and recovered when they had passed over them. Thomson had the right ideas and tackled the subject systematically but unfortunately materials and

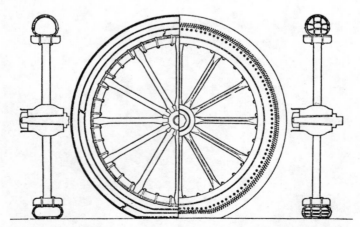

Figure 20.1 Drawings of two pneumatic tyres from R W Thomson's patent of 1845. The right-hand tyre has a multicomponent tube. By permission of Elsevier Science Publishers BV.

production methods were not available to make the tyre a commercial success.

Thomas Hancock in 1846 used solid rubber tyres about 1.5 in wide and 1.25 in thick either attached to a metal hoop or sprung on to the wheel and held in position by flanges. In 1852 a builder of invalid carriages encircled wheels with india-rubber, but the tyres could only be used on good roads and at low speeds or they were soon shed. Uriah Scott improved matters by vulcanising the inner part of the tyre sufficiently for it to be bolted to the wheel and joining to it an outer annulus of soft, pliable rubber. Mulliner secured a solid rubber tyre to the wheel by clamping it between lipped flanges attached to the felloes.

Bicycles were so popular and the advantages of spring or rubber tyres so obvious that hundreds of patents for all sorts of tyres were taken out and by 1870 solid rubber tyres for bicycles were fairly common.

Dunlop

A major advance occurred in 1888 when J B Dunlop of Belfast, a veterinary surgeon of Scottish descent, patented a pneumatic tyre made of textile material and rubber, and containing an inflatable rubber inner tube. He was unaware at the time of Thomson's earlier patent. He noticed one day that the wheels of his ten year old son's tricycle were leaving deep tracks in soft ground. He considered that larger, lighter tyres should sink in less, thus reducing the rolling friction and he had the inspiration of using a hollow tube filled with air as a tyre. He made a circle of wood, connected a tube of soft rubber to its circumference and inflated the tube with a football pump. The wheel rolled further than a solid-tyred wheel. He put 2 in pneumatic tyres on the driving wheels of the tricycle and a good adult rider had difficulty in keeping up with the boy. This led to further experiments on a tricycle which were so encouraging that a local Belfast firm built some bicycles with forks big enough to hold pneumatic tyres. Dunlop tried to interest people in his tyres without much success, but in May 1889 he got William Hume to ride in the Queen's College sports races using his tyres. Hume defeated two of the du Cros brothers who were so impressed that their father, Harvey du Cros, became interested. The older du Cros was a very remarkable and able man and among other things President of the Irish Cyclists Association. When he was approached by Dunlop and his friends he set up the Pneumatic Tyre and Booth's Cycle Agency Ltd with himself in effective command and started making tyres. The tyres were considered clumsy and ugly by the public and repairing them was a business 'which resembles an operation for appendicitis with complications' but prejudice against them was eventually overcome and this was helped by a remarkable series of racing successes. Bicycle racing was a very popular sport and the Irish riders on pneumatic tyres swept all before them.

Arguments and controversies also gave the new tyres enormous publicity, and testing the tyres under racing conditions speeded their development. In 1890, however, Thomson's patent was rediscovered and Dunlop's became invalid. du Cros retrieved the situation by buying Welch's patent for detachable tyres in 1891 (see below), eight months after the original patent application. Charles Kingston Welch was a fertile and successful inventor; amongst many other things he built a tandem tricycle for his wife and himself. Bartlett patented the beaded-edge method of attaching the tyre to the wheel rim within weeks of Welch's patent and du Cros acquired this too in 1896. In the same year the firm was reconstituted and became the Dunlop Pneumatic Tyre Co.

The firm from the start did very well financially. Other people tried to jump on the bandwagon and there was a boom in bicycle tyres and a little later in bicycles. The Welch and Bartlett patents expired in 1904 but by then Dunlops were well entrenched and had developed improved production methods which put them in a very strong position. The lawyers incidentally did very well out of the pneumatic tyre, the Dunlop Co. alone initiated 1350 actions for infringements of their patents up to 1904 and they in turn had any number of actions brought against them. The financial manipulations of du Cros's firms were very interesting to say the least; they were also extraordinarily profitable for du Cros and his friends.

The motor cars of the early nineties, however, ran on solid rubber tyres which were reasonably satisfactory up to about 20 mph so far as the ride was concerned, but even at this low speed they were difficult to attach securely. The tyre tended to stretch because of the centrifugal force, which was considerable at speed because of the weight of the tyres, and so tended to leave the wheel. To reduce the effect of thermal expansion caused by frictional heating it was recommended that the tyre should have a gap of 0.25 in between its ends (it was not continuous). Solid tyres were also very noisy and very expensive.

A few early cars had metal tyres, which says a lot for our ancestors' bottoms.

Solid tyres died out on cars in the early years of this century, but the starkly utilitarian Trojan of the twenties had them—pneumatics were an optional extra! Solid tyres were used on trucks until much later times.

Curiously enough solid tyres are still in use today on forklift and similar industrial trucks, for solid tyres can carry very heavy loads on small-diameter wheels and these trucks move sufficiently slowly to give the driver a reasonable ride.

Soon motor cars could reach respectable speeds, and as driving was a sport rather than a means of transport cars were driven hard and, as roads were bad, they had to be fitted with pneumatic tyres if they were not to be intolerably uncomfortable.

The Michelin brothers

Although the Dunlop Co. had founded the bicycle tyre industry, the credit for developing the pneumatic tyre for the automobile goes to the Michelin brothers. The way in which their firm first became involved in the rubber industry and then in tyres is curious.

A niece of Macintosh, the man who invented a way of water proofing cloth by covering it with rubber, married the part owner of a small factory near Paris making agricultural machinery. Madame Daubree, as she then became, made some rubber balls to amuse her children, and from this small beginning the firm went on to make a wide range of rubber products. The firm did not do very well in the early 1880s and André and Edouard Michelin, grandsons of the other part owner of the original factory, and who now owned 90 % of the company's shares, took over its management and with the financial help of an aunt got the firm going again.

One day an English cyclist arrived at the factory on top of a bullock cart complete with bicycle and asked for his punctured Dunlop tyre to be repaired. This was such a long and complicated business that the Michelin brothers were appalled and set out to make a better tyre. In 1891 they patented a tyre with a separate inner tube, the tyre being attached to the wheel by a steel band and bolts. The brothers demonstrated their tyres in long distance races, bicycles fitted with their tyres generally winning because Michelin tyres could be changed much faster when punctured than the tyres of rival makers. In the early nineties André Michelin made pneumatic tyres for horse-drawn vehicles, beginning with his own Phaeton. The tyres were necessarily thick and bulky and spoiled the light graceful appearance of a fashionable equipage. Though not unpopular in France, they were little used in the UK and America.

Michelin made a series of very careful measurements in 1895–6 comparing the rolling resistance of pneumatic tyres with that of solid rubber and solid steel tyres on different types of road surfaces and at various speeds. He found, for example, that if the resistance of the pneumatic tyre was taken as 100 units on a given surface the resistance of steel and rubber tyres was 123 and 127 units, respectively, on a good dry surface, and 150 and 143 on a bad surface. The pneumatic tyre gave an obviously better and quieter ride, but to make a more objective comparison Michelin mounted a pencil at the end of the axle and constrained the latter to move in a circle. The pencil trace on a piece of card showed that a steel tyre jumped over an obstacle and rebounded again. The rubber tyre on the other hand absorbed, or to use Michelin's phrase 'drunk in' the obstacle and the pencil trace showed that the axle moved slowly and smoothly up and down. The idea of the tyre drinking in the obstacle was the inspiration behind M Bibendum.

Next Michelin tried pneumatic tyres on motor cars. One of the first cars fitted with pneumatic tyres was a Peugeot which the Michelin brothers entered in the Paris–Bordeaux–Paris race in 1895. The tyres were similar to those that Michelin was supplying to the Paris cabs. All the twenty two spare inner tubes carried were used, and innumerable punctures and bursts were repaired at the roadside, but the Peugeot did finish the course though it was disqualified because outside help had been used to repair broken wheel spokes. The vehicle incidentally was called 'L'eclair'—'Lightning'—not because of its speed but because it proceeded by a series of zig-zags.

However Michelin persevered, and half the entrants in the 1896 Paris to Marseilles race used their pneumatic tyres. All the tyres discarded in the race were recovered and studied, and the reasons for their failure determined. Eventually their painstaking work paid off until the advantages of the pneumatic tyre were so much greater than their disadvantages that the pneumatics were generally adopted.

Some troubles arose because at first manufacturers based their tyres on bicycle practice. They used small section, highly elastic tyres, and if they were underinflated, which was likely, for it was hard work pumping them up, they rolled and lateral stability was affected. In addition, bicycle tyre construction was unsuitable; bicycle tyres were made by laying up strips of solutionised canvas by hand, adding solutionised rubber tread and finally vulcanising the lot. Patchy vulcanising could cause trouble and it was difficult to make the tyre uniform. The Doughty process replaced the handmade method; here the tyre was vulcanised, tread and all, between male and female moulds under considerable pressure.

The tyres, both pneumatic and solid, did not last many miles and for many years the cost of tyres was the biggest running cost the motorist had to face. In 1900 C S Rolls considered tyres cost him 2d per mile (a chauffeur would have cost him about 30 shillings a week) though admittedly he was driving one of the largest and fastest cars of the time. A small light car in the early 1900s might manage 2000 miles on one set of tyres if driven slowly, but anything over 1000 miles was considered good and a journey of a hundred miles without a puncture was remarkable.

Punctures were the bane of the early motorist's life and must have taken much of the pleasure out of driving. It was not that tyres were so bad but that the roads were dreadful, and a quick means of replacing a flat tyre was not available. Punctures were often caused by the cover being gouged or damaged, water could then get in and rot the fabric so that the inner tube would no longer be supported. Nails lost from horses' hooves were only too plentiful and caused many punctures.

Construction

In Dunlop's first pneumatic tyres the casing was made from 'Gents Yacht Sail Cloth', probably cotton or flax, but this had the disadvantage that the yarns chafed one another where they crossed and eventually broke and the tyre burst. Later the cover was built up out of several layers of canvas or fabric, and the outer casing covered with a protective layer of rubber which was considerably thickened at the tread which was also reinforced with two or three layers of canvas. Because they had so many wrappings and layers these early Dunlop tyres were called 'Mummy' tyres (figure 20.2).

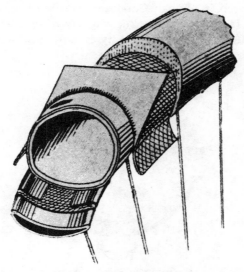

Figure 20.2 Dunlop Mummy bicycle tyre of 1888.

In 1907 three layers of light cloth were used for voiturettes, three of heavy cloth for light cars, and four or five layers of heavy cloth for large, heavy cars. Square woven cloth cut on the bias was considered very suitable. By 1910 a tyre life of 3 to 4000 miles could be expected.

Bicycle experience had shown that if threads were radial the tyre was resilient but rolled badly, if they were at an angle they tended to be in tension when the wheel was accelerating, and the tyre was more rigid and kept a better line. By having a second set of threads at another angle again, the tyre was more rigid during braking. The tyre thus had a crossply construction.

The warp and weft threads of the canvas casing chafed when the tyre flexed, but this was overcome by making the warp threads into cords by twisting threads together and virtually eliminating the weft threads. The cords were anchored at the edges of the tyre and impregnated with rubber to form plies and the casing was built up from these plies. The cords were inclined at an angle to the radius and alternate plies ran in opposite directions. Palmer patented cord construction in 1892 and car tyres were being constructed from two layers of cord in the early years of the century although it took a considerable time (presumably for patents to expire) before cord construction was generally adopted. It was not until the early twenties that the major manufacturers were all using cord construction.

Tubeless tyres were used on bicycles but they never became popular on cars. Inner tube and casing were integral (figure 20.3) and on cars the latter was generally very thick. Tubeless tyres, however, were liable to puncture and difficult to repair, though Goodyear tyres worked well on very light American cars. In 1900 a Locomobile was driven from Land's End to John O'Groats on tubeless (single-tube) tyres without, incredibly, a single puncture—or so the driver claimed.

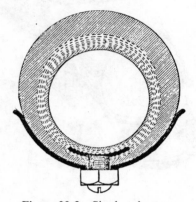

Figure 20.3 Single-tube tyre.

Strictly speaking these were single-tube, rather than tubeless, tyres. The true tubeless bicycle tyre had a normal cover which contained a sheet of rubber, a flap of which projected from one edge and pressed against the rubber on the other edge to form an airtight seal.

It was not until more than forty years later that tubeless tyres reappeared.

Rubber

Christopher Columbus in his second voyage (1493–6) noticed natives of Haiti playing with balls made from the gum of a tree, and Mexican games in which rubber balls were used were described in 1535. de la Condamine visited Peru in 1735 with a survey party which measured the arc of the meridian at the equator (the metre was eventually based on this measurement) and when he returned to Paris he brought rubber to the attention of the scientific world. Joseph Priestley popularised its use as an eraser and this lead to its name of 'India rubber'. Macintosh found that naphtha, a product of the newly founded gas industry, was an excellent solvent for rubber and in the 1820s invented his waterproof fabric—two layers of textile fabric with a layer of rubber in between. Thomas Hancock, brother of the Walter Hancock of the steam carriages, discovered how to process and masticate rubber, and tried to make the properties of rubber less temperature dependent so that it was not too stiff at low temperatures or too soft at higher temperatures. His efforts were stimulated by samples of a much superior rubber compound brought from the USA by Stephen Moulton in 1841, and in 1843 he patented the vulcanisation process in which rubber is interacted with sulphur. The American samples were probably made by Charles Goodyear, or someone using his ideas, for Goodyear had already hit upon the idea of vulcanisation, though his patent application was a little later than Hancock's. Incidentally, Walter Hancock made the machinery for processing the rubber for his brother before going on to build his steam carriages. Hancock thought it advisable to plant rubber trees rather than rely on wild ones growing in remote parts of the world and through his interest Sir John Hooker, Director of Kew Gardens, became interested and H A Wickham (1841–1923) was commissioned in 1876 to collect seeds in Brazil. The seeds were highly perishable so Wickham had to collect a large number and get them to the hot houses of Kew as quickly as possible which entailed smuggling them out of Brazil so as to save delay or even confiscation of the seeds. Only a small proportion of the seeds germinated but their progeny was sent to Ceylon and Singapore and from these a vast plantation industry eventually grew.

Para rubber was preferred for the early tyres with the rubber being vulcanised with sulphur. Various catalysts and other compounding agents were used to assist manufacture and increase the durability of the rubber. In 1910 it was found that carbon black could be a very effective reinforcing filler which increased the life of the tyre, but a considerable time elapsed before it was in general use.

Attachment of tyre to wheel

Dunlop's early tyres were attached to the wheel rim by a wrapping which passed between the wire spokes. This wrapping had to be undone before a repair could be made. Tubeless tyres could be cemented to the wheel.

For detachable tyres two methods were invented in 1890 for keeping the tyre on the wheel. W E Bartlett, an American who was Managing Director of the North British Rubber Company, patented the clincher beaded-edge tyre in 1890, though apparently T B Jeffrey had invented the clincher rim earlier. A thickened edge, bead, or clinch, usually of hard rubber, moulded into the form of a hook locked into a corresponding loop in the wheel rim so that the tyre was securely held when the tube was inflated. Correspondingly, if pressure was lost the tyre could leave the wheel.

Didier in France patented a clincher tyre about the same time as Bartlett; he sold it to Michelin for £800 and they used it on their tyres (figure 20.4).

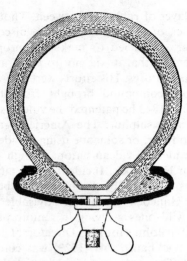

Figure 20.4 Michelin clipper tyre showing beads and security nut.

The second type of attachment was invented by Welch. He used an inextensible wire instead of a bead, the wires being embedded in the two outer edges of the cover (figure 20.5). In order that the tyre could be attached to the wheel the middle part of the wheel rim was recessed so that its circumference there was less than that of the wires. A wire could then be pushed into the recess and pulled over the wheel rim on the opposite side of the wheel. The stiff wires were later replaced by coils of thinner wires and eventually coils of fine wires twisted together were used to give more

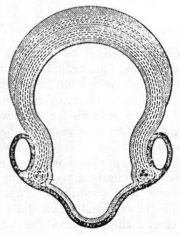

Figure 20.5 Welch recessed rim and inextensible beads.

flexibility and make it easier to remove and replace the tyre. The Welch method worked well on bicycle tyres but Dunlop found that wired-on tyres could not be made in quantity sufficiently accurately for motor cars. Indeed in 1901 the Dunlop Co. had to import under licence the Clipper-Michelin tyre and in 1903 they acquired the Bartlett patent from the North British Rubber Co. at a cost variously reported as £120 000 and £200 000. Both the Bartlett and Welch patents expired in 1904.

The Dunlop-Welch patent indirectly encouraged the development of the beaded-edge type tyre which was not so well protected by patents and therefore more widely used.

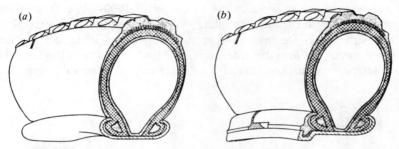

Figure 20.6 (*a*), Beaded-edge tyre; (*b*), straight-sided tyre with detachable flange. By permission of Elsevier Science Publishers BV.

Beaded-edge (BE) tyres (figure 20.6(*a*)) were always hard to fit and generally too tight at first and after a while too slack. They were likely to roll off on corners if not fully inflated. Stresses at the rim clinch were very high and could cause chafing, and it was fatally easy to nip an inner tube. Inner tubes incidentally were coloured red to make it easier to see whether they had been pinched. The beads, if all went well, held the tyre on to the wheel but if not sufficiently inflated the tyre could creep round the wheel. To prevent this from happening, and to stop the wheel shedding the tyre in the event of a puncture, security bolts (figure 20.4) were used to secure the tyre casing to the wheel rim and these were used until the twenties. Three or more bolts were normally used and sometimes the valve was shaped to form an additional security bolt; indeed on some tyres the valve was the only means of preventing the tyre from moving round the rim.

The BE or clincher tyre was gradually superseded in the USA by wired-on (WO) or straight-sided (SS) tyres. Steel wire and not hardened rubber was used in the beads. As the wire was inextensible one side of the wheel rim had to be removed to change the tyre (figure 20.6(*b*)) and so, unless rust caused detachable flange, rim and tyre to stick together, the tyre could be easily removed and replaced. In the straight-sided tyre the rim was flat but grooved on its outside edge. A flange was sprung into the groove which was split in one place so that when the tyre was deflated the flange could be removed. On small cars the rim was made in one piece and was flexible enough for the tyre to be coaxed on to it.

By 1916 the clincher tyre was used in the USA only on Fords; the balloon tyre superseded it completely. Presumably in the Old World the relatively well-to-do owner had his chauffeur or a mechanic to replace his tyres, whereas in the New World the owner generally wrestled with the tyre by himself and wanted something more manageable.

Valves

Several types of valve were in use in the early 1900s. The Woods valve (figure 20.7(*a*)) was the most popular; a rubber tube fitted over a metal tube drilled with holes. During inflation the rubber was lifted off the holes allowing air to pass into the inner tube, and back pressure in the latter kept the holes covered at other times. To deflate the tube the valve was partly

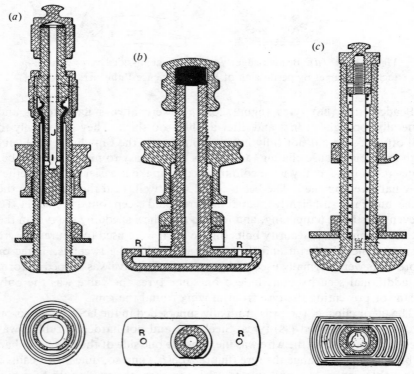

Figure 20.7 (*a*), Wood's valve—air passed through centre J of plug to holes I covered with rubber tube. (*b*), Welch's valve—holes R were covered by inner tube itself. (*c*), Sangster's valve—air passing down valve, unseated plug C.

taken to pieces. In Welch's valve (figure 20.7(*b*)) the inner tube normally covered holes in a box-like contrivance on the valve, but within the inner tube inflation pressure lifted the tube off the box. Sangster's valve simply used a conical rubber plug seating in the metal body and held in place by a spring (figure 20.7(*c*)). The Lucas valve was apparently similar to Sangster's.

In the Schrader valve (figure 20.8) a spring-loaded conical plug seated in a conical bore in the body of the valve and was held in place by the pressure within the tyre. Depressing a rod above the valve released air and allowed the pressure in the tyre to be measured. This type of valve, with modifications, is substantially that used today.

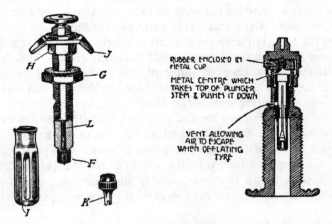

Figure 20.8 Schrader valve—air unseated plug, and plug could be depressed to release air.

Patterned treads

The early cars had smooth tyres and indeed these were mandatory until 1903 in order to minimise road damage and as only the rear wheels were braked it is not surprising that 'sideslips' were frequent, but as Dr Lanchester wrote, 'Whatever the state of the roads it is BAD DRIVING to navigate your car sideways'. Just why the rear end swung round when the brakes locked the rear wheels was not generally understood. Even Hele-Shaw as late as 1908 did not quite seem to understand why, though he made experiments braking front and rear wheels of a car on a concrete floor smeared with so much grease that he could barely keep upright on it. Lanchester gave a correct but rather complicated explanation, and George Darwin, a son of Charles Darwin, and Dr Burton gave a complete dynamical explanation in 1904.

Tread patterns had been tried on bicycle tyres as early as 1892, but with existing manufacturing methods it was difficult to incorporate a pattern, and on cars the back tyre spoon brakes would soon have worn off any pattern in any case.

By about 1906 spoon brakes were a thing of the past, manufacturing processes had improved and most automobile pneumatic tyres carried a patterned tread, but for many years the soft rubber tread wore so rapidly that the tyre sooner or later became smooth. Dunlop started moulding transverse grooves into the treads, the grooves being semi-circular in shape, 0.75 in wide and at 1 in intervals; some modern tractor tyres have similar patterns. The pattern was effective at first but the edges soon wore making it ineffective.

In 1908 Palmer tyres carried broad ribs running lengthwise to counteract sideslips. It was realised that the pattern should have sharp edges to penetrate water or mud on the road surface, and that longitudinal patterns counteracted sideslip and transverse patterns skidding.

In 1904 the legal requirement that all projections on the tyres had to be of the same material as the tyre itself was withdrawn. This led to the introduction of a number of different ways of reducing skidding and sideslip. Some tyres had steel wires embedded in them so that bits of metal came through as the tyre wore away. Leather strips carrying steel studs and fastened to the wheel rims were also used and Michelin made tyres in which metal-studded covers were vulcanised into the tread. Some tyres had steel plates attached round their periphery and Parson's non-skid chain was a zig-zag chain loosely attached to the wheel. Studded tyres and chains are still used in snow today. However, most of these devices, though they were reasonably effective, increased the wear of the tyres and played havoc with the roads.

By the end of the period a great number of tread patterns were in use—parallel ribs, cross bars, zig-zag bars, round-, square- or diamond-shaped blocks, etc.

Developments Between the Wars

By the early twenties all the major tyre manufacturers were using cord construction.

Tyres with canvas casings had to be small in diameter relative to the wheel diameter in order not to crease the fabric, and pressures therefore had to be high to carry the load; cross sectional areas of 3 to 5 in^2 were usual and the inflation pressure varied between 50 and 120 lb in^{-2} depending on the weight of the car.

Balloon tyres

It had been realised for a long time that low pressure, large section tyres (figure 20.9) would give a better ride but they became possible only when cord casings replaced canvas ones. When they came in about 1924, their adoption was about as big a step forward as when high-pressure pneumatics displaced solid tyres.

The balloon tyre had a cushioning effect and absorbed small shocks and lessened bigger ones. Minor rattles were less likely to develop and shock and vibration at the steering wheel was reduced. As a result a car could be driven on French roads at speeds of 30 to 35 mph whereas with high-pressure tyres speeds were only 15 to 18 mph for the same quality of ride. Because of the greater contact area the adhesion was increased and therefore the risk of skidding reduced.

Figure 20.9 Relative proportions of low-pressure tyre and the high-pressure tyre which it replaced. By permission of The Council of the Institution of Mechanical Engineers (from *Proceedings of the Institution of Automobile Engineers*).

Low pressures meant the end of the beaded-edge tyre. Stretching of the beads when fitting the tyre was not a serious problem when high pressures were used but with balloon tyres there might not be sufficient pressure to keep the tyre on the wheel. Consequently the much better idea of Welch was used, for the tyre could easily and reliably be fitted in the drop well rim. Michelin also adopted a recessed well, but the well was eccentric and had its maximum depth at the valve, and tapered off on each direction until opposite the valve the well disappeared.

The major drawback of balloon tyres was that they made steering heavier and accentuated wheel wobble. They also absorbed more power than the high-pressure type. The outside diameter of the tyre could not be increased to make room for the thicker tyre and so the wheel diameter had to be reduced, and this in turn limited the size of the brake drum (front wheel brakes were just coming in) which could lead to the brakes overheating.

The large car manufacturers did not wholeheartedly adopt balloon tyres at first because of their width, which meant increasing the track width, or decreasing the seating width; and because the steering, etc, had to be strengthened and the steering geometry altered. There were other troubles; the first covers were extremely large, thin and flexible and the pressure very low indeed, as low as 10 lb in^{-2}, but they were found to be too sensitive to inflation pressures, so sections were reduced and pressures increased. In the USA wheel rims were at first too narrow. The low pressures meant the tyres were not securely held to the wheel, and as the casing was thinner they did not withstand the manhandling to put them into position as well.

Once the low-pressure balloon tyre was adopted there was a trend for many years for wheel rim diameter to decrease and rim widths to increase. Tyre widths increased but not quite so rapidly. The cross sectional area of the tyre accordingly increased and at the same time inflation pressures were progressively reduced. There was also some reduction in overall tyre diameter. Those changes improved the ride, lowered the centre of gravity of the car and reduced the moments on the axles. They were gradual, because improvements in manufacturing techniques for tyres were continuous and because the changes were to some extent dependent upon improvements in the design of suspension and steering.

For typical medium-sized cars the outer and inner tyre diameters decreased from 31.9 to 28.4 in and 23 to 16 in between 1924 and 1940 whilst the tyre width increased correspondingly from 3.9 to 6.2 in. The inflation pressure fell from 50 to 28 lb in^{-2}. For comparison the tyres of similar cars in 1980 had outer and inner diameters of 22 and 11 in, a width of 6 in and an inflation pressure of 26 lb in^{-2}.

Tyres continued to be made by plying up. Each ply consisted of small cords held in place by widely spaced cross threads and by 1939 four plies were normal. Additional part plies or breakers were incorporated under the tread and there was similar reinforcement around the bead wire. The plies were separated by thin rubber sheet. Once assembled, together with the wall and tread rubber, the whole tyre was vulcanised in a special mould.

Improvements in technology and materials

There were improvements in rubber technology, plasticisers and antioxidants were used, and carbon black came into general use. By 1930 a tyre life of 15 to 25 000 miles was not unusual whereas before 1914, 3 to 4000 miles was normal. Though this increase was partly due to better construction and materials, the introduction of four-wheel brakes and, particularly, the making of better roads had a lot to do with the longer life obtained. The resulting longer life meant tread patterns lasted much longer. There was still a great variety of tread patterns (figure 20.10) and some extravagant claims were made about their effectiveness. Sipes were apparently invented at Brooklands in the twenties. Transverse razor cuts were made in the tread, these cuts or sipes opened up under load when in the contact area between tyre and road and threw the water clear when they moved out of the contact area. In the early thirties the National Physical Laboratory investigated the skidding characteristics of tyres on road surfaces. This work was taken over by the Road Research Laboratory (later the Transport and Road Research Laboratory) and still continues.

The road is half the tyre/road system and work was and is still needed to determine the optimum macrostructure and microstructure the road should have for good adhesion under all conditions.

As improvements in design made cars quieter there was pressure for the tyre to be made quieter too. This was achieved by spacing the tread elements slightly unevenly so that resonance effects did not occur, and by making sure the elements were well supported.

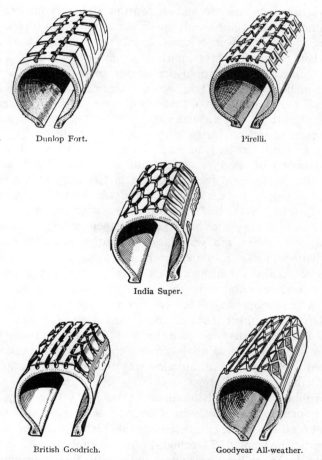

Figure 20.10 Tread patterns of various tyres. By permission of Haymarket Publishing Co (from *Autocar* and *Autocar Handbook*).

Lower tyre pressures and the softer springing made possible by independent front suspensions caused camber angles and wheel attitudes to change, particularly during cornering. This could affect the handling of the car in unexpected ways, and led to the systematic study of tyre behaviour. Indeed, until the thirties it was not realised how much the tyre contributed to the stability and steering of the vehicle, and how complicated the suspension/tyre interaction could be.

Developments Since 1945

After 1945 tyre design became more and more a science rather than an art. The interaction between tyre and suspension and the contribution of the tyre to comfort was studied with much work going into investigating and improving cornering power. Likewise treads were improved to increase the adhesion between road and tyre, particularly in wet conditions. The tyre has become a complicated and sophisticated part of the vehicle.

Instead of being circular, tyre sections became wider and wider, that is lower and lower in profile, to put more and more rubber in contact with the road. Another trend was to make the tyres stiffer to resist deformation during cornering and yet for them to retain their resilience.

There were considerable changes in the materials used and in the design and construction of the tyre. Buna rubber (butadiene polymerised with natrium, that is, sodium) was developed in Germany in the thirties and Germany had perforce to use this during the war. About 1940 du Pont made a synthetic rubber 'Neoprene' based on chloroprene and because of the loss of Malaya in 1942 the Allies had to eke out their supply of natural rubber with synthetic rubber and very large quantities of GR-S (Government Rubber Synthetic) were produced in the USA.

It was soon realised that the synthetic rubbers had a great potential and since the war a large number of other synthetic rubbers have been developed and indeed some modern tyres have no natural rubber in their composition. Styrene butadiene rubber (SBR) based on the original German formulation continues to be used. It was found that if oil was incorporated the rubber had better wet grip. Increasing the styrene content also had the same effect and indeed rubber of this formulation was used by shoe manufacturers in the late forties.

Other rubbers such as polybutadiene rubber had very good wear properties but poor wet road adhesion. However, with a number of different rubbers at his disposal the tyre designer could blend them together to obtain the properties he wanted. Thus high-hysteresis rubbers give very good wet grip but add to rolling resistance and so to fuel consumption. This drawback can be largely overcome by using high-hysteresis rubber in the tread alone, but a recent development has been to use a rubber that has high hysteresis at the higher frequencies that occur in the tyre–road interactions, but low losses at the lower frequencies associated with the rolling resistance of the tyre.

Special rubbers have also been developed for specific purposes, for example, butyl rubber which, being highly impermeable, is used for lining tubeless tyres. Until the Second World War plies were made of long staple cotton, rayon was then introduced, not to save cotton but to save rubber as less rubber was needed with rayon plies. Rayon bonds well with rubber and remains in wide use. Nylon and other polymers are also used.

Radial tyres

A major change in tyre design and construction took place in the fifties when radial tyres were introduced. The idea of radials was, however, very old, Gray and Sloper having taken out a patent in 1913 in which the plies were radial and a flexible inextensible belt ran under the tread, and under the outside of the ply.

In the later radial-ply tyre the plies were again radial but the inextensible belt or rigid breaker was outside the plies (figures 20.11, 20.12). The idea was that the radial plies should deform easily in the radial direction and so give a good ride, whereas the rigid breaker, though flexible, was very rigid fore and aft and so should give good cornering. The breaker also reduced distortion of the tread and kept a larger area of rubber in contact with the road, thereby increasing the adhesion. The lack of distortion and the support of the breaker also give the radial tyre a much better life than an equivalent crossply tyre. The suspension engineer liked the radial tyre because it was relatively insensitive to changes in camber and so did not make such demands on independent front suspensions as the crossply type.

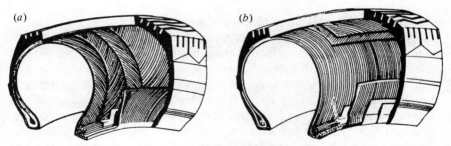

Figure 20.11 Construction of (*a*) crossply and (*b*) radial-ply tyres. By permission of Elsevier Science Publishers BV.

The disadvantages of radials were that they were more expensive to make than crossplies, they were rather harsh and noisy and made the steering heavier at low speeds. They could also behave disconcertingly in some manoeuvres, such as when going fast through an S bend.

The radial tyre was considered in the later forties as a means of increasing tyre life for the economy-minded French motorist. They were produced by

Michelin who used rayon casing and steel wires. In the early fifties Pirelli introduced the all-textile Cinturato tyre. Car manufacturers began to design cars, such as the Rover 2000, to match the radial tyre.

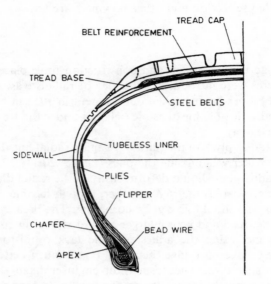

Figure 20.12 Steel-belted radial-ply tyre of the eighties. By permission of SP Tyres UK Ltd.

Eventually radials became practically standard in Europe, and some of the tyre manufacturers became victims of their own success; radials lasted so much longer than crossplies that the replacement market contracted and firms got into financial difficulties. It took much longer for radial tyres to become standard in the USA; tyre manufacturers did not want anything to do with them because they could not be built on existing equipment and enormous investment was needed for retooling. There was pressure, however, from the car manufacturers for lower profile tyres to improve handling of the vehicle and in the mid-sixties the belted-bias tyre had crossply construction and in addition a rigid belt under the tread as in the radial. In the late sixties attempts were made to use polyester cords and fibre glass belts, but difficulty was experienced with quality control and interaction between plies and belt led to overheating and other troubles. In 1971 GM decided that in future they would change to steel-belted radials and in 1974 Ford specified similar tyres for most of their range. The American tyre industry could not at first cope with the demand and tyres were imported from Europe. There was a hiccup in 1975 when Firestone radials gave trouble and lawyers and politicians had a field day but Firestone brought in a new thoroughly tested and reliable tyre and extricated themselves from a tricky position.

Tubeless tyres

Tubeless tyres reappeared in the USA about 1948, in Europe about 1953 and now are almost universal. Spokes were no longer used on rims which could therefore be made airtight, and the inside of the tyre was covered with an impervious layer of rubber. Tubeless tyres reduce the chance of a blowout and they are quick to mount, and this was an attraction to the vehicle manufacturer as their use saved time on the assembly line.

Punctures and blowouts have become rare events—in the mid-seventies about one in 18 000 miles—and so the spare tyre may not be needed for years at a time. The vehicle manufacturer would like to dispense with it, as it is either difficult to house or takes up too much space, and adds to the original cost of the vehicle. This has led to the development of 'get you home' tyres by several manufacturers but no set of four is yet cheaper than the conventional four plus spare.

Chapter 21
Bodywork

Developments to 1914

The bodywork and seating arrangements on the first cars were naturally fairly simple. The designer concentrated on the engine and chassis and then added seats on the vehicle as best he could. There was, however, a flourishing coach building industry and before long it seemed natural for the engineer to build the chassis and then hand it over to the coach builder to mount a body on it. Relationships between the two were apparently rather strained, for the more conservative coach builder wanted nothing to do with motor cars, and the others were not very willing to adapt their designs and methods to the motor car. Indeed in the 1900s many manufacturers were more or less forced to become their own coach builders. This did not go down well, for the car people were accused of poaching skilled men from the carriage builder. Austin did not help matters when he flatly told them that they did not know their own business.

In many cases coach builders tried to adapt existing carriage styles; the results were often very freakish, but sometimes quite delightful, particularly in the smaller cars. By building his own bodywork the manufacturer could get exactly what he wanted and did not have to accept unsuitable or shoddy work that would give him a bad name.

A short, high carriage looked well behind a horse so that the carriage builder wanted as short a wheel base as possible; he considered it impossible to put a nice looking body on a long chassis or to make a long body that would not sag. But a motor car with a short wheelbase was unstable and prone to sideslip. The early wheels were large in diameter which resulted in the driver and passengers being up in the air, but after all they did not have to look over a horse and so the coachwork was lowered. The adoption of the Système Panhard and mounting the engine in front of the passengers and not under them also helped, as did the use of pneumatic tyres.

The major criticism of the coach builder, however, was that the bodywork was far too heavy, so that a chassis which was quite satisfactory when carrying a phaeton body was grossly underpowered when fitted out as a limousine. In the mid-1900s the body of a large touring car on a chassis weighing 15 cwt could weigh 7 cwt, a small landaulette 6 to 8 cwt, and a large limousine more than half a ton. Though a heavy body impaired performance it generally improved the ride, for the suspension could be designed for the one constant weight, the designer knowing that the weight of the vehicle changed relatively little, full or empty.

Bodies were made of wood. English ash was preferred for frames, though it was difficult to obtain, and mahogany for the best panelling, otherwise sapele or acajou. The wood was steamed and bent to shape and the bodywork was built up by skilled craftsmen. For many years they used rule of thumb and experience to decide the dimensions and spacing, etc, of the frame members. The panels were mounted and the body was then painted and received quite an incredible number of coats, being rubbed down between each coat. It is not surprising that as late as 1913 a man year of work went into making the body of a not particularly relatively expensive car. Aluminium panelling was also used and after some preliminary difficulties was successful; it did not soak up such quantities of paint as wood, and could be formed into complex shapes.

Wooden panels could not be given curvature in two directions and so to make a body with fashionable curves wooden blocks had to be glued together and shaped. This was not very satisfactory, whereas aluminium or steel panels could be beaten into shape. Wood panelling was still used for flat surfaces, and surfaces with simple curvature, and indeed was sometimes used on doors, even in the late twenties. Mouldings masked the join of panel and frame.

So far as the seating arrangements and bodywork were concerned the manufacturer and the coach builder had to satisfy several different types of owner. The racing driver was not much of a problem as he did not mind how stark his vehicle was in appearance and, being a man of iron, he was not particularly worried about the seat. The overwhelming proportion of car owners were sporting motorists who motored for pleasure and wanted something smart in appearance and with some measure of comfort.

There was also a market for small, cheap vehicles with single-cylinder engines, and rather larger vehicles for the professional man. Finally, there was the relatively small proportion of owners who wanted and could afford a horseless carriage—a vehicle with all the comfort, luxury even, of the best contemporary carriages but without the horse.

The sporting motorist drove an open car, a tourer, and the bodywork of the open car evolved in a fairly straightforward manner.

Open cars

On the larger cars of the nineties there were generally two pairs of transverse bench seats. The seats could face one another in a *vis-à-vis* arrangement, or the front seat face forward and the rear backwards both sharing the same back, the *dos-à-dos* arrangement, or both seats could face forward and a motor car with the latter arrangement was called a double phaeton. The rear wheel was generally larger than the front and the engine was often under the back seat so the latter was way up in the air. On some cars there was a little dickey seat behind the car. This originally carried the mechanic, and when the owner could venture out without him, the owner's mother-in-law, or so says tradition.

There was a dashboard at the front of the vehicle to give driver and passenger some protection from mud thrown up from the road. By the end of the nineties mudguards were fitted. Early mudguards were simple curved strips of painted wood or of leather, and were attached to the body by brackets, with a considerable gap between mudguards and body. The rear ends of the rear mudguards were later for a while turned up and made flat, and wicker baskets for carrying the odd bit of luggage were sometimes fitted on to the flat part.

Platform steps enabled the occupants to clamber aboard but the steps were soon replaced by a running board that was generally continuous from front to rear mudguard. It was quite a time before the gap between running board and body was screened by a metal valance. The working bits of a railway locomotive could be seen, so why should not drive shaft, brake rods, and so on, of a car not also be visible?

Very soon a fabric valance was buttoned to the inside edge of the front guard and along its bottom edge to the frame which gave the occupants much more protection. By about 1910 the rear mudguard was curved downwards to follow the wheel round and also made wider. Lips were formed along the edges of the mudguards to try to stop mud from creeping round their outer edges. By the end of the first decade of the century the valance was made of metal, the mudguards were wider and curved in cross section (on the more expensive cars) and the steps had become running boards.

The smaller cars had a single transverse seat and these were called voiturettes. Three wheelers were in vogue for a while, largely because of the Leon Bollées.

The Système Panhard became popular and eventually displaced other arrangements. The engine was moved to the front of the vehicle and housed under an elementary bonnet and front and rear seats faced forwards. The driver and his passenger incidentally just climbed aboard as there were no doors. Side panelling joined the gap between the front and rear seats and as the rear wheel took up a lot of space, particularly when it was larger than the front wheel, the easiest place to put a door for the rear passenger was at the back of the vehicle. The rear door also made things easier with chain-driven cars.

And so the tonneau body was evolved. It derived ultimately from the Governess cart. (Tonneau incidentally is French for a tub or cask.) There was a seat in each rear corner and usually a tip-up seat attached to the door itself though other arrangements could be used. The rear seats were often above the level of the front seats, possibly to give the passengers more leg room, and a better view (and more dust!). The tonneau became fashionable about 1898 and was in vogue until 1905 or so. Considerable development occurred over these years for, as one owner claimed, the first tonneaus were designed for 4 ft dwarfs whereas the last were meant for 7 ft giants! The tonneau was never popular in the USA. The disadvantages of the rear entrance were that passengers had to step down on to the road instead of the footpath, and that it was difficult to attach hoods and screens to protect the passengers from the weather.

Then some sensible character who knew what he wanted demanded side doors on the tonneau being built for him, and refused to buy the car without them and so this was done. Troubles were experienced with the body twisting and making the doors gape or jam, but this was largely overcome by bracing and stiffening the chassis. The side doors also accelerated the trend to longer wheelbases and smaller rear wheels. This vehicle was called a side entrance tonneau or simply a phaeton. The phaeton was originally a small open light four wheeler generally driven by the master rather than by his coachman, and therefore at a fast and possibly dangerous pace. Phaeton was the mythological character who recklessly drove the chariot of the Sun too close to the Earth. A particularly baroque form of the phaeton was called the Roi des Belges Phaeton and took its name from Leopold II. Apparently when the king was deciding which coachwork to use on his new 1903 Mercedes, his mistress put two easy chairs together in line ahead and suggested the designer should copy their shape. Hence the curving form. The panels had, in cross section, a curve outwards at the top and a turn under at the bottom.

A dashboard was mounted behind the engine on the early front-engined cars and this dash was particularly noticeable on cars with high seats and low engines. There was quite a gap between dash or bonnet and the front seat. It was not until 1910 and later that the dash was replaced by a curved cowl or scuttle mounted behind the bonnet, and extending backwards lessened the gap between driver and bodywork and gave more protection to the driver's legs and feet. The original dashboard became a bulkhead and the instrument panel confusingly called the dashboard. The radiator and water tank of the Canstatt Daimler of 1901 were constructed in one unit which was faired into the bonnet. Practically everybody else except Renault followed suit and for many years most manufacturers maintained a distinctive radiator style shape to enable their cars to be readily identified. Rolls-Royce and Daimler still do. Renault persevered with their arrangement of the radiator behind the engine and with their distinctive bonnet until the late thirties.

Racing cars needed larger and larger engines and therefore larger, longer bonnets, which could give them quite attractive lines. This style was copied on touring and sporting cars, and the bonnets made larger.

For a number of years the front seats were quite open and the driver had no door or side protection. Later a fabric screen was sometimes fitted between the dashboard and the back of the front seat and this gave way to fixed wooden panelling together with a front door (figure 21.1) but this was not general practice until the First World War. The level of the radiator was also raised during these years and the angle of the scuttle (the strip connecting bonnet and body) reduced making the profile of the car smooth except where it was interrupted by the tops of the seats. These changes were accompanied by the introduction of the torpedo body (figure 21.2). This was flush sided, the body did not bulge out in plan at the seats, and instead of abrupt breaks where bonnet joined body, body and bonnet contour were continuous. A number of people claimed to have originated this type of body including Messrs Rheims and Auscher of Paris when they built Jenatzy's 'Jamais Contente'.

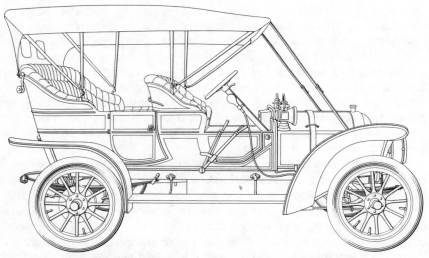

Figure 21.1 Oldsmobile touring car 1907.

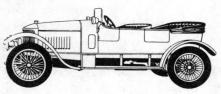

Figure 21.2 Vauxhall Prince Henry 1913. By permission of Unwin and Hyman (from A Bird 1967 *Early Motor Cars*).

Smaller cars

While the larger open vehicle was evolving into torpedo touring cars, the smaller cars developed along two main lines. First there were scaled-down versions of the tourers, with two seats only, that is, coupés. In the early years the bonnet was stuck out in front and the transverse seat was mounted on a platform. If the seat was right back near the rear wheel the back of the seat could be curved to give the car an attractive shape, but as wheelbases increased the seat merely sat on the platform which projected behind it and there was no attempt to put any body work at all behind the seat. Then someone put a shallow tool locker on the platform or a fairing of sorts and gradually this built up to form the normal boot. Similarly over the years panelling and doors were fitted between scuttle and seat, and the radiator level raised so that bonnet, doors and boot could be made to blend to give the torpedo style (figure 21.3). The process was not complete until after the First World War with some makes.

Figure 21.3 Morris Oxford 1912. By permission of Unwin Hyman Limited (from A Bird 1967 *Early Motor Cars*).

The second type of light car comprised the runabouts or voiturettes and these led to cycle cars which had their heyday just before and after the war. The runabout simply had a transverse seat and an engine mounted on the chassis, with floor boards filling the gap between seat and engine. The cycle cars owed more to the quadricycle and tricycle than to the carriage and were very cheap and spartan. Their nakedness was partially covered by bits of painted tin but nevertheless, simple though their bodywork was, their lines were generally rather rakish.

Windscreens

The driver had little protection in touring and sporting cars. Some of the smaller vehicles had windscreens, but not many of the larger ones. Windscreens were rare before 1905 or so; they could be dangerous in wet or dirty weather when the screen would become obscured and the driver could not see where he was going. One way to overcome this was to divide the screen into two parts and tilt the top half slightly backwards so that the driver could see the road ahead. Mechanical windscreen wipers, incidentally, were used in the USA about 1916. At first the windscreens were carried in heavy frames, for the forces on large windscreens could be considerable, but the frames interfered with vision so later 'Clear vision' types without frames were used.

Windscreens of safety glass were exhibited in 1906 and consisted of two thin sheets of plate glass with a thin sheet of celluloid cemented between them. The safety glass was not a commercial success but in 1910 improved cements changed the situation and the Triplex Co. was launched. Although the windscreen did not shatter the celluloid tended to discolour with age.

In 1927 Henry Ford was cut by glass in an accident; an enterprising salesman pointed out the virtues of Triplex and Ford was so impressed that thereafter all Ford cars were fitted with safety glass.

At first the windscreen was on the dash and so a long way from the driver, but later it was placed on the scuttle when it was nearer the driver and gave him more protection. A large windscreen created a lot of drag, slowed the car and introduced very large stresses on mountings and scuttle. A windscreen in the now conventional position was probably not of much use to the rear passengers in any case.

However, by 1910 or so windscreens had become fairly general. One sheet of glass could be used or two joined at the centre line. A common arrangement was to have two narrower sheets, one above the other, and the top could be rotated and locked in any position the driver required. The rear passengers could also have their own windscreen, which was usually demountable so that it could be stowed away. For protection against the rain many open cars had Cape Cart hoods. These were large, unwieldy looking folding contraptions of fabric and wood. They looked charming on

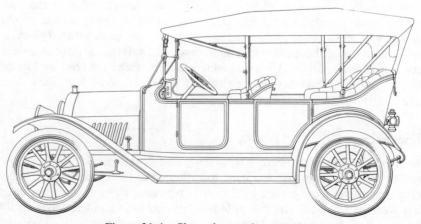

Figure 21.4 Chevrolet touring car 1914.

little voiturettes but had such an area that, with the hoods up, small cars would have been blown backwards in a high wind.

The normal touring car (figure 21.4) had a folding hood and detachable sidescreens to keep the rain out. The hood could project in front of the windscreen to help keep it dry. Some cars had a detachable canopy and a few like the 1901 Lanchester and 1903 Cadillac had detachable hard tops so that they could be converted from tourers to enclosed carriages. Storing the hard top and putting it on and off must have been problems.

Enclosed cars

As late as 1912 97 % of the cars made in that year in the UK were tourers. Enclosed cars were generally luxurious and could only be afforded by the rich.

The brougham, popularised by Lord Brougham in early-Victorian times, was a four-wheeled closed carriage without a perch, the absence of which enabled the floor to be lowered making side entrance easier. The early enclosed horseless carriages were based on the brougham and were more likely to be powered by electric motors and batteries than by petrol engines, for electric motors gave a smoother, quieter ride and there were no fumes, but as better engines became available the electrically powered carriage died out. The Edwardian enclosed cars (figure 21.5) were generally stately town carriages or limousines. They were very large and had to be high so that ladies and gentlemen could enter the carriage and seat themselves with minimum disturbance to their headgear. The chauffeur sat out in front

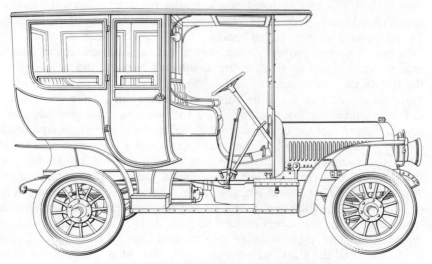

Figure 21.5   Cadillac Limousine 1907.

under a shallow roof or awning. The origin of the name limousine is rather curious; a coarse woollen cloak was made in Limousin and this, because of its weight and water repellent qualities, was used as a carriage rug and then later as an awning to protect the driver, so that a motor car with an enclosed body for the passengers and an awning for the driver became known as a limousine. Later the body was extended to enclose the driver as well as the passengers, but an internal glass partition kept the chauffeur in his place. Usually two occasional seats were fitted, with their backs to the driver, which could be tipped up and folded away into the back of the partition when not in use. If the division could be removed for the owner or his friends to drive the car it was called a sedanca de ville. A large car with a division between driver and passengers is still called a limousine.

The Berline was originally a smaller, lighter type of coach with the body hung on leather straps, the driver sitting in front and grooms standing behind but the motor Berline was a large, luxurious limousine. Later it denoted a limousine with windows all round including the rear corners. In the USA the Berline had, in addition, a moveable window in the glass partition behind the driver.

The sedan or saloon car, unlike the limousine, carried driver and passengers all in one enclosed body and generally held four people. It was many years, however, before saloon bodies became popular. The wealthy owner did not expect to drive his own vehicle or to sit too close to his chauffeur; he had not when his carriage was drawn by horses so why should he now? Indeed he rather suffered for this, for he sat well back, almost over the back axle, whereas the chauffeur sat near the centre of gravity and had a better ride. Sedans naturally came into widespread use earlier in the USA than in the UK, where enclosed cars were so expensive that only the wealthy could afford them. Even so saloons were rare in the USA until the twenties. Two-door saloons were called coaches in the USA between the wars.

Sedan cars were probably so named from the enclosed sedan (Italian sedere = to sit) chair invented in the seventeenth century, and 'saloon' from the Italian salone = a hall.

An enclosed body could be stuffy and unpleasant in hot weather and although some had roof ventilators, so following carriage practice, it was soon arranged that the back part of the roof of the passenger compartment could be folded back so that the passengers could get a breath of fresh air when they wanted. The chauffeur was often protected by a sliding or folding roof. This type of vehicle was called a motor landau or landaulette. Just before the First World War the landaulette was the most popular form of closed car. The folding part of the hood was made of patent leather. The landaulette could be further divided into the quarter and the three-quarter types, having an extra window behind that on the rear door. The D landaulette had the glass front curved and D-shaped in section or made up of three sheets of glass, the two side ones at an angle. The passengers had a better view out of the car with these arrangements. The original landau was named after the town where it presumably originated, and was a four-wheeled vehicle with suspended bodywork and a double folding top. The interior seats were transverse and faced one another and one half of the top folded behind each seat. The driver had a higher seat at the front.

The motor cabriolet had a completely folding top and generally two fixed seats, two folding seats and an exposed driving seat. The touring car could also have a folding top, not unlike the 'all weather' tourer that became popular in the twenties, which had canvas and celluloid curtains that could be clipped on to the sides. Today's convertibles with their folding tops and glass windows are direct descendants of the cabriolet.

The name 'cabriolet' derives ultimately from the Latin *capriolis*—the goat. The Italians had a two-wheeled vehicle with long springy shafts which capered like a goat and so they called it a curricolo. The vehicle went to France where it was called a cabriole. The vehicle plied for hire in France where they were known as 'cabs'. They eventually arrived in England, somewhere acquiring an extra 't' to become cabriolets.

Coupés were cut-down versions of large vehicles and so were, and still are, two seaters. Early coupés could be chauffeur driven with the chauffeur sitting out in front.

The shooting brake was a saloon car without a boot but with luggage space behind the rear seat and accessible by a rear door. The car had a long wheelbase and could carry a lot of luggage. Breaks were four-wheeled carriages with two facing seats placed length-wise for six or more passengers, with raised seats at front and back for driver and grooms. They were found to be very useful for breaking in horses because of their strength and weight and high driving seats and so acquired their name. Again their strength made them suitable for shooting parties. For some curious reason when the name was applied to the motor car it became a shooting 'brake', not 'break'. Nowadays the term 'estate car' or 'station-wagon' is used, even if the 'estate' is a suburban semi. They were called 'woodies' in the USA because the body had a wooden frame.

Developments Between the Wars

Tourers

The tourer was still the most popular type of car immediately after the war but by the end of the decade it had been displaced by the saloon car—only 5 % of UK cars were tourers by 1930.

The touring body was not very different in appearance from the torpedo body of ten or fifteen years earlier and pre-war trends continued. The radiator was high and narrow and instead of the scuttle fairing upwards to the windscreen, bonnet and scuttle were in one straight line and the cant line of the bonnet was continued in one straight line along the top of the body side. There was a general trend to lower the chassis and so reduce the overall height of the body and make it appear less top heavy. The appearance of low build was accentuated by using, for example, a well marked line between body and sill. The body had some degree of turn-under to enable wider seats to be used. The rear seat was more or less over the back axle and there was also turn-under under the rear seat. The rear turn-under rather spoilt the profile of the car, which otherwise could be attractive, at least when the hood was down. The mudguards and running board became more integrated with the body and the running board would continue to carry the spare wheel, the tool box, the battery, spare petrol cans and luggage. The lazy-tongs expanding holders were very effective. Luggage trunks could also be carried on racks which slid or folded under the rear of the body when not in use.

Though the driver now had protection he did not always have his own offside door, particularly on smaller cars. Indeed, the 1926 Morris Cowley had no doors at all on the driver's side of the car.

Doors could be hinged at the front or back; sometimes the front door was hinged at its front end and the rear door at its rear end. The safer arrangement of hinging both doors at their front ends was not generally adopted in the USA until after the Second World War.

Horizontally divided windscreens were popular but when reliable electric motors became available single piece screens became usual. Windscreens V-shaped in plan were used at times to reduce wind resistance, and Kissel introduced curved windscreens in 1914 and Arrol-Johnston in 1922.

For shelter the car still had a Cape hood. This hood was made of waterproof twill and mounted on a pair of body irons at the rear and carried forward on rods. Its framework was of ash or some other timber that could be easily bent by steaming, though metal tubing of brass or japanned D-sectioned steel was also used. The metal tubing kept its shape better but it was not so easy to attach the fabric to it. The rods could be folded backwards on the lazy-tongs principle so that in theory one man could quickly extend or fold the hood. When the hood was to be concealed in the down position the body had to be curved out to carry a horseshoe-shaped channel to enclose the hood. If the hood was not to be concealed an envelope was provided to slip on it for protection from dust. The 'all-weather' hood had, in addition to the hood, side curtains of stiff twill in which celluloid windows were inserted. There were generally four curtains on each side, numbers one and three being carried on rods on the doors so that the latter could be opened. The driver's screen had a flap at the bottom so he could put his arm out to make hand signals. When not required the curtains were stored in a compartment situated, for example, behind the front seat. Sometimes the side windows could be attached securely to the body alone so that the car could be driven with the windows up and the hood down. Collapsible roofs, however, tended to crease and look shabby after some use, and to sag between the supports. Celluloid windows eventually crazed and turned yellow.

Small cars

Open-bodied two seaters were powered by 4 to 8 hp engines and had to be cheap to buy and to run. The wooden framing was simple and the panelling was normally sheet metal as this could be beaten or pressed to have curvature in two directions. The boot behind could carry luggage, the spare tyre (under a false floor), or a dickey or rumble seat for extra passengers. The spare tyre, however, took up a lot of space so it was often carried on the step, and as it got in the way of the door, on the offside. If the boot was small it had a single lid hinged at the front, larger boots would have two lids meeting in the centre. Dickey seats were meant for occasional use only as they were generally uncomfortable and unprotected. There were two main types. In one the seat board was fixed to the floor of the boot and the padded rear lid when swung back formed the backrest and was held in place by knuckle joints, the front half-lid being folded up and over. In the other type of dickey the front lid rotated backwards on a hinge until it was upside down on the floor to form a seat, the rear half-lid again forming the backrest. Folding seats of metal struts were also used. The passenger got into the dickey by a step tread or by a small door in the side of the boot and hinged from the bottom. Cape hoods were fitted and Victoria heads on more expensive vehicles.

The post-war Triumph 1800 of 1946, incidentally, not only had a rumble seat but the rumble seat passengers had their own windscreen.

Various ways, besides using dickey seats, were used to try to fit more seating space in these small cars. The front seat could not be made wide enough for three people because of trouble with the controls, and without spoiling the appearance of the car. However, small rear seats could be put in. The clover leaf arrangement was popular, it had a small seat at the back, but some vehicles had a full width back seat, or two staggered seats facing one another. Access to these seats was obtained through a gang-way between the front seats, or by folding back the nearside front seat. To make things easier the door was often widened. These four seaters were called 'Occasional Fours'.

Enclosed cars

In the early twenties enclosed cars could cost three times as much as tourers and so could be afforded only by the rich. The customer could still specify the type of body he wanted and as these were hand built by skilled craftsmen there was a bewildering number of body styles; limousines, broughams (now limousines but without a window behind the rear door window), various types of landaulette, cabriolet and coupé. Upholstery and trimmings and finish could be luxurious. Of the various types the limousine was the leading style of town carriage of the twenties, though closely followed by the three-quarter landaulette.

A lot of thought went into giving cars the best proportions, for example, the rear light (window) of the limousine was made longer than the rear door light because the surrounding panelling made it appear smaller than it was. The roof line rose a couple of inches towards the rear, and the elbow or mid-line was straight until it rose near the rear end. The rear corners were curved and, as on the touring car, the body had a pronounced turn-under at the rear and often on the sides under the passenger compartment. Doors were made neater in appearance by using flush panels which extended a little out of the door proper and engaged in corresponding recesses in the bodywork, and by using concealed door hinges. Simpler 'plant on' locks came in about 1923. In the early wooden-framed doors the glass light was pulled up by a strap in much the same way as the door windows of railway carriages were operated until recently. In the twenties the glass was fitted in velvet-lined channels boxed into pillars and raised and lowered by finger pulls or window handles through spring-loaded mechanisms.

The Depression put an end to many of the high-class carriage builders who either went out of business or were taken over. The Second World War finished off most of the remaining firms. In their day, however, they set fashions which more mundane people tried to emulate.

Construction

In the early twenties the normal body had a wooden frame (figure 21.6), metal panelled on the outside and trimmed and upholstered inside. A great deal of handwork was required, increasing the cost of the car. In addition the body was heavy which affected the performance of the vehicle. The body was not very stiff; joints and panels tended to rub against each other and to work loose, and all sorts of squeaks and rattles could develop which were generally difficult to track down and eliminate. Paintwork could also crack because of flexure of the panels.

In 1923 Weymann in France introduced a quite different type of body. The framework was made of wood in the normal way but no wooden member joined any other, the connections being made by metal strips and angle pieces. There was practically no panelling and instead leather or leather fabric compounded of rubber and leather was used. This was backed by wadding and canvas. The resulting body was free from squeaks and rattles and could be rapidly built, but it was costly and water and damp could cause the fabric to deteriorate.

Some manufacturers made normal bodies but covered the body with fabric instead of cellulose, these were also sometimes called Weymann bodies. Weymann construction was quite fashionable in the late twenties but had practically disappeared by 1932. One critic observed that Weymann bodies were remarkably free from drumming, a major problem with steel bodies, but better suited to those people who preferred to live in a tent rather than in a house.

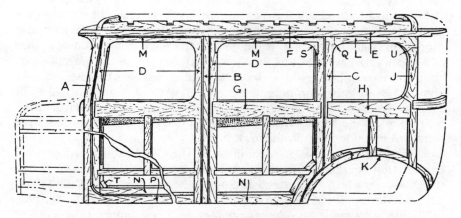

Front standing pillar *A*, middle standing pillar *B*, hind standing pillar *C*, door pillar *D*, cant rail *E*, cant rail make-up piece *F*, door rail *G*, elbow *H*, side-light pillar *J*, wheel-arch *K*, side-light make-up piece *L*, door top *M*, door bottom *N*, bottom side *P*. The lower front portion of the front door, which is broken away, represents a door opening above the bottom side in contrast to the other door which is "cut through"

Figure 21.6 Side framework of a four-door saloon.

Another patent system of body building of the twenties was that of Gordon in England. The body shell was built up from a framework covered with continuous sheets of plywood, the door and window apertures being cut out. The continuity of the sheets meant that there were no joints to work loose and squeak, and the door openings were lined with special mohair to make the doors draught and rattle proof. The floor and seats were mounted on the chassis frame as with the Weymann body, so the body shell was lightly loaded and was carried on the frame at three points on rubber insulators, consequently twisting of the frame did not affect the body.

All-metal bodies

Some all-metal bodies had been built at the turn of the century and a trickle made at later times, one example being the 1911 BSA. Edward Budd worked in the USA for a firm making steel railway carriages before he set up with Joseph Ledwinka to make steel bodies for motor cars. Their first big contract was for Dodge bodies in 1914 and volume production of all metal bodies in the USA began in 1916 when Budd supplied bodies for an open tourer. Twelve hundred pressings were needed for each body. In 1919 Budd supplied four-door saloon bodies to Dodge, but the first all-metal bodies to have a great deal of impact were those on Hudson cars. Again the enterprising Budd organisation was responsible.

In 1921 the Hudson Motor Co. introduced an all-metal enclosed car—the Essex 'coach'—at only 25 % above the price of the Essex tourer and a couple of years later reduced the price further still so that the coach cost less than the tourer had cost in 1921. This Essex 4 was replaced by the Essex 6 in 1924 and again the coach was very competitively priced. These cars had a tremendous effect. They showed that closed cars, with their comfort and protection from the weather, could be sold at a price comparable with that of an open car. The Essex was not a particularly cheap car but the makers of cheaper cars soon followed its lead. In 1920 closed cars represented only about 17 % of American output, but this grew to 43 % in 1924, 72 % in 1926 and 85 % in 1927. Even Ford had to put closed bodies on an increasing proportion of the Model T. Similarly in the Old World touring cars were displaced by saloon cars in the mass market.

Citroën were the first large company in the Old World to adopt steel construction but the first volume production was at Coventry in 1927 when the Pressed Steel Co. was established under the auspices of G W Budd and Co. and Morris Motors. The first bodies were those for the Morris Oxford, but bodies were made for other companies; Morris withdrew from Pressed Steel after a few years, competitors did not like buying from a firm in which he had a major interest. Other firms soon began turning out steel bodies. As a result demand for specially built coachwork fell rapidly. So cheaply could steel bodies be produced that a good copy of a coach-built saloon costing £350 could be built at one tenth of its price. At first 500 or so separate panels had to be assembled into larger assemblies which were again put together to make the complete body. Panels were joined by riveting and welding while held in jigs, and had to be die pressed so that they were sufficiently accurately made to enable immediate assembly. As manufacturers became more expert, and suitable ductile steels, better welding techniques and larger presses became available, larger units were made so that in 1930 only about 120 parts were required. Progress too depended on developments in the steel industry, thus strip mills for the continuous rolling of hot steel were operating successfully in the early twenties and sheet steel was cold rolled by 1930. By the early thirties bodies were being assembled from four major components, the two side panels, rear panel, scuttle and windscreen frame (and roof). These were held in a jig and spot or flash welded. It was later found, however, that it was easier to maintain tolerances using a number of smaller parts than one big part, and less metal was also required.

All-metal construction had its drawbacks. Metal bodies did not absorb much noise and could also give rise to resonance effects such as drumming, so that ways had to be worked out to damp the panels by stiffening them suitably or by using sound-absorbent materials. Corrosion was (and still is!) a problem. Metal bodies made the mass-produced cars of the thirties cheap but also in some cases nasty—they could be tinny, noisy contraptions.

Many manufacturers continued to use composite construction, that is, a wooden framework clad with metal panels and indeed this was still used on some cars in the late forties. Generally these cars were not made in sufficient numbers to warrant the capital outlay for equipment to make all the body components out of metal. Even so, composite bodies could be built relatively cheaply in some volume by the extensive use of jigs which cut down greatly on the amount of skilled labour required. It was still practicable to use a variety of body styles on one chassis, and an incredible range of bodies was available; Austin, for example, listed 52 different bodies in their 1934 catalogue.

Some manufacturers, and not just manufacturers of luxury cars, made chassis only and turned them over to specialist body builders who mass produced middle-range bodies; some manufacturers made bodies for only part of their range. Some big groups, particularly in the USA, had specialist firms within the group to make their bodies. Luxury cars were made in far too small numbers to justify the cost of large presses and dies and so the metal coach work continued to be shaped by skilled panel-beaters.

Painting

Even in the early thirties some vehicles were still painted in the traditional manner. First the surface was cleaned and slightly roughened to help key the first coat, the priming coat of lead oxide, turpentine, and hard drying varnish for metal surfaces, or white lead, linseed oil, turpentine and lampblack for wooden parts. Then came two to six coats of filling-up to make good small imperfections on the surface. These could be applied with a putty knife or with a brush. Stopping to hide screwheads, etc, then followed. Next the surfaces were smoothed by rubbing down with a wet artificial pumice stone, or waterproof abrasive papers. The surfaces were now ready for the ground colour coats; the number applied depended upon the colour. These coats were smoothed down using moistened felt pads dipped in pumice. Abrasive paper was also used and so were strips of cuttle-fish bone, so the process became known as 'fishing down' as well as 'flatting'. The flatting operation was intended to remove brush marks as well as other minor irregularities. Then came a coat of glossy colour which was flatted. Finally, several coats of varnish were applied.

The various filling up, colour and glaze paints were based on white lead, gold size, linseed oil, turpentine, appropriate pigments and various varnishes. The varnishes themselves were mixtures of resin or gum, linseed oil, turpentine, litharge, etc.

The paint work had to be treated with care, it scratched easily and needed careful washing down and leathering with a chamois leather after exposure to the elements.

Enamels were also used; these were mixtures of ground pigment and varnish and had more body so fewer coats were used, for example, after priming and filling one undercoat of enamel plus one coat of finishing enamel would be applied. The enamels were oven dried and quite durable. Unfortunately only black pigments were satisfactory and so enamel paints were widely used on mudguards in particular. Henry Ford said that his customers could have any colour they liked provided it was black!

With mass production painting had become an intolerable production bottleneck. Cars that took a day to build could take literally weeks to paint so that enormous amounts of space were taken up by cars waiting for their paint to dry. In 1920 a DuPont chemist in the USA trying to find uses for wartime chemicals showed the possibility of developing synthetic paints and GM and DuPont collaborated in an attempt to make a fast drying synthetic paint. After some work the chemists came up hopefully with a product they thought would reduce the total painting time from weeks to days. Kettering was then head of GM Research and was not satisfied—he wanted the painting time cut to hours, not days! So the chemists went back to the laboratory and finally produced a nitrocellulose lacquer which dried in minutes and yet was very durable and could be made up in many colours. This Duco lacquer was used by GM on the 1924 Oakland True Blue (it was painted blue!) and very soon by the industry generally. Nitrocellulose paints were used until the late fifties when they were displaced by acrylic resins.

The cellulose lacquers were complex formulations of pyroxylin (cellulose nitrate) solvents, thinners, and plasticisers and had to be applied by spraying, which was a relatively unskilled job. Even when using cellulose, painting was a protracted job. On the better cars the surfaces had to be cleaned, primed and filled. Then an undercoat was sprayed on, then a couple of cellulose coats, rubbing down between each, and then a final coat of finishing cellulose. After the last coat was rubbed down the whole body might be polished but this could take five or six hours, so polishing was omitted on cheaper cars. As late as the fifties a luxury coach-built body could be given as many as twenty two coats, these would include priming, oil filler, cellulose filler and cellulose colour coats.

Styling

The front end of the car changed drastically during the thirties. At the beginning of the decade the high and narrow radiator was generally exposed and the dumb irons, suspension, and chassis cross members all very visible, for the wings did not come very far down in front of the wheels and there was no apron. Bumper bars were coming in and they did not make things any tidier. Hitherto it had been thought that, for the car to look right, the radiator should be immediately above the front axle, but to put more weight on the front axle and to increase the passenger space the engine was pushed slowly further and further forwards. The radiator was placed behind a wire mesh which protected the radiator proper from flying stones, and later radiator slats were used to form either a vertical or horizontal grille. The radiator could be as much as a foot behind the grille. Radiators with such false fronts soon became standard. The traditional honeycomb radiator was in any case being replaced by cheaper tubular cores which were more unsightly than the honeycomb.

Many UK designers, more conservative than the Americans, made the false front resemble the radiator it displaced and surrounded the grille by a more or less rectangular frame. The disappearance of distinguishing radiator shapes in the USA was hastened because the Americans liked to copy the radiator of the most popular car of the year. About 1933 there was a vogue in the USA for V-shaped radiators which terminated in a point at the bottom, and other curious shaped radiators were also used. Soon grilles appeared set in very narrow frames which followed patterns determined by the bonnet lines rather than by the radiator block and about 1938 the Lincoln Zephyr set a new fashion by lowering the radiator block and placing two inlet grilles low down in the front of the bonnet so that the appearance of the nose of the car owed nothing to the shape of the radiator itself. The surround of the grille could be merely a beaded edge.

Year by year the wings came further and further down until, in front view, wing bottoms and apron formed one continuous line so that the suspension and steering were completely hidden. At the same time the wings or rather the mudguard proper became more and more crowned. The valance inside the mudguard extended down to the bottom line of the bonnet but this made the lower part of the valance and bonnet rather inaccessible and difficult to clean. This was overcome by moving the join between wing and bonnet further up the bonnet and so doing away with the valance and mudguards as such, and using a highly crowned curving wing. Instead of opening the bonnet from either side on hinges along its mid-line, the bonnet now had to be hinged either at its front or back, alligator fashion, to gain access to the engine.

The semi-flared wing was popular on sports cars as it could make the car look very attractive indeed but it was difficult to keep the underneath of the wing clean and its finish was soon spoiled by flying stones. Long swept wings with deep outer valances which merged into swept running boards were used on some of the larger more expensive cars. The rear wings were a restrained version of the front wings. On streamlined cars the rear wheels could be completely tucked in, panels being detached to give access to the wheels; in less extreme examples the wheel was covered in, but still had a rudimentary wing. A 1935 Mercedes had even the front wheel covered in but this could reduce the wheel lock although the Praga Super Piccolo overcame this problem by hinging the side panel so that it moved with the wheel.

The paintwork of the swept running board was protected by rubber-filled chromium-plated metal strips, or by moulded rubber sheeting. However, running boards of all sorts were being discarded, for vehicles were by now generally low enough to step directly in to, and this allowed the body to be made wider. Some protection was lost and the sides of the car needed more frequent washing. Space also had to be found elsewhere for the battery, spare wheel, extra fuel cans, tool box and extra luggage which could clutter up the running board, though some of these had begun to disappear within the car in the previous decade.

The line of the running board was sometimes marked by a moulding, and this was still done on some cars in the seventies, or by a strip of chrome-plated metal. The running board was omitted more frequently on the cheaper mass-produced cars.

Head lamps were also drastically transformed during the thirties. In the late twenties they were set fairly high and generally braced by a crossbar which passed in front of the radiator. Each lamp was in a more or less streamlined shell and quite separate from wing or bonnet, but as the wings became longer and larger they tended to swallow the lamp. The lamp could be half-buried in the wing, as in the 1937 Lancia or the Willys. A further stage was to have the lamps buried in the wing but nevertheless conspicuous because of surrounding chromium plating or grilles. The ultimate treatment was that used on the Cord in which the headlamps retracted into the wings when not in use so that the wings were smooth and clean, and were wound out when required, rather like the retractable undercarriages of aircraft.

With regard to the body itself open cars in the thirties were generally limited to expensive convertibles and sports cars. Some of these cars were the finest, so far as appearance was concerned, that have ever been produced. Their proportions were so well balanced that they looked right from all angles, not only in profile. Over the decade their lines became cleaner and more graceful. The appearance of the saloon car, however, changed considerably.

Panels could not be given much curvature at first and so the all-metal saloon had a square, boxy appearance which was made much more apparent by its high roof line. The rear end turned under. By the early thirties edges and corners were being rounded and the profile of the roof was slightly curved with the rear part a little higher than the front.

The roof height was progressively lowered and so the floor also had to be lowered if the passenger was not to get into the car 'like a rabbit into a burrow'. By about 1938 the roof height of a small car could be 63 in and the floor 13 in from the ground. The main reason for the development of the hypoid gear (p 268) was to lower the prop shaft and make the transmission tunnel less obvious. Low floors eventually did away with the need for running boards. The rear seat passengers profited by these changes because the rear seats were moved progressively forward until they were well within the wheelbase and therefore gave a more comfortable ride. The turnunder at the back gradually disappeared and instead the rear-end came down at an angle and tended to sweep out at the bottom (figure 21.7).

Figure 21.7 Trends in car design: Vauxhall 14 hp saloon cars of 1933 and 1939. By permission of Reed Business Publishing Ltd (from *The Automobile Engineer*).

Luggage was first carried on the running board or on steel slats mounted over the petrol tank. Next a special trunk rather like a cabin trunk with the top opening was attached to the rear of the car. About 1925 the body was sometimes extended backwards to form a tail compartment with a door or lid. This compartment or boot was later faired-in but for many years it looked like an afterthought and it was not until after the Second World War that the boot was made an integral part of the body on some makes.

The spare wheel was also moved from the running board to the rear end of the car. It could be merely bolted to the back, or bolted in a shallow recess. With a boot available it could be mounted on the inside of the boot lid or simply bolted to the floor of the boot, in some cases having a shallow compartment to itself. The overall shape and appearance of the car, however, was greatly influenced by the fashion in the mid-thirties for 'streamlining'.

Streamlining

The importance of air resistance at higher speeds was soon realised and racing cars like Jenatzy's 'Jamais Contente' and the Baker Electric 999 were shaped to reduce air resistance. After the First World War there was a resurgence of interest in streamlining, largely as a result of wartime work on aerodynamics, and Jaray, who had been involved in the design of Zeppelins, Rumpler (figure 21.8) and others worked on streamlined cars. It was demonstrated that air resistance at high speeds, and therefore the power required to propel the vehicle, could be much reduced by suitably shaping the body. Interest continued and work went into reducing the air resistance and increasing the aerodynamic stability of record breaking cars and racing cars, but streamlining did not have much practical impact until the thirties.

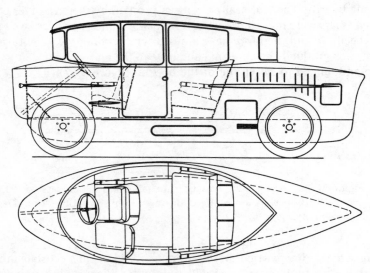

Figure 21.8 Rumpler enclosed touring car. By permission of Reed Business Publishing Ltd (from *The Automobile Engineer*).

The time was ripe because all-steel bodies were becoming popular and steels and presses were available to make bodies of complex shapes. There was also the possibility of making bodies of distinctive appearance which would appeal to the motorist. Well streamlined shapes, however, were wasteful of space and it was thought that they might be considered ugly by conservative buyers. However, by about 1934 'streamlined' vehicles were produced in the USA, the sales people hoping that streamlining the vehicles might make them sell better, even though it might have little effect on their performance. The idea took and for a year or so there was a craze for streamlining. The Chrysler Airflow of 1934 was one of the first in the USA. It had a well curved radiator shield and the headlamps were buried in the sides of the bonnet. The public thought it ugly and the car was not a success. In the UK streamlining was at first applied only to special coachwork, though it was used on the Hillman Aero Minx which was a cheap production car, for the British public were not so receptive to new ideas and novelty as in other countries. It was soon popular on the Continent and in France, in particular, some of the specialist coach makers went to extremes and in the middle of the decade turned out some extraordinarily ugly bodies. Indeed some of the 1933 French vehicles were so extreme that the courage of the manufacturers was more to be admired than the appearance of their vehicles. The more curious freaks disappeared within a few years.

For effective streamlining the bonnet has to be curved, the windscreen raked and very shallow, and the rear curved to prevent an extensive low-pressure space from building up behind the vehicle. The resultant overall profile is ugly. The wheels were sometimes within the vehicle, but often streamlined 'parabolic' wings were used and these too were ugly, particularly when compared with the long swept wings used on contemporary cars of more conservative styling. The shape of the VW Beetle (figure 21.9) was reminiscent of the 1935 styles. Indeed streamlining seemed to last longer in Germany than elsewhere, possibly because the new Autobahnen permitted high enough speeds to show the advantages of streamlining. In moderate doses, however, streamlining gave the vehicle a cleaner line, even if it did not improve its performance, and could increase the amount of luggage space. It could also reduce wind noise at higher speeds.

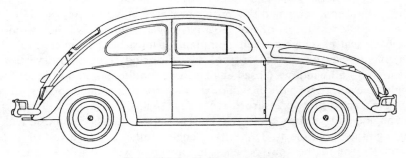

Figure 21.9 Volkswagen Beetle.

The basic principles of styling were well accepted—to give an impression of speed, horizontal lines should predominate over vertical, and the major breaks or changes in the horizontal lines and the front and rear mudguards should have something in common. Consequently, as streamlining was relatively ineffective at speeds of less than 50 mph it was important to give the vehicle lines that made it look fast rather than give it a shape that would actually make it go faster, at least as far as production saloons were concerned. The colour scheme was also important for if two colours were used they could be arranged so as to apparently alter the shape of the body.

Consequently, although aerodynamically streamlined vehicles continued to appear many bodies were designed to combine a measure of streamlining with an attractive appearance. So, although the extreme forms of streamlining lasted only a few years and many cars were built at the time with no attention to streamlining, it did have a profound effect on body shape; cars lost their boxy appearance, the roof line was curved, the waistline was no longer straight and the line of the rear end sloped downwards and outwards.

There were drawbacks; designers liked a high waistline despite wanting a low roof line so windows were small and the dashboard was too high, particularly on smaller cars, so that a small driver had to peer through the spokes of the steering wheel. The driver could not see immediately ahead because of the high bonnet, and his nearside visibility was poor and as a result he tended to drive on the crown of the road because of the difficulty of judging where his nearside wings were.

In 1930 toughened glass was introduced for windscreens—the glass was cooled rapidly in an air blast so that hard skins of glass in compression were formed on either side while the inside glass was in tension. If the glass was fractured it broke up into smaller fragments, the edges of which were not sharp.

Ventilation and heating

The closed saloon body had one drawback—it was not easy to ventilate. The open tourer never had this problem, it was only too well ventilated. If in a closed car the driver or front seat passenger wound down their windows the back seat passengers would complain. It must be remembered too that for many years on the majority of cars the windscreen could swivel and open. With all windows closed engine fumes could leak in and indeed this could be even worse with the windows open for they tended to suck air out and the resulting low pressure inside the car sucked the fumes in. Heat from the engine and gearbox could also be a nuisance. Fumes could also leak in through the floor boards. To keep out engine fumes and heat rubber gaiters were put round the foot pedals and openings were sealed off. Unitary construction was eventually a great help for the scuttle was steel sheet, and fumes from the exhaust were kept out by a sheet metal pan under the body.

There was an advance in 1933 when the Fisher-bodied cars of General Motors were fitted with swivelling lights, generally quarter-lights. These could be adjusted to divert air into the vehicle as well as to extract it and they were so successful that swivelling front quarter-lights were fitted on many cars until they were replaced by eye ball ventilators in the seventies.

Air was also brought into the body from intakes in the bonnet, etc. A favourite position was on the scuttle just in front of the window and an intake here had the advantage that it could be easily opened or closed by a lid operated by the driver. The trouble with some of these intakes was that they also let in water.

There was an enthusiasm from about 1927 for sunshine roofs which became so popular during the thirties that even luxury saloons were fitted with them. A lot of ingenuity went, not altogether satisfactorily, into making the attachments inconspicuous and making the panels slide easily and yet not let water into the car.

Soon the driver wanted not only fresh air but also warmed air in cold weather. He wanted what the salesmen of the day called air conditioning.

For many years passengers had relied on hot water bottles to keep their feet warm, a practice derived from trains. Foot warmers were also used which were heated by the exhaust gases, and closed cars in the USA had been heated by the exhaust before 1914. In the mid-twenties some more expensive cars ducted the air from an intake over a heat exchanger heated by water from the radiator, and then into the interior of the car, boosting the air flow by a fan, the driver being able to control things as he required.

The UK lagged behind other countries, particularly the USA, in the provision of heating systems. Prior to 1938 no UK car had a heater as original equipment and on only a few expensive cars was it available as an optional extra. In 1939 certain models had heaters, but these merely heated and circulated the air already inside the car; the heater itself was under the seat or under the scuttle though later it was installed in the engine compartment. After the Second World War heaters took in fresh air and heated it, with ducting and ports so arranged that warm air could also be directed to the windscreen if desired. Heat was derived from the radiator and the fan powered by electricity. Heating remained an optional extra on many UK cars for years after the war. Units to cool the air in hot weather were first used in 1938 on a Nash and were next used on the more expensive Packards of 1941. There is not much market for such things in the UK!

Developments Since 1945

It was easier to change the bodywork of a car and call it a post-war model than to design and tool up for a completely new vehicle, and so a number of new bodies appeared on pre-war chassis in the late forties. Many of the bodies had bulbous, rounded contours reminiscent of American designs and of the streamlined shapes of the mid-thirties but they were unattractive and indeed could look remarkably ugly, particularly on smaller chassis.

A few cars like the Triumph Mayflower and Renown went to the other extreme and had razor lines.

A number of cars used composite construction—Triumphs had wooden frames and metal panels on a conventional chassis frame but most cars of the first true post-war generation used unitary-type construction to some extent but carrying the engine and front suspension on a frame carried forward from the body.

Unitary construction

The Austin 7 of 1951 was one of the first British cars with true chassisless construction. The 7 had side members extending the whole length of the frame with two additional side members at the front to carry the front and side loads, and three main cross members. The various components did not merely give added strength; the rear cross member, for example, carried the rear spring shackles, the next the front eyes of the springs and the heel board, the third cross member was also the toe board and transferred vertical loads from the suspension out to the side members, and the front cross member formed part of the front apron and radiator support.

By the early fifties many cheap and medium-priced cars had full unitary construction. At first the underfloor frame rather resembled a chassis frame, but soon pressed-steel platforms were used (figure 21.10). Other cars used frames with boxed or open side members and various types of diagonal bracing between the frames. Some sports cars and small-volume quality cars continued using tubular members.

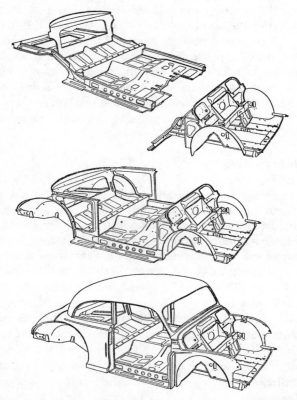

Figure 21.10 Chassisless construction of Morris Minor. By permission of Haymarket Publishing Co (from *Autocar* and *Autocar Handbook*).

By 1955 the number of new British cars with separate frames were in the minority. Daimler cars had cruciform frames, for example, and the Riley Pathfinder had box section side members with a half-cruciform frame at the front. The Bristol 405 had a composite body of aluminium alloy panelling on ash.

The general unitary arrangement was simple: typically the floor pressing was welded to box sections, side sills and front and rear transverse members, the former carrying the front seat and the latter forming the heel board of the rear seat. Box extensions at the rear connected to wheel arches which carried quarter panels and the lower parts of the rear end structure. At the front the floor panel joined the dash and toe board panels and the base structure was completed by wheel arch side panels which extended from the A post to the front bulkhead. Larger cars had extra longitudinal box section members inboard of the sills to give more support to the transverse members.

As the designer obtained more experience cars tended to become lighter whilst still retaining their strength and stiffness. Stress analysis carried over from the aircraft industry helped.

By the end of the fifties unitary construction was so general on all but the most expensive cars in the UK that the Triumph Herald of 1959 was considered quite remarkable because it had not only a separate chassis but the panels were also bolted on and not welded in place. This was done partly, however, to enable subassemblies to be farmed out and to save jig and welding costs, and also to permit future styling changes.

The 1965 Rolls-Royce Silver Ghost had a unitary body in order to reduce its external dimensions but at the same time increase its internal space, and it was indeed 4.25 in lower and 3.75 in narrower than its predecessors. The most difficult problem experienced in developing the new Rolls-Royce body was minimising road noise; to achieve this subframes were used and attached to the structure by resilient mountings.

Preventing the transmission of noise to the occupants of the car was probably the biggest headache for the body designer with the use of subframes on resilient mountings becoming fairly general. Differing types of rubber were employed for different positions; for example, one type was used to help damp out road noise, and another type to damp out engine vibrations. The use of subframes also simplified production on the assembly line and made repairs in service easier.

The second problem with unitary bodies was corrosion; if the body was not very carefully designed water could get into vulnerable places and cause 'inside out' corrosion. Box sections and the like were particularly at risk. Corrosion was aggravated by the increasing use of salt on roads in wintry weather. Even in 1970 many authorities claimed that corrosion could be greatly reduced at almost negligible cost by improved design and manufacturing techniques. Pre-war chassis frames had been protected by heavy coatings of high-temperature stove enamels, body surfaces had been rinsed in dilute phosphoric acid and small components dipped in the acid. After the war improved phosphate treatments permitted spraying. The phosphated surfaces were dried and a priming coat applied and cured in an oven; this coat was intended to protect the internal as well as external surfaces.

Then came a mid-coat and finally a high-gloss top coat on the visible surfaces, and these were baked together.

In the fifties dipping was introduced for the priming coats, the body being partially or completely immersed in the bath, or spitted and rotated in it, so that the primer could reach the inside of box members and other awkward places. There was a further advance in the sixties with the introduction of electrophoretic primers. The body was cleaned, dipped in a phosphating bath, and then immersed in the priming bath and current passed between the body and electrodes in the bath which by electrophoresis, that is, the motion of charged particles (of paint), followed by electrodeposition, caused the paint to be relatively uniformly deposited over the whole body surface, both visible and invisible, and gave good covering of edges. This type of priming was rapidly accepted and was used on 90 % of vehicles produced worldwide in 1983.

The priming coat contained chromates as well as pigments. It was followed by sealer and colour coats, the former of alkyd or epoxy ester and the latter of alkyd or acrylic paints.

In the USA chassis frames with separate bodies continued to be manufactured for many years. It was easier to put different bodies on the chassis, particularly of convertibles which would not be stiff enough without a frame, and easier to implement the annual styling changes. Also in America the greater weight, and therefore cost, of the car was tolerated and there was no objection to the larger engine required, and the resulting increased fuel consumption. The Corvair of the late fifties had one of the first unitary bodies. Unitary construction was used on some other compacts for a time but after a few years there was a tendency to revert to separate chassis. The Chevy II compromised by having a bolt-on front end which permitted styling changes. Indeed this arrangement became not uncommon. However, by the end of the sixties unitary bodies were becoming much more normal and in 1969 all Chrysler bodies were unitary with subframes.

Safety measures and safety legislation brought about improvements like anti-burst locks, padded fascias, seat belt anchorages, high impact strength windshields and bodies designed to absorb some of the energy of impact during a collision. Volvo and Rover in the sixties made the passenger compartment rigid and it was intended to over-ride the engine as the front part of the car collapsed progressively.

Plastics

A plastic body was considered, according to Hitler, for the VW in 1938, and Ford a couple of years later built an experimental and much publicised car to show that such a body could be made. The polymer used was derived from soya beans. More practically, plastics were used before the war on instrument panels, steering wheels and for interior fittings. They have been used more and more over the years, and in many parts of the car, but so far as the body is concerned only for fittings and trim. They have not been used, at least not on volume-produced cars, on the body itself, except occasionally in areas where structural strength is not important, such as boot and bonnet lids, bumpers, and more recently, spoilers. Plastic bodies, however, do not need such expensive tooling as steel ones and so glass-reinforced plastics (GRP) have been used on cars produced in small numbers, like the Reliants, and sports cars (though a large number, by European standards, of the 1953 Chevrolet Corvette roadster was made). Plastic bodies were used on kit cars and other specials that were built on chassis frames.

Styling

Most American cars appeared large, ostentatious and vulgar to Old World eyes. Until the end of the thirties, however, American, UK and Continental cars were similar in appearance but after the war they diverged more and more but even so many of the American styling innovations were adopted, though in a modified form, on the other side of the Atlantic. The leaders in styling, however, in the fifties and sixties were the Italians—they designed vehicles with clean and simple lines and relied on subtle balancing of proportions. The Mustang of 1964 showed a strong Continental influence so far as appearance was concerned and it was very successful indeed.

French cars showed the national characteristics—cars like the remarkable Citroën DS19 Space Ship of 1954 (figure 21.11) and some early but utilitarian Renaults were made, and sold very well in addition to cars of more conventional appearance.

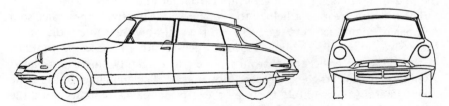

Figure 21.11   Citroën DS19 of 1954.

During the 1945–70 period the roofline of the car was lowered and more and more glass appeared above the waistline but the biggest changes were in the treatment of the four corners of the car. The wheels were tucked within the width of the car and the wings no longer masked the wheels and, except insofar as they carried front and rear lights, they were non-functional and could be manipulated at the stylist's whim.

The styling at front and rear ends in the forties and early fifties generally continued pre-war trends. The front wings were well marked and often extended backwards in a convex curve to the rear wings which were given matching lines. Footboards had disappeared completely. The next step in many cases was to extend the top line of the front wing backwards and almost horizontally and merge the wing into the front door. In other cars the line of the wing curved down but in a concave instead of a convex curve. Wing and bonnet merged into one another and so the headlamps had to be buried in the wing. There were of course many exceptions, the front wing of the A55 of 1957, for example, extended right back to the rear bumper, there was no external rear wheel arch and the top half of the rear wheel was hidden.

As time went on the front wing extended further and further backwards and slightly downwards, swallowing up the rear wing. Semi-circular cut-outs were left for the rear wheels to make wheel changing easier. Next the bonnet was sloped downwards towards the front of the car so that the headlight and front wings were higher than the front of the bonnet. This also improved visibility somewhat, but made life cramped for the carburettor. The headlamps were sometimes accentuated by bright plated hoods and the waistline was marked by a strip of metal or by moulding which ran from about the mid-point of the headlights to the rear wing. From the late fifties there was a trend to use twin headlights side by side on larger cars, particularly in the USA. Often the headlamps—single or twin—were mounted within the radiator grille. Wrap-round windscreens were introduced about 1956 and soon their use was universal, though at first curious makeshifts such as cranked pillars had to be used to fit such screens. Wrap-round windscreens also caused problems with wipers. Curved windscreens had appeared in 1949. All windows tended to increase in area in the mid-fifties making the interior appear larger and lighter.

The radiator treatment helped to give a car a distinctive appearance so a number of manufacturers tried to keep recognisably the same radiator on successive models. From about 1950 the open-mouth grille became popular; in the USA the grille and surround were extravagantly chromium-plated giving American cars the celebrated 'dollar grin'. The main lines of the grille were generally horizontal, the major exception being the cars with the traditional radiator in dummy form. There was a trend also for the grille to become narrower and narrower; on some cars it was almost invisible below the front bumper.

The pre-war trend for the rear wheel mudguard and wings to disappear continued so that soon all that was left were mouldings round the wheel cut-outs front and rear. The luggage boot was faired in at the rear; about 1950 the rear panel was made slab-sided and extended backwards in a line with the waist to make a rear wing. The rear wing became very fashionable and instead of continuing back it was swept upwards and outwards to form a fin. This fashion passed and the fin reverted to a wing but now the boot lid was tilted until it was almost horizontal and its rear surface was vertical so that the boot occupied almost all of the space between the wings. The rear wings or fins eventually carried rear parking, braking and turning lights, and on some cars reversing lights which were very convenient for the purpose.

In the forties the rear window was merely a sort of rectangular porthole but when larger expanses of glass and of curved glass became available the rear window eventually extended from the waistline to the roof and across the whole of the rear of the body, with a consequent improvement in rear visibility. The rear of the car was first curved but with a glass screen the glass was straight in side elevation and at a small angle to the vertical, but generally curved in plan view to give the same wrap-round effect at the rear as at the front of the car. The stylist began experimenting on the angle of the rear window and in a couple of cars the window had a reverse rake but in general the angle was made flatter and flatter in the fastback style. This meant, however, that the window came back further and further over the boot; one way to reduce this was to tilt up the front end of the boot, so that by the end of the sixties there was often only a slight discontinuity caused by their slightly different angles to separate window and boot in profile, and in some cars the boot continued the line of the window. To overcome the difficulty of opening the boot the whole of the back formed a door which lifted upward, and the car became a 'hatchback'. This arrangement was used on the Renault 16 of 1965 and the Austin Maxi of 1968. It later became quite popular. If the boot extended beyond the rear window the car was a 'notchback'.

The spare wheel was still a nuisance. It could be placed in the boot under a false floor or in the top of the boot or on one side of it. Sometimes it was mounted underneath the boot and lowered directly on to the ground. On the Bristol 406 the wheel slid into a compartment ahead of the front door, in the Simca Vidette into a rear wheel fin and on rear-engined cars the tyre could be put under the bonnet—this was even done on some cars with front engines so that the tyre was kept nice and warm even in cold weather. On some cars—the Nash Metropolitan and the Rover 3500 for example—the designer gave up and simply mounted the wheel externally on the boot.

Work on the aerodynamics of high-speed and racing cars and in particular on their stability at high speeds resulted in some odd-looking vehicles. Some of this influenced production cars but again things like spoilers and air dams and wedge profiles were generally cosmetic rather than effective. Some cars like the Citroën DS19 and the Saab were shaped to reduce their drag coefficients, and streamlining has been taken more seriously in the last decade to help reduce fuel consumption, and 'slippery' shapes have become fashionable.

Chapter 22
Lighting

Acetylene Lamps

The coach driver had a high seat, he could rely on the help of his horses and so he needed only a simple lighting system. The early motorists, however, soon found that ordinary carriage lamps fitted with candles were far too weak, oil lamps were also inadequate, and by the turn of the century special acetylene head lamps had been developed for the motor car. These lamps permitted speeds of up to 25 mph whereas carriage lamps had limited speeds at night to 8–10 mph. Until the First World War, and in some cases until later, most cars had paraffin side and tail lamps and acetylene head lamps. Incidentally only one (offside) tail lamp was required in the UK until 1953. Often battery-operated electric lights were used to illuminate the dashboard and to light the interiors of closed cars.

At first acetylene was generated on the vehicle itself by the action of water on calcium carbide. In the drip generator the water passed from a reservoir through a needle valve controlled by the driver to the carbide chamber (figure 22.1). The carbide in the diving bell type of generator was completely immersed in water which passed up through the bottom of the chamber; the resulting buildup of gas pressure prevented the water from wetting more carbide unless the gas tap was turned on. The carbide was coated to slow down its rate of reaction. The generators were messy and smelly so that 'bottled acetylene' became popular—acetone was carried in a porous medium in the reservoir, and as enormous quantities of acetylene will dissolve in acetone the gas did not have to be stored under pressure.

Acetylene lamps were powerful and were made more effective still by fitting parabolic reflectors. They could dazzle oncoming traffic and a number of ways were tried to reduce dazzle. One arrangement was to put a series of horizontal metal slats in front of the lamp with their upper surfaces blackened and their lower surfaces polished. In another system light travelling forward passed through a lens to give a divergent beam while a reflector behind the lamp together with a lens gave a strong parallel beam which could be cut off by a shutter in front of the reflector and operated by the driver.

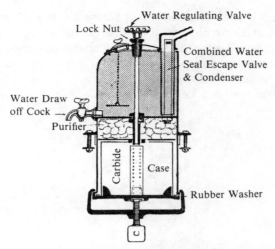

Figure 22.1 Acetylene generator.

Electric Lighting

Electric lighting was hardly practicable in the early years of the century because carbon filaments were weak and dry cells expensive to use. Accumulators could be used but were affected by vibration; they were rapidly exhausted, and they could not be recharged in use because the voltage of the contemporary dynamos varied with engine speed. By 1910 lamps with tungsten filaments were available, accumulators (now called batteries) and suitable dynamos with automatic cutouts had been developed, and electric lighting first appeared on the 1911 Cadillac. It was in general use in the USA by about 1915 and in the Old World after the war.

Dazzle

Electric lighting made the dazzle problem worse, particularly in the USA where street lighting in cities was often bad. Electric lights were, however, easier to control, so far as dazzle was concerned, than acetylene lamps. One expedient was merely to dim the lamps with a resistance in circuit with the battery. A less hazardous arrangement was to swivel the headlamps on a transverse shaft; later a cam could be fitted in addition to rotate the nearside light to illuminate the gutter. An anti-dazzle bulb was invented in the early twenties in which a metal screen was positioned below the axial filament so that only the top half of the metal reflector was used, a second bulb giving the driving beam. This arrangement led to the double-filament bulb (figure 22.2); the main filament was at the focus of the reflector and gave the normal driving beam, and the second filament was slightly ahead of the main filament and was fitted with a shield to cut off the light that would otherwise be reflected upwards out of the lamp. Other arrangements were to use a reflector split horizontally with the top half shifted forwards relative to the lower half and a line filament used, or to use a V-shaped filament with only its point at the focus of the reflector.

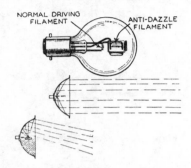

Figure 22.2 Double-filament bulb.

The double-filament lamp could get out of adjustment so that the US manufacturers collaborated in the late thirties to produce standardised sealed units in which the components were securely mounted, and the unit itself could be accurately aimed. When dipped the beams from both lamps were deflected downwards and to the nearside of the car. Many UK designers, however, preferred to use a mechanical dip and switch method arrangement—the offside head lamp was switched off and the nearside reflector tilted to deflect its beam to the left and downwards. Reflector and bulb were moved pneumatically (figure 22.3) or by a solenoid.

Until the late thirties the headlamps were mounted in separate units which were originally fairly functional but gradually became more attractive and streamlined in appearance.

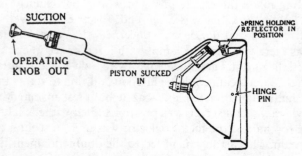

Figure 22.3 Mechanical dipping arrangement.

Developments Since 1945

After the Second World War designers in the UK concentrated on obtaining a brighter, more penetrating beam and compromised with the best dipped arrangement they could manage. Two-filament lights were widely used, with the beams of both nearside and offside lights deflected downwards and to the left when dipped.

Aluminised steel reflectors displaced the traditional silver-plated brass reflector by about 1950, and from about 1960 sealed units began to displace separate reflector and bulb assemblies in the UK. They were made by melting together the rim of a glass lens and an aluminised glass reflector, the latter containing an accurately focused filament unit. The air was then extracted and replaced by an inert gas. If the filament failed the whole unit had to be replaced.

As more and more cars could reach high speeds consideration was given to how the light should best be distributed to give good penetration and enable distant objects to be picked out when the car was travelling at high speeds and yet to illuminate the foreground so that the driver could see his position on the road not only when driving on full beam but also when the lights were dipped. It was also essential that the dipped lights did not dazzle oncoming drivers.

In the mid-sixties Continental manufacturers dropped their system of a flat cutoff and adopted the US/UK asymmetrical dipped-beam system. The flat cutoff gave little dazzle but the hood beneath the dipping filament blocked off a considerable amount of light. The French, however, stuck to their yellow head lights.

After the war the head lights were no longer mounted as separate units but were recessed in the front wings and when the latter became more and more vestigial the head lights were merged into the front end of the car. The lights were circular and still obviously head lights.

Cadillac pioneered the use of four head lights in the USA in the late fifties, and the four-headlamp set-up appeared in the UK in the following decade. Typically the outer head lights had double filaments and could be dipped in the usual way, and the inner head lights were used for driving only. Two rectangular lenses looked neater than four circular ones, and eventually most cars used this arrangement, though the advantage of the volume production of the standard sealed head light was lost. Rectangular (high aspect) and irregularly shaped head lights also blended in better with the aerodynamically efficient shapes of the last decade. Head lights and turning lights could also be incorporated in one plastic housing. Turning and braking lights had been housed at the rear in one housing on a number of cars for some time.

It is difficult to control the curvature of steel pressings of complicated shapes accurately, and in the past few years reflectors have been made from aluminised thermo-setting polyesters (dough moulding compounds). These compounds can be accurately moulded to complex shapes, and so different areas of the reflector given different curvatures and therefore focal lengths. Main and dipped beams can therefore give higher outputs and more closely defined beam patterns. Similarly, lenses have been moulded from polycarbonate polymers and then given a protective layer.

One great advance was the introduction of tungsten halide lights in the sixties. They used quartz bulbs containing halogen, the quartz bulb permitting higher gas pressures and therefore less evaporation of the tungsten filament and so less blackening of the bulb. The halogen reacted with evaporated tungsten, and when the halide moved back to the vicinity of the filament it decomposed to deposit tungsten back on the filament. The latter could be run at higher temperatures and therefore greater brightness.

Bibliography

Periodicals

The Autocar 1896 to date (London: Haymarket Publishing Ltd)
Automobile Engineer 1910–72 (London: Iliffe and Sons Ltd)
Automotive Engineer 1975 to date (London: Institution of Mechanical Engineers)
Journal of Automotive Engineering 1970–75 (London: Institution of Mechanical Engineers)
Motor 1903 to date (London: Haymarket Publishing Ltd)
Proceedings of the Institution of Automobile Engineers 1907–47
Proceedings of the Institution of Mechanical Engineers (Automobile Division) 1947–69
Proceedings of the Institution of Mechanical Engineers Part D 1982 to date
Journal and Transactions of the Society of Automotive Engineers, USA

Books

W Worby Beaumont 1896 *Mechanical Road Carriages* 1st edn (2nd edn 1906) (London: Constable and Co. Ltd)
A Bird and F Hutton Smith 1963 *The Veteran Motor Car Pocketbook* (London: B T Batsford Ltd)
C F Caunter 1968 *The Light Car* 1st edn (2nd edn 1970) (London: HMSO)
H O Duncan *World on Wheels*
A Harding (ed) 1966 *Classic Car Profiles* (Profile Publications Ltd)
A N Harmsworth 1902 *Motors and Motor Driving* (London: Longmans Green and Co)
P N Hasluck 1903 *The Automobile* 1st edn (2nd edn 1903)
A W Judge 1925 *Motor Manuals* 1st edn (to 7th in 1965) (London: Chapman & Hall Ltd)
Land Transport: II, Mechanical Road Vehicles 1936 (London: Science Museum)
C Maxcy and A Silberston 1958 *The Motor Industry* (London: Allen & Unwin Ltd)
E Molloy and G H Lanchester 1956 *Automobile Engineers' Reference Book* (George Newnes Ltd)
Motor Vehicles 1975 1st edn (London: HMSO) (followed by later editions)
K Newton and N Steads (and T K Garrett) 1929 *The Motor Vehicle* 1st edn (to 11th in 1989) (London: Butterworth Scientific Ltd)
W Plowden 1971 *The Motor Car and Politics* (London: The Bodley Head Ltd)

The Autocar Handbook 1906 (London: Iliffe & Sons Ltd) 1st edn (and succeeding editions)
The Motor Manual 1903 (London: Temple Press Limited) 1st edn (and succeeding editions)

Index

Absorbers, shock, 288, 292, 303
Ackermann R, 313
Action, knee-, 295
Accumulators (batteries), 160, 174
Additives
  antifreeze, 202
  oil, 189, 197
Adhesion, tyre–road, 337, 383
Alternators, 176
Amalgamations
  UK, 63
  USA, 67
Apperson, 32
Austin H, 31, 53, 63, 388
Axles,
  back, 268
  Collinge, 314
  fully floating, 269
  Lemoine, 314
  location of, 271
  semifloating, 269
  stub, 314

Balancer, harmonic, 85
Barnett W, 14
Bartlett W E, 370, 375
Batteries, see Accumulators
Bearings
  ball, 365
  big-end, 110
  roller, 365
  thin-wall, 123
  wheel, 365
Beau de Rochas A, 15
Bendix V, 343
Bentley W O, 107, 196
Benz K, 15
Berline, 396
Bicycles, 38
Block, cylinder, 95, 106, 122

Bodies, 388
  all-metal, 401
  construction of, 400
  steel, 51, 401
  torpedo, 392
  unitary, 281, 411
  Weymann, 400
Bollée A, 13, 229
Bourbeau R, 42
Brake, shooting, 397
Brakes, 38, 333
  antilock systems, 353
  band, 333, 343
  disc, 60, 349
  drum, 334
  duo-servo, 343
  front-wheel, 50, 321, 337, 338
  Girling system, 344
  hydraulic, 335, 340, 342, 348
  servos, 342, 349
  spoon, 333
  transmission, 334, 336
Bremner F W, 29
Brooklands, 41, 54
Brougham, 395
Brown S, 13, 21
Budd E, 401
Buick D D, 46
Butler E, 29

Cabriolet, 397
Camshaft, 104, 105
Cars
  cost of early, 42
  rear-engined, 61, 275
  running expenses, 40
Carburation, 37, 128, 137
Carburettors
  Benz, 128
  constant-vacuum, 135, 139, 145

INDEX

Carburettors (*contd*)
 Daimler, 129
 downdraught, 140
 multijet, 136, 139
 multiventuri, 138
 spray, 131
 surface, 128
 wick, 129
 in USA, 137, 143
Carriages, 388
Cayley G, 358
Cecil W, 13
Chevrolet L, 47
Chains, transmission, 37
Chamber, combustion, 95, 99, 101, 116, 119
Charge, stratified, 14, 118
Choke (strangler), 141
 automatic, 142
Christie J W, 272
Citroën A, 34, 52, 267, 281
Classification, SAE oil, 190
Clerk D, 93, 99, 123
Clutches
 automatic, 222
 band, 220
 cone, 215
 diaphragm spring, 225
 facings, 221, 227
 free-wheel, 243
 multiplate, 218
 other designs, 220
 plate, 217, 221
 scroll, 220
 stops, 216
Coaches, 3
Coachbuilders, 388
Coils, 166
 Ruhmkorff, 159
Column, steering, 318
 collapsible, 328
Compensation, brake, 340
Competition, Japanese, 70
Contact breaker, 164
 transistorised, 166
Converter
 three-way catalytic, 116
 torque, 254
Coolant, 202
 circulation of, 203
Cooling
 air, 206

Benz, 200
 pressurised, 202
Corrosion, 403, 413
 and cylinder wear, 191
Coupé, 397
Coupling, fluid, 223
Crankcase, 96, 106
Crankshaft, 98, 110, 123
Cross R C, 92
Cugnot N, 6
Current, generation and control, 172
Cut-out, automatic, 172
Cycle cars, 42, 87

Daimler G W, 15, 18, 78
Dance C, 9
Darwin E, 8
Darwin G, 379
Dashboard, 391
Davis F W, 330
Davis T M and W, 360
Dazzle, 419
de Dion A, 12, 26
Delamere-Deboutville E, 21
De Lavaud, 293
Delco system, 163, 178
Depression
 in UK, 53
 in USA, 57
de Rivaz I, 7, 13, 21
Diesel R, 125
Disc, brake, 354
Drive
 camshaft, 105
 four-wheel, 275
 friction, 231
 front wheel, 272
 speedometer, 211
Drums, brake, 354
Dubonnet A, 298
du Cros H, 369
Dunlop J B, 39, 369
Durant W C, 46, 56
Duryea C E and F, 32
Dynamo, 172
Dynamotor, 177

Edge S F, 39, 41
Elliot O, 5
Emissions
 control, 146
 legislation, 114

INDEX

Engineering, badge, 112
Engines
  aircraft (1914–18), 43
  Benz, 75
  compression ignition, 125
  Daimler, 77
  de Dion Bouton, 81
  diesel, 125
  early internal combustion, 13
  Ford Model T, 87
  four-stroke, 14
  gas, 14
  Lanchester, 84
  lean burn, 59, 119
  Mors, 82
  multicylinder, 83
  Otto, 14
  OHC, 120
  OHV, 90
  sleeve valve, 91
  stratified charge, 118
  traction, 11
  two-stroke, 93
  Wankel, 59, 114

Fans, 205
Farish, 13
Filter, oil, 186, 194
Forces, out of balance, 81, 83
Ford H, 32, 34, 44, 46, 56, 57, 67, 236, 281, 394
Fottinger H, 223
Frame, chassis,
  cruciform, 283
  ladder, 281
  perimeter, 283
  subframe, 279
Free-wheel devices, 243
Friction materials, 354
Frood H, 355
Fuel, 156
  injection, 152
  requirements, 158

Galton D, 337
Garrett G, 12
Gauges
  Bourdon, 212
  oil-pressure, 212
  petrol, 213
  radiator temperature, 212
Gears differential, 265

Gearboxes, 37, 232, 250, 256
  automatic, 246, 254, 259
  constant-mesh, 234
  design features, 237, 249, 256
  epicyclic, 236
  noise, 238, 241
  preselector, 244
  requirements, 239
  semiautomatic, 246, 253
  sliding gear, 232
  synchromesh, 247, 250
  Wilson, 245
  in USA, 239
Gearchanging, 35
Gears, hypoid, 268
Gibbs, 11
Girling A H, 344
Governors, 94
  hit and miss, 37, 94
Guinness K L, 169
Gurney G, 9

Hancock T, 369, 375
Hancock W, 9, 157, 375
Haynes E, 32
Head, cylinder
  F-type, 90
  L-type, 90
  T-type, 88
  turbulent, 100
Head, steering
  Elliot, 315
  Lemoine, 315
Heating, 411
Hele-Shaw H S, 219, 379
Henry E, 239
Hewetson H, 29
Hill F, 266
Hitler A, 275, 414
Hock J, 15, 131
Hood, Cape, 398, 399

Industry, automobile
  Japanese, 69
  UK, 29, 38, 51, 61
  USA, 32, 44, 55, 67
Injection, fuel (petrol), 152
  throttle-body, 154
Ignition, 159
  coil, 160, 164
  capacitance discharge, 168
  Delco system, 163

INDEX

Ignition (contd)
  electric, 159
  hot-tube, 36, 78, 159
  magneto, 161
  timing, 163
  transistorised, 166
Inventors, early, 21

James W H, 11
Jaray P, 407
Jeffrey T B, 375
Jellinek E, 35
Johansen H, 21
Johnston G, 31
Joints, universal, 262
  for FWD, 273

Kaiser H, 67
Kettering C, 57, 163, 177, 295, 404
Knock, 50, 100, 156
Krebs, 132, 288

Lamps
  acetylene, 418
  double-filament, 420
  electric, 419
  four headlamps, 406, 421
  sealed, 420, 421
  tungsten halide 422
Lanchester F, 31, 34, 85, 91, 95, 98, 130, 185, 233, 236, 267, 287, 317, 333, 349, 379
Lanchester G, 265
Landau, 396
Landaulette, 396
Lawson H J, 30, 31, 39
Leckensburger G, 313
Ledwinka J, 281, 401
Legislation, 41, 58, 60, 68, 335
Leland H M, 34, 44, 163, 177
Lenoir J, 14, 21, 168
Levassor E, 22, 232
Lighting, electric, 419
Limousine, 396
Linkages, steering, 325, 341
Lord L, 63
Loughead M, 342
Lubricants
  extreme pressure, 198, 199
  SAE classification, 184, 190
Lubrication, 181
  camshaft, 192

dry-sump, 196
forced-feed, 185
gravity-feed, 184
sleeve-valve, 194
splash, 183
systems, 192, 195, 199
trough, 184
Lubricators, sight feed, 184, 207

McPherson E A, 282, 304
Magneto, 36, 161, 162
  high-tension, 162
  in the twenties, 164
  low-tension, 162
Manifold, inlet, 146
Marcus S, 22, 161
Martin P, 244
Maxim H, 32
Maybach W, 17, 18, 20, 21, 35, 131
Michelin A and E, 355, 371
Microprocessors, 59
Midgley T, 50, 100, 157
Morris W, 53, 63
Motor, starter, 178
  pre-engaged, 180
Moulton S, 375
Mountings, engine, 110
Murdock W, 7
Murray B, 31

Nader R, 68, 275, 308
Nasmyth J, 9
Number, octane, 50, 58, 62, 100, 157, 158

Oil
  compound, 182
  mineral, 181, 188
  multigrade, 196
Olds R E, 44
Olley M, 295, 327
Otto N, 14, 18, 161
Overdrive, 251

Packard J W, 32
Painting, 403, 414
Panel, instrument, 207
Papin D, 13
Patent, *Selden*, 46
Patents, 75
Pecquer O, 266

INDEX 429

Perrot H, 337
Petrol, see Fuel
Peugeot A, 27
Phaeton, 391
Pistons, 97, 122
  aluminium, 97, 107
  slap, 107
Plastics, 414
Plunkett T J, 356
Pomeroy L, 87
Porsche F, 275, 299
Plug, sparking, 168
  aluminium oxide, 171
  mica, 169
  porcelain, 169
  surface discharge, 171
Production
  early methods, 33
  mass, 34, 52
Pressure, oil, 186, 193
Pugh J W, 360
Pumps
  oil, 193, 195
  water, 203

Races, classic road, 24
Radiators
  cellular, 201
  film-type, 201
  honeycomb, 201
  multitube, 200
  single-tube, 200
Rating
  octane, see Number, octane
  RAC, 97
Ratio
  compression, 95, 158
  stroke-bore, 96, 107
Red Flag Act, 11
Refining, 156
Renault L, 29
Renouf P L, 337
Reynolds O, 186
Ricardo H, 50, 100, 118, 127, 157
Road locomotives, steam, 6
Roberts R, 266
Rod, connecting, 109
Roesch G, 103
Roger E, 28
Roll, body, 300
Rolls C S, 372
Royce H, 132

Rubber
  natural, 374
  synthetic, 384
Rumpler, 407
Runabouts, 45, 393
Russell J S, 9, 10

Saloon, 396
Scott W, 369
Seat
  dickey, 399
  rumble, 399
Sedan, 396
Serpollet L, 12
Shutters, radiator, 205
Slide-slip, 337, 388
Simms F R, 30, 161
Smog, 69
Sloan A P, 57
Sludging, 196
Solomon D, 30
Speedometer
  centrifugal, 209
  chronometric, 210
  electro-magnetic, 210
Sprags, 333
Springs,
  coil, 285
  leaf, 284, 287
  types of, 284, 290
Squish, 100, 118
Starkey J, 359
Starley J K, 39
Street, 13
Suspensions, 284, 289, 301, 309
  active, 311
  air, 305
  de Dion, 307
  independent front, 293
  independent rear, 302
  rear, 307
  rubber, 306
  unconventional, 304
Steering, 38, 321, 328
  Ackermann, 313, 330
  Benz, 312
  centre-pivot, 312
  centre-point, 316
  Daimler, 312
  four-wheel, 332
  irreversible, 317
  powered, 330

Steering gear
  epicyclic, 317
  proprietary, 324
  rack and pinion, 317, 318
  worm and nut, 318
  worm and sector, 319, 320, 322
  worm and wheel, 322
Strangler, see Choke
Streamlining, 407
'Stretching', 112
Styling, body, 404, 409, 415
Supercharging, 123, 151
Switch, starter, 180

Tappets, 89, 102
Tetraethyl lead, 50, 100, 157
Thermostats, 203
Thomson R W, 368
Thousand Mile Trial, 40
Tiller, 317
Timing, automatic, 164
Timkens H, 365
Tizard H, 50
Tjaarda J, 282
Tonneau, 390
Tourer, 42, 395, 397
Towers B, 185
Transmissions, 37
  belt, 228, 256
  chain, 37
  Hotchkiss, 271
Tread, tyre, 379
Trevithick R, 8
Tube, torque, 271
Turbine, gas, 59
Turbocharging, 123
Turbulence, 99, 119
Tyres, 38, 384
  attachment to wheels, 375
  balloon, 380
  beaded edge, 375
  belted-bias, 386
  clincher, 375
  construction of, 372, 382
  cord, 374
  crossply, 373
  development by Michelin, 371
  get you home, 387
  low-pressure, 324

Mummy, 373
pneumatic, early, 368
radial, 385
solid rubber, 370
treads, 379
tubeless, 374, 387
valves, 378
wired on, 377

Valves
  exhaust, 88
  desmodromic, 102
  inlet, 36, 88
  mechanisms, 88, 101, 119
  overhead, 91, 101, 103
  overhead cam, 104, 120
  pushrod-operated, 103, 119
  poppet, 36, 89
  rotary, 36, 92
  sleeve, 36, 91, 105
  timing, 90
  tyre, 378

Varlo C, 365
Ventilation, 410
Viscosity index, 190
  improvers, 197
Vibration, engine, 37
Voiturettes, 42, 393

Welch C K, 370, 376
Wheels, 356
  artillery, 356
  auxiliary, 360
  detachable, 360
  detachable rims, 360
  disc, 359
  light alloy, 364
  Magna, 362
  Rudge-Whitworth, 360
  Sankey, 357
  spare, 417
  wire, 358
Wickham H A, 375
Williams C G, 191
Windscreens, 393, 398
Winton A, 32
Wright L W, 14